High-Power Piezoelectrics and Loss Mechanisms

High-Power Piezoelectrics and Loss Mechanisms

Kenji Uchino

CRC Press
Taylor & Francis Group
Boca Raton London New York

CRC Press is an imprint of the
Taylor & Francis Group, an **informa** business

First edition published 2020
by CRC Press
6000 Broken Sound Parkway NW, Suite 300, Boca Raton, FL 33487-2742

and by CRC Press
2 Park Square, Milton Park, Abingdon, Oxon, OX14 4RN

CRC Press is an imprint of Taylor & Francis Group, LLC

ISBN: 978-0-367-54069-2 (hbk)
ISBN: 978-1-003-08751-9 (ebk)

Typeset in Times
by codeMantra

Contents

List of Symbols

D	Electric displacement
E	Electric field
P	Dielectric polarization
P_S	Spontaneous polarization
α	Ionic polarizability
γ	Lorentz factor
μ	Dipole moment
ε_0	Dielectric permittivity of Free Space
ε	Dielectric permittivity
ε (ε_r)	Relative permittivity or dielectric constant (assumed for ferroelectrics: $\varepsilon \approx \chi = \varepsilon - 1$)
κ	Inverse dielectric constant
χ	Dielectric susceptibility
C	Curie–Weiss constant
T_0	Curie–Weiss temperature
T_C	Curie temperature (phase transition temperature)
G	Gibbs free energy
A	Helmholtz free energy
F	Landau free energy density
x	Strain
x_s	Spontaneous strain
X	Stress
s	Elastic compliance
c	Elastic stiffness
v	Sound velocity
d	Piezoelectric charge coefficient
h	Inverse piezoelectric charge coefficient
g	Piezoelectric voltage coefficient
M, Q	Electrostrictive coefficients
k	Electromechanical coupling factor
η	Energy transmission coefficient
Y	Young's modulus
μ	Friction constant
$\tan \delta$ ($\tan \delta'$)	Extensive (intensive) dielectric loss
$\tan \phi$ ($\tan \phi'$)	Extensive (intensive) elastic loss
$\tan \theta$ ($\tan \theta'$)	Extensive (intensive) piezoelectric loss

Preface

As one of the pioneers of "Piezoelectric Actuators" (I authored the first book titled this in 1984), I have been contributing to various product commercialization in these 45 years, including million-selling devices, micro-ultrasonic motors for smart-phone camera modules by Samsung Electromechanics (Korea), piezoelectric transformers for backlight inverters by Apple laptops (USA), multilayer PZT actuators for diesel injection valves by Denso Corporation (Japan), and piezoelectric energy harvesting modules for Programable Air-Burst Munition (25 mm φ caliber) by the US Army. During the development period on "piezoelectric actuators and transformers", I found that the "bottleneck" for the device miniaturization is "heat generation" under a high-power drive condition. When I started in the early 1980s, the PZT ceramics could generate only less than 10 W mechanical power per cm^3 volume. When higher power level was input, the ceramic specimen generated a significant heat higher than 50°C, which I could not touch without burning the finger. Thus, in parallel to the piezo-actuator developments, I have been developing various high-power density piezo-ceramic materials with the loss mechanism clarification. Now, our high-power density piezo-ceramics can generate 40 W/cm^3, high enough to miniaturize the device down to a quarter of the previous device size. I spent almost 35 years for establishing so-called "HiPoCS™" (High-Power Piezoelectric Characterization System), which have been transferred worldwide to encourage the following researchers. Hence, I considered that it is time to organize a textbook based on the previous studies, including my "personal" materials development philosophy, in order to stimulate younger generation to reach to the dreaming energy density up to 100 W/cm^3 in the future. Increasing the efficiency and saving the energy and space (compactness) are one of the important approaches in this 21st-century "sustainable society".

This textbook introduces the theoretical background of piezoelectrics, electromechanical phenomenology, loss mechanisms, practical materials, device designs, drive and characterization techniques, typical applications, and looks forward to the future perspectives in this field. Though the discovery of piezoelectricity is relatively old (Jacque and Pierre Curie brothers in 1880), since the high-power actuator development is relatively new and interdisciplinary, it is difficult to cover all the recent studies in a limited-page book. Therefore, I focused important and basic ideas to understand how to design and develop the high-power piezoelectric materials and devices. Many of the studies cited in this textbook are intentionally from our group's lab notes in order to keep the consistency of our development philosophy. Thus, this book is NOT an overall review of this area, remaining the reader's further search for other scholars' approaches.

Let me introduce the contents. Chapter 1 introduces the overall "Background of High-Power Piezoelectrics". Following two chapters, Chapter 2 "Foundations of Piezoelectrics" and Chapter 3 "Fundamentals of Losses", summarized the basic prerequisite knowledge on the piezoelectrics and losses for junior researchers, without referring to other textbooks. Thus, senior researchers with these knowledge may skip these chapters. Chapter 4 "Piezoelectric Loss Phenomenology" is the starting chapter of the main content. Based on the phenomenology, we construct "Equivalent Circuits with Piezo Losses" in Chapter 5. Though the applicable area is limited to the equivalent circuit analysis, it is occasionally useful. We also point out a problem on the present IEEE Standard document on the loss determination. Chapter 6 treats "Heat Generation in Piezoelectrics", where we also provide the insight how the "thermal" measurement can predict the "mechanical performances". Chapter 7 "High-Power Piezo Characterization System" (HiPoCS) is one of the highlights of this book, in which I introduce our characterization system's history in these 30 years, and persuade the reader how the "vibration velocity constant" admittance spectrum measurement and the improved "Burst Mode Method" are essential. After discussing "Drive Schemes of Piezoelectric Transducers" in Chapter 8, from the high-power driving circuit viewpoint, we will discuss another highlight, Chapter 9 "Loss Mechanisms in Piezoelectrics". Microscopic (unit cell base) and semi-macroscopic (domain wall dynamics) models are presented in order to develop high-power density piezoelectrics. In Chapter 10 "Practical High-Power Piezoelectric Devices", we describe up-to-date problems in practical device manufacturing, such as electrode selection in multilayer piezoelectric transducers. Finally, Chapter 11 "Concluding Remarks and Future Perspectives" addresses the importance of three development stages: (1) low-loss piezoelectric materials (Chapter 9), (2) device designing, such as multilayer manufacturing of hard PZT ceramics (Chapter 10), and (3) drive system/power supply schemes (Chapter 8), in order to achieve practical high-power piezoelectric components and devices.

This textbook was written for graduate students, university researchers, and industry engineers studying or working in the fields of piezoelectric actuators, transducers, and energy harvesting systems. This textbook is designed for self-learning by the reader by himself/herself aided by the availability of:

- Chapter Essentials – Summary for your quick memory recovery.
- Check Points – Answers are provided in the book Appendix.
- Example Problems – To enhance the reader's understanding with full detailed solutions.
- Chapter Problems – For the final exam or further consideration.
- Quick E-Answer from the Author via e-mail (kenjiuchino@psu.edu) – Any questions are welcome.

Since this is the first edition, critical review and content/typo corrections on this book are highly appreciated. Send the information directed to Kenji Uchino at 135 Energy and The Environment Laboratory, The Pennsylvania State University, University Park, PA, 16802-4800. E-mail: KenjiUchino@PSU.EDU

For the reader who needs detailed information on ferroelectrics, smart piezoelectric actuators, and sensors, *Ferroelectric Devices 2nd Edition* (2010) and *Micromechatronics 2nd Edition* (2019) authored by K. Uchino, published by CRC Press, are recommended. Further, *FEM and Micromechatronics with ATILA Software* published by CRC Press (2008) is a perfect tool for practical device designing, with three losses in simulation. ATILA FEM software code is available from Micromechatronics Inc., State College <www.mmech.com>.

Even though I am the sole author of this book, it nevertheless includes the contributions of many others. I express my sincere gratitude to my former visiting researchers, associates, and graduate students who worked on "high-power piezoelectrics" in the ICAT research center; Doctors Seiji Hirose, Sadayuki Takahashi, Hideaki Aburatani, Seok-Jin Yoon, Alfredo Vazquez Carazo, Yun-Han Chen, Jungho Ryu, Jiehui Zheng, Uma Belegundu, Shashank Priya, Hyeoungwoo Kim, Shuxiang Dong, Seyit O. Ural, Burhanettin Koc, Serra Cagatay, Gareth J. Knowles, Yongkang Gao, Seung-Ho Park, Sahn Nahm, Yuan Zhuang, Safakcan Tuncdemir, Aditiya Rajapurkar, A. Erkan Gurdal, Menglun Tao, Husain N. Shekhani, Weijia Shi, Mariyam Majzoubi, Xiaoxiao Dong, Tao Yuan, Minkyu Choi, Anushka Bansal, Hossein Daneshpajooh, and Yoonsang Park. I am also indebted to the continuous research support funds from the US Office of Naval Research Code 332 during 1991–2020 without any intermission through the grants N00014-96-1-1173, N00014-99-1-0754, N00014-08-1-0912, N00014-12-1-1044, and N00014-17-1-2088.

Finally, my best appreciation goes to my wife, Michiko, who constantly encourages me on my activities.

Kenji Uchino

December, 2019 at State College, PA
Kenji Uchino

MATLAB® is a registered trademark of The MathWorks, Inc. For product information, please contact:

The MathWorks, Inc.
3 Apple Hill Drive
Natick, MA 01760-2098 USA
Tel: 508-647-7000
Fax: 508-647-7001
E-mail: info@mathworks.com
Web: www.mathworks.com

Author

Kenji Uchino, one of the pioneers in piezoelectric actuators and electrooptic displays, is the director of International Center for Actuators and Transducers (ICAT) and professor of Electrical Engineering and Materials Science and Engineering, also currently a Distinguished Honors faculty member at Schreyer Honors College at the Pennsylvania State University. He has been a university professor for 45 years so far, including 18 years in Japanese universities. He has also been a company executive (president or vice president) for 21 years in four companies, most recently the founder and senior vice president of Micromechatronics Inc., a spin-off company from the above ICAT, where he was elaborating to commercialize the ICAT-invented piezo-actuators and transducers. He was a Government officer for 7 years in both Japan and the United States, recently a "Navy Ambassador to Japan" (2010–2014) for assisting the rescue program from the Big Northern Japan Earthquake. He is currently teaching "Ferroelectric Devices", "Micromechatronics", "FEM Application for Smart Materials", and "Entrepreneurship for Engineers" for the Engineering and Business School graduate students.

After being awarded his PhD degree from Tokyo Institute of Technology, Japan, Uchino became research associate/assistant professor in Physical Electronics Department at this university. Then, he joined Sophia University, Japan, as an associate professor in Physics in 1985. He was then recruited to Penn State in 1991 under a strong request from the US Navy community. He was also involved with Space Shuttle Utilizing Committee in NASDA, Japan, during 1986–1988, and was the vice president of NF Electronic Instruments, USA, during 1992–1994. He has his additional Master's degree in Business and Administration from St. Francis University, PA. He has been consulting more than 120 Japanese, US, and European industries to commercialize the piezoelectric actuators and electrooptic devices. He was the chairman of Smart Actuator/Sensor Study Committee partly sponsored by the Japanese Government, MITI (1987–2014). He is editor-in-chief of J. Insight-Material Science, PiscoMed Publishing, and Associate Editor in Chief of J. Actuators, MDPI, also was associate editor for *Journal of Materials Technology* (Matrice Technology) and the editorial board member for *Journal of Ferroelectrics* (Gordon & Breach) and *Journal of Electroceramics* (Kluwer Academic). He served as an administrative committee member for IEEE, Ultrasonics, Ferroelectrics, Frequency Control Society (1998–2000), and secretary of American Ceramic Society, Electronics Division (2002–2003).

His research interests are in solid-state physics—especially dielectrics, ferroelectrics, and piezoelectrics, including basic research on materials, device designing, and fabrication processes, as well as development of solid-state actuators and displays for precision positioners, ultrasonic motors, projection-type TVs, etc. He has authored 582 papers, 78 books, and 33 patents in the piezoelectric actuator and optical device area. He is a fellow of American Ceramic Society from 1997, fellow of IEEE since 2012, and senior member of National Academy of Inventors since 2019, also a recipient of 31 awards, including Wilhelm R. Buessem Award from the Center for Dielectrics and Piezoelectrics, The Penn State University (2019), Distinguished Lecturer of the IEEE UFFC Society (2018), International Ceramic Award from Global Academy of Ceramics (2016), IEEE-UFFC Ferroelectrics Recognition Award (2013), Inventor Award from Center for Energy Harvesting Materials and Systems, Virginia Tech (2011), Premier Research Award from The Penn State Engineering Alumni Society (2011), the Japanese Society of Applied Electromagnetics and Mechanics Award on Outstanding Academic Book (2008), the SPIE Smart Product Implementation Award (2007), R&D 100 Award (2007), ASME Adaptive Structures Prize (2005), Outstanding Research Award from Penn State Engineering Society (1996), Academic Scholarship from Nissan

Motors Scientific Foundation (1990), and the Best Paper Award from Japanese Society of Oil/Air Pressure Control (1987).

In addition to his academic carrier, Uchino is an honorary member of KERAMOS (National Professional Ceramic Engineering Fraternity) and obtained the Best Movie Memorial Award as the director/producer in the Japan Scientific Movie Festival (1989) on several educational video tapes about "Dynamical Optical Observation of Ferroelectric Domains" and "Ceramic Actuators".

Prerequisite Knowledge Check

Studying of "High-Power Piezoelectrics" assumes certain basic knowledge. Answer the following questions by yourself prior to referring to the answers on the next page.

Q1 Provide definitions for the *elastic stiffness, c,* and *elastic compliance, s,* using stress (X)–strain (x) equations.

Q2 Sketch a *shear stress* (X_4) by arrows and the corresponding *shear strain* (x_4)/*deformation* on the square material depicted below.

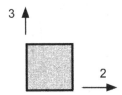

Q3 Describe an equation for the *velocity of sound, v,* in a material with mass density, ρ, and elastic compliance, *s*.

Q4 Given a rod of length, *L*, made of a material through which sound travels with a velocity, *v*, describe an equation for the *fundamental extensional resonance frequency, f_R*.

Q5 When two solid materials are contacted, and moved along the contact plane, friction force is introduced. How do you describe the friction force *F* in terms of the force *N* normal to the contact plane and the friction constant μ?

Q6 Provide the capacitance, *C*, of a capacitor with area, *S*, and electrode gap, *t*, filled with a material of *relative permittivity, ε_r*.

Q7 Describe an equation for the *resonance frequency* of the circuit pictured below:

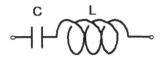

Q8 Given a power supply with an internal impedance, Z_0, what is the optimum circuit impedance, Z_1, required for maximum power transfer?

Q9 Calculate the polarization *P* of a material with dipole density N (m^{-3}) of the unit cell dipole moment $q \cdot u$ (C·m).

Q10 Provide the polarization, *P*, induced in a *piezoelectric* with a piezoelectric strain coefficient, *d*, when it is subjected to an external stress, *X*.

ANSWERS

[60% or better score is expected.]

Q1 $X = c \cdot x, x = s \cdot X$
> [*Note*: c stands for "stiffness" and s stands for "compliance".]

Q2 $x_4 = 2 x_{23} = 2\phi$
> [*Note*: Radian measure is generally preferred. This shear stress is not equivalent to the diagonal extensional stress.]

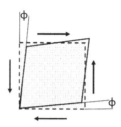

Q3 $v = 1/\sqrt{\rho s}$

Q4 $f = v/2L$

Q5 $F = \mu \cdot N$

Q6 $C = \varepsilon_0 \varepsilon_r (S/t)$
> [*Note*: Do not forget ε_0]

Q7 $f = 1/2\pi\sqrt{LC}$

Q8 $Z_1 = Z_0$
> [*Note*: The current and voltage associated with Z_1 are $V/(Z_0 + Z_1)$ and $[Z_1/(Z_0 + Z_1)]V$, respectively, the product of which yields the power. The maximum power transfer occurs when $Z_0/\sqrt{Z_1} = \sqrt{Z_1}$ (when impedance is resistive).]

Q9 $P = Nqu$
> [*Note*: The unit of the polarization is given by C m^{-2}, equivalent to the charge density on the surface.]

Q10 $P = d \cdot X$
> [*Note*: This is called the *direct piezoelectric effect*.]

1 Background of High-Power Piezoelectrics

ABSTRACT

"Background of High-Power Piezoelectrics" describes the necessity of loss mechanism clarification in order to develop high-energy density piezoelectric components for the portable electronic devices. Keys to high-power piezoelectric developments include: (1) low-loss piezoelectric materials, (2) multilayer technology, and (3) high-power supply/drive circuit. This book focuses on high-power density "actuator" developments, not on low-power "sensor" applications. In this text book, we starts from the loss phenomenology, and propose typical high-power characterization methods. Then, current high-power piezoelectric materials and drive/operation schemes are introduced. In the latter part of this book, semi-microscopic loss mechanisms are discussed, based on the domain dynamics models, followed by some know-how on commercialization. Note that the "high-power piezoelectrics" in this textbook does not mean high absolute power devices (kW, MW), but "high-power density", 10–100 MW/m^3 in practice.

1.1 RESEARCH MOTIVATION FOR HIGH-POWER PIEZOELECTRICS

1.1.1 PIEZOELECTRIC ACTUATORS/MOTORS COMMERCIALIZATION TREND

Piezoelectric actuators have been commercialized in various smart structures and systems such as precision positioners, adaptive mechanical dampers, and miniature ultrasonic motors (USMs). Piezoelectric ceramics are forming a new field between electronic and structural ceramics. Application fields are classified into three categories: positioners, motors, and vibration suppressors. Recently, piezoelectric transformers and energy harvesting systems have been moving in similar high-power piezo-device categories. The manufacturing precision of optical instruments, such as lasers and cameras, and the positioning accuracy for fabricating semiconductor chips, which must be adjusted using solid-state actuators, is of the order of 10 nm. From the market research result for 80 Japanese component industries in 1992, tiny motors in the range of 5–8 mm are highly required in the early 21st century for office and factory automation equipment, and the conventional electro-magnetic (EM) motors are rather difficult to produce with sufficient energy efficiency. The author's group collected power vs. efficiency data for 2,000 commercial EM motors, as plotted in a logarithmic scale in Figure 1.1. The trend slope seems to be 2/3, which can be explained by the scaling law. Power level is proportional to the actuator size $\propto L^3$; while the efficiency is determined by the dissipation of the Joule heat generated by the coil wire resistance under the same current (i.e., the same magnetic field level) $\propto L^2$. Thus, the slope 2/3 is expected. The data on piezoelectric USMs are also inserted with almost horizontal line (size-insensitive efficiency), because the loss is mostly originated from the material's intrinsic losses. The two lines crossing data correspond to 30 W power/30% efficiency. Figure 1.2 compares size difference among EM and piezoelectric components: (1) 50 mW-micromotors for a mobile phone silent alarm, and (2) 30 W-compact transformers for a laptop backlight inverter. You can recognize roughly 20:1 volume/weight ratio of the piezo-device against the EM type in the same power level components.

The advantages of piezoelectric devices over EM types are summarized:

a. More suitable to miniaturization – Since the stored energy density is larger than that of an EM type, 1/10 smaller in volume and weight can be achieved.
b. No EM noise generation – Since magnetic shielding is not necessary, we can keep a compact design.

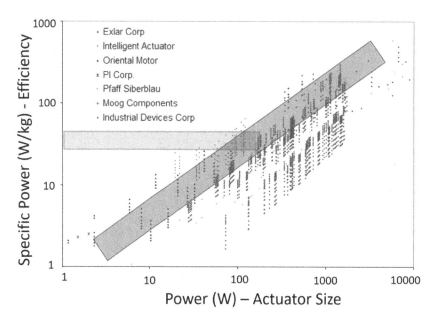

FIGURE 1.1 Power vs. efficiency for various commercial EM motors and piezoelectric USMs. The slope 2/3 is originated from Joule loss.

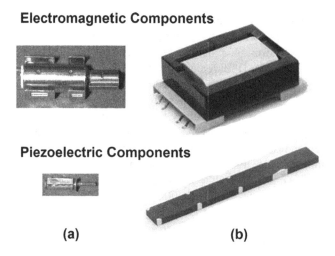

FIGURE 1.2 Size comparison among EM and piezoelectric components: (a) 50 mW-micromotor for a mobile phone silent alarm; (b) 30 W-transformer for a laptop backlight inverter.

    c. Higher efficiency – Since the efficiency is insensitive to the size, the piezo-device is effective in the power range lower than 30 W.

    d. Non-flammable – The piezo-device is safer for the overload or the short-circuit at the output terminal.

Figure 1.3 shows the piezoelectric device market estimated from multiple sources.[1] The current (2017) revenue is around $40 billion. Actuator/piezo-generator (energy harvesting) is the largest category, followed by transducer/sensor/accelerometer/piezo-transformer. Then, resonator/acoustic device/USM category chases.

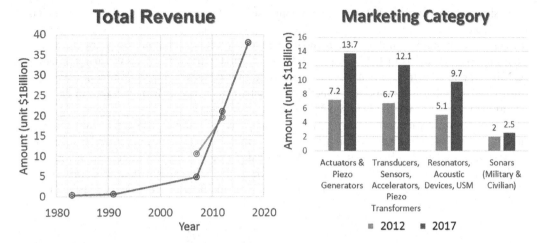

FIGURE 1.3 Piezoelectric device market trends.[1]

1.1.2 MULTILAYER TRANSDUCER INVENTION

In order to enhance the handling energy level of a piezoelectric device, we need to increase the capacitance, because the input electro static energy U under pseudo-DC condition is expressed by

$$U = (1/2)CV^2 \tag{1.1}$$

or the *apparent power P* under sinusoidal AC voltage V_{RMS} is given by the product of voltage and current ($V \cdot I$), and the admittance Y of the device, that is, $Y = j\omega C$, under a voltage constant condition:

$$P = \omega C V_{RMS}^2. \tag{1.2}$$

Thus, in a limited compact size, or even in a smaller size, to realize a high-power device, an increase in the capacitance C is essential. Thus, the author's group invented a cofired multilayer (ML) piezo-actuator design to primarily increase the capacitance, in addition to reduce a practically applying voltage and the resulting current increase in a compact piezo-device. When we adopt the n-layer ML structure, instead of a single-layer actuator with the same total sample size, we can expect the capacitance increase by a factor of n^2, since each layer capacitance increases by n times, and n layers are parallelly connected:

$$C = n \cdot \varepsilon_0 \varepsilon \left[S / \left(t_{\text{total}} / n \right) \right] = n^2 \cdot \varepsilon_0 \varepsilon \left(S / t_{\text{total}} \right). \tag{1.3}$$

To achieve a low-driving voltage and a high-operating current, device miniaturization, and high-energy density of the devices, ferroelectric ceramic ML structures have been investigated intensively for capacitor and actuator applications. Note that the ML design is also used to reduce the electrical impedance. Key words for the future trend will be "finer" and "hybridization". Layers thinner than 3 μm, which is currently used in ML capacitors, can also be introduced in actuator devices technologically instead of the present 60–80 μm thick sheets. A typical actuator with 10 mm length exhibits 10 μm displacement (strain of 0.1%) with a fundamental resonance frequency around 100–150 kHz. Nonuniform configurations or hetero-structures of the materials, layer thickness, or the electrode pattern can be adopted for practical devices.

There are two techniques for making multilayered ceramic devices: the *cut-and-bond method* and the *tape-casting method*. The tape-casting method has been widely used for ML capacitor fabrication, and requires expensive fabrication facilities and sophisticated techniques, but is suitable for mass production of more than hundred thousand pieces per month.

As shown in Figure 1.4, a ML structure is composed of alternate ferroelectric ceramic and internal electrode layers (here, *interdigital electrode pattern*) fabricated by *cofiring*. An adjacent pair of electrodes composes a unit displacement element, which is connected in parallel by the external electrode up to hundreds of layers. Figure 1.5 shows a flowchart of the manufacturing process of the ML ceramic actuators. *Green sheets* are prepared in two steps: slip preparation of the ceramic powder and a doctor blade process. Sometimes so-called "templates" (BaTiO$_3$ flakes in PZT powers) are added to enhance the textured ML devices, as in Figure 1.5. The *slip* is made by mixing the ceramic powder with *solvent*, *deflocculant*, *binder*, and *plasticizer*. The slip is cast into a film under a special straight blade, called *doctor blade*, whose distance above the carrier determines the cast film thickness. After drying, the film, called a *green sheet*, has the elastic flexibility of synthetic leather. The volume fraction of the ceramic in the polymer matrix at this point is about 50%. The green sheet is then cut into an appropriate size, and internal electrodes are printed using

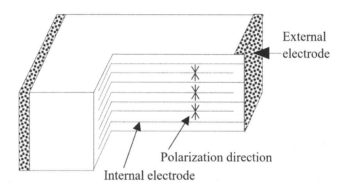

FIGURE 1.4   Structure of an ML actuator.

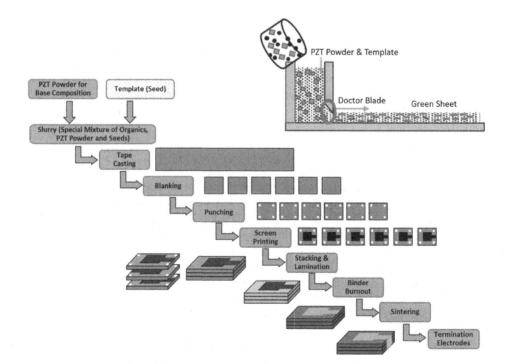

FIGURE 1.5   Manufacturing process of the ML ceramic actuators.

Ni (for BT-based capacitors), silver–palladium, platinum, or Cu ink (for Pb(Zr, Ti)O₃ or PZT-based transducers). Several tens to hundreds of such layers are then laminated and pressed using a hot press. After cutting into small chips, the green bodies are sintered at around 1,000°C (for Cu, Ag)–1,200°C (for Ag/Pd, Pt) in a furnace, taking special care to control binder evaporation around 500°C. The sintered chips are then polished, externally electroded, lead wires are attached, and finally the chips are coated with a water-proof spray. Special care must be taken for the electrode material's conductivity in the ML structures; due to a small impedance of the piezoelectric transducer itself, the resistance of the electrode and lead wire occasionally provides larger loss as a total device, such as in piezoelectric transformers, as described in Chapter 10.

1.1.3 POWER SUPPLY DEVELOPMENT

When Uchino et al. invented cofired ML piezoelectric actuators in the late 1970s, the impedance measurement for characterizing the ML was the initial problem, because the conventional impedance analyzer could not supply sufficient voltage (max only 30 V) nor current (max less than 0.2 A) with the output impedance 50 Ω. Uchino considered this problem for determining the specifications for a new piezo-actuator drive product/power supply as a Vice President at NF Corporation, Yokohama, Japan.

First, the ML actuator has 100 ceramic layers, each 100 μm thick with an area of $5 \times 5\,mm^2$. The relative permittivity is 10,000. Using Eq. (1.3), C = $(100)(8.854 \times 10^{-12}\,F/m)(10,000)$ $[(5 \times 5 \times 10^{-6}\,m^2)/(100 \times 10^{-6}\,m)] = 2.21 \times 10^{-6}$ (F). ML actuators have a *capacitance of higher than* 1 μF.

Second, the resonance frequency of the actuator is calculated. Assuming it has a density $\rho = 7.9$ g/cm³, elastic compliance $s_{33}{}^D = 13 \times 10^{-12}\,m^2/N$ and the ML 10 mm long in total (the electrode weight load is ignored for this calculation),

$$f_R = \frac{1}{2L\sqrt{\rho s_{33}^D}}$$

$$= \frac{1}{2[100]\left[100 \times 10^{-6}\,(m)\right]\sqrt{\left[7.9 \times 10^3\,\left(kg/m^3\right)\right]\left[13 \times 10^{-12}\,\left(m^2/N\right)\right]}} = 156\,(kHz) \quad (1.4)$$

The response speed of the power supply must be greater than the actuator's resonance frequency.

Third, we determine the current required. If 60 V is to be applied to the actuator as quickly as possible (up to the resonance period t_R), the relationship between the actuator voltage and the charging current is given by:

$$I = Q/t_R = C V f_R$$

$$= \left[2.21 \times 10^{-6}\,(F)\right] [60\,(V)]\left[156 \times 10^3\,(Hz)\right] = 21\,(A) \quad (1.5)$$

Ideally, a significant current is required from the power supply, even if just for a relatively short period (6 μs). The *apparent power* is estimated to be [60 (V) × 21 (A)]. So, we see that more than 1 (kW) is needed for the quickest (i.e., resonance period) drive. [The *real power* consideration is conducted, taking into account the phase lag of the current from the applied voltage in Chapter 8.]

Fourth, the *cut-off frequency* (1/RC) of the drive circuit, where R is the output impedance of the drive circuit and C is the piezo-actuator's capacitance, must also be higher than the ML mechanical resonance frequency. Assuming $\omega_R = 2\pi f_R = 1/RC$, R can be calculated as

$$R = 1/\left[2\pi f_R C\right]$$

$$= 1/\left[2\pi[156 \times 103\,(Hz)]\left[2.21 \times 10^{-6}\,(F)\right]\right] = 0.46\,[\Omega] \quad (1.6)$$

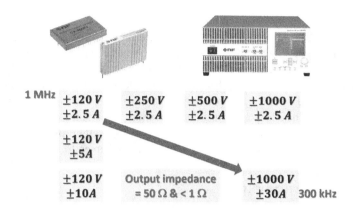

FIGURE 1.6   Power supply development strategy (NF Corporation, Yokohama, Japan).

The output impedance of the power supply should be less than 1 Ω. In conclusion, the power supply specifications for the ML resonance measurement should be better than the followings:

- Maximum voltage: 200 (V)
- Maximum current: 10 (A)
- Frequency range: 0–500 (kHz)
- Output impedance: <1 (Ω).

You may use a conventional impedance analyzer, on which you should recognize that maximum voltage is only 30 V and maximum current less than 0.2 A with the output impedance 50 Ω. Therefore, your measured admittance value on the ML device at its resonance is one order of magnitude smaller than the expected value, because the peak current cannot be supplied from the analyzer power system. Thus, through NF Corporation, Japan, Uchino's team developed and commercialized new power supplies for driving ML USMs with the specs: 300 V/10 A and output impedance less than 0.2 Ω, which have practically accelerated the progress in the piezoelectric actuators, in particular, in high-power applications. The development strategy of the power supply (NF Corporation, Yokohama, Japan) is illustrated in Figure 1.6.

1.2 RESEARCH MOTIVATION FOR LOSS MECHANISMS

1.2.1 Device Miniaturization

"Micromachine" field started in the 1980s. We initially searched the moving principle of the various insects (this may be the original "bio-mimetic" engineering period). A large insect like a cockroach (~cm) moves with their six legs. With reducing the body size down to ~mm like a flea, it does not use six legs, but only two legs for jumping. For a much smaller body size ~100 μm like paramecium and ameba, they use the body cilia or surface movement. The key factor is a ratio (surface area $\propto L^2$)/(volume or weight $\propto L^3$) to change the moving fashion. With reducing the body weight, the friction force obtained by the small leg contact area is not sufficient for the body propulsion. Thus, the full surface friction is used for a micro-mechanism or micro-robot. The piezoelectric USMs using surface friction have been developed along with this development strategy in general. "The Micromachine Contests" were frequently organized in Japan during that period. The contestant machine needs to fulfill the following criteria: (1) size: less than $10 \times 10 \times 10\,mm^3$, (2) battle field: $1 \times 1\,m^2$ curved steel plate with a center hill height of 150 mm, (3) running route of grid-pattern narrow passages: in addition to "Start" and "Goal" 2 cm circles, the machine needs to pass additional three points accurately, (4) time trial race, basically. The propulsion mechanism did not matter;

EM rotary, linear or step motors, piezoelectric or magnetic inchworms, jumping/hopping machines, USMs, micro engines, and so on. The author's team (as one of the organizers) attended with one of the smallest four-wheel-drive vehicles ($7 \times 7 \times 7\,mm^3$) composed with two USMs, which could climb up and down even on a human index finger [Figure 1.7: unpublished]. However, due to the worldwide economic recession, this event stopped unfortunately in the middle of 1990s. Roughly after one generation (25 years), the second "Micromachine" boost has occurred recently (i.e., engineering "Renaissance"), because of the necessity in the medical (e.g., drug delivery and internal surgery with catheters), precision machinery, and optics areas.

As we discussed with Figure 1.1, conventional EM motor designs cannot exhibit sufficient energy efficiency in a micromotor category. Though the 30 W level EM motor still shows an efficiency ~30%, the motor in a wrist watch (less than 100 mW) has the efficiency less than 1%, in fact. The main reason of this exponential decrease in the efficiency is originated from the Joule heat in the thin coil wire (usually Cu wire). Thus, piezoelectric USMs, whose efficiency is insensitive to size (around 30%), are superior to the conventional EM devices when motors of millimeter size are required. Many EM machines failed to climb up the hill in the "Micromachine Contest". In a word, the piezo-motors are superior to the EM motors in the micromotor category less than 30 W.

With accelerating the commercialization of piezoelectric actuators and transducers for portable equipment applications, we identified the bottleneck of the piezoelectric devices; that is, significant heat generation limits the maximum power density. Though the problem is much smaller than the EM motors and transformers, the piezo-ceramic devices become "ceramic heater" with increasing the input/output power significantly. Figure 1.8 shows ring-dot-type piezoelectric transformers with Cu (left) and Ag (right) electrodes (a) and the heat generation profile under operation. A 20°C temperature rise is easily observed with increasing the vibration velocity or power level, which is called "maximum vibration velocity" in our community, taking into account the safety to human (e.g., human finger is burned on a 50°C electronic component!). The current maximum handling power of a well-known hard PZT is only around 10 W/cm^3. We desire 100 W/cm^3 or higher power density for further miniaturization of devices (such as piezo micro electro-mechanical system (MEMS) actuators) without losing efficiency. Taking into account the practically required minimum power levels, 30–100 mW for charging electricity into a battery, 10–20 mW for soaking blood from a

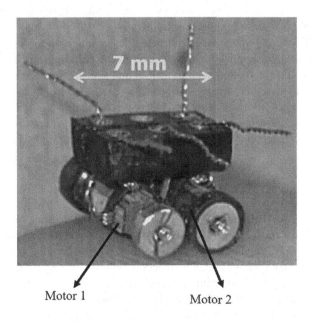

Motor 1 Motor 2

FIGURE 1.7 One of the smallest 4-wheel-drive vehicles with piezo-motors

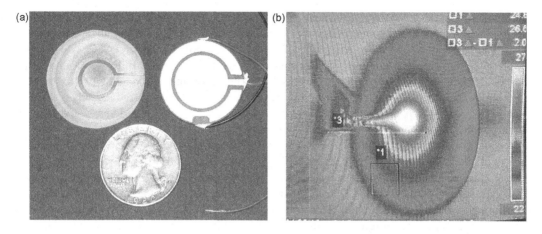

FIGURE 1.8 Ring-dot-type piezoelectric transformers (30 W level) (a), and the heat generation profile under operation (b).

human vessel, or 1–3 mW for sending electronic signal, minimum 1 mW handling is necessary. Accordingly, the MEMS devices with less than 1 µm thin PZT (current commercial base) films are useless from the actuator application viewpoint (1 µW level is usually called "sensing"). At least, 10–30 µm thick films should be used with minimum $3 \times 3\,\text{mm}^2$ device area, as long as we use the current existing materials.

Thus, aiming at wide commercialization, the main research focus seems to be shifting from the "real parameters" such as larger polarization and displacement, to the "imaginary parameters" such as polarization/displacement hysteresis, heat generation, and mechanical quality factor, which are originated from three loss factors (dielectric, elastic, and piezoelectric losses) in piezoelectrics. Reducing hysteresis and heat generation and increasing the mechanical quality factor to amplify the resonance displacement are the primary target. However, since the resonance displacement is provided by a product of the mechanical quality factor Q_M and the real parameter (piezoelectric constant) d, and the Q_M value is usually reduced with increasing d in the material, the development strategy of the material is highly complicated.

On the other hand, loss and hysteresis in piezoelectrics exhibit both merits and demerits in general. For positioning actuator applications, hysteresis in the field-induced strain causes a serious problem, and for resonance actuation such as USMs and transformers, loss generates significant heat in the piezoelectric material. Further, in consideration of the resonant strain amplified in proportion to a mechanical quality factor, low (intensive) mechanical loss (i.e., high Q_m) materials are preferred for USMs. In contrast, for force sensors and acoustic transducers, a low mechanical quality factor Q_m (which corresponds to high mechanical loss) is essential to widen a frequency range for receiving signals.

1.2.2 PREVIOUS STUDIES IN PIEZOELECTRIC LOSSES

As a historical "bible", K. H. Haerdtl wrote a review article on electrical and mechanical losses in ferroelectric ceramics in the early 1980s.[2] Losses are considered to consist of four portions: (1) domain wall motion, (2) fundamental lattice portion, which should also occur in domain-free monocrystals, (3) microstructure portion, which occurs typically in polycrystalline samples, and (4) conductivity portion in highly ohmic samples. However, in the typical piezoelectric ceramic case, the loss due to the domain wall motion exceeds the other three contributions significantly.

Interesting experimental results have been reported concerned with the relationship between dielectric and mechanical losses in piezo-ceramics.[3] These results are shown in Figure 1.9, in which

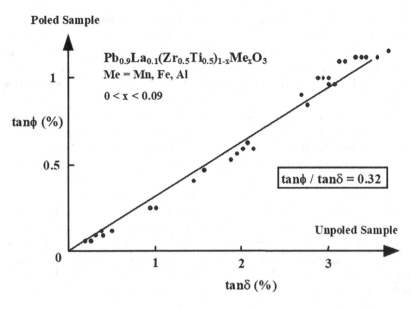

FIGURE 1.9   The correlation between mechanical loss, tan ϕ, and dielectric loss, tan δ, for a series of piezo-electric ceramics with composition $Pb_{0.9}La_{0.1}(Zr_{0.5}Ti_{0.5})_{1-x}Me_xO_3$, where Me represents the dopant ions Mn, Fe, or Al, and x covers a range between 0 and 0.09.[3] [f = 520 kHz].

mechanical loss, tan ϕ, is plotted with respect to the dielectric loss, tan δ, for a series of piezoelectric ceramics with compositions, $Pb_{0.9}La_{0.1}(Zr_{0.5}Ti_{0.5})_{1-x}Me_xO_3$, where Me represents the dopant ions Mn, Fe, or Al, and x covers a range between 0 and 0.09. The mechanical losses were measured on poled ceramic disks 0.4 mm thick and 5 mm in diameter, at their respective radial resonance frequencies (around 520 kHz). Since it was not possible to measure the dielectric losses on poled ceramics at the resonance frequency then, due to the occurrence of very strong electromechanical interactions, the measurements were made on depoled specimens at approximately the same frequencies (now possible; refer to Chapter 7). The Mn-doped ceramics showed dielectric losses below 1%, while Fe doping produced losses of 1%–2%, and Al doping increased the loss to over 3%. The linear relationship:

$$\tan\phi = 0.32\tan\delta \tag{1.7}$$

was found to apply to these nominal compositions.

Assuming that 90° domain wall movements constitute the primary loss mechanism for these materials, the proportionality constant was defined in terms of the relevant physical parameters, such that:

$$\tan\phi = \left[m\, x_S^2\, \varepsilon_0\, \varepsilon^X / P_S^2\, S^E \right]\tan\delta \tag{1.8}$$

where m is a constant assigned a value between 0.7 and 0.8, dependent on the crystalline structure, x_S the spontaneous strain, ε^X the dielectric constant at constant stress, P_S the spontaneous polarization, and s^E the elastic compliance at constant electric field. Taking into account the relationship:

$$x_S = Q\, P_S^2 \tag{1.9}$$

$$d = 2Q\, \varepsilon_0\, \varepsilon^X\, P_S, \tag{1.10}$$

where Q is the electrostrictive coefficient, the proportionality constant in Eq. (1.8) is found to be also proportional to the square of the electromechanical coupling factor, k^2. However, note that since the experiments were carried out on samples with different polarization states (i.e., unpoled and poled), the correlation is not significant from a theoretical viewpoint. In principle, only the dielectric loss associated with poled samples should be considered. Refer to Example Problem 9.6 on the dielectric loss difference.

Not many systematic and comprehensive studies of the loss mechanisms in piezoelectrics have been reported, particularly in high-electric field and high-power density ranges. Although part of the formulas of this textbook was described by T. Ikeda in his textbook[4], the piezoelectric losses, which have been found in our investigations to play an important role, were neglected then. Equivalent circuit analysis is a simple method to consider the electrical behavior, taking into account losses.[5–13] The Mason's equivalent circuit was derived from the piezoelectric constitutive and motion equations; therefore, the electromechanical conversion process is clearly described. Though there exist three fundamental losses in the piezoelectric materials (dielectric, elastic, and piezoelectric losses), equivalent circuits so far reported included the mechanical/elastic loss factor (and dielectric loss, occasionally) on the basis of the Mason's model.[14–16] D. Damjanovic proposed an equivalent circuit to present the influence of the piezoelectric loss[17], while the coupled loss formulas are insufficient to measure the losses in piezoelectric materials.

1.2.3 DILEMMA IN THE PRESENT IEEE STANDARD

The major problem is found even in the present IEEE Standard on Piezoelectricity, ANSI/IEEE Std. 176-1987[18], which many engineers are still using as the standard. Figure 1.10 shows an equivalent circuit for a k_{31}-type piezoelectric resonator, for example, proposed by the IEEE Standard on Piezoelectricity, ANSI/IEEE Std. 176-1987. Only one loss parameter R_1 results in the same mechanical quality factor Q_m for both resonance and antiresonance peaks in its admittance/impedance spectrum, as suggested in the following forms for the k_{31} mode:

$$s_{11}^E = 1/\left(4\rho f_r^2 l^2\right) \tag{1.11}$$

$$K_{31}^2 = \frac{\pi}{2}\frac{f_{ar}}{f_r}\tan\left(\frac{\pi}{2}\frac{\Delta f}{f_r}\right) \quad \text{where,}\ K_{31}^2 = \frac{k_{31}^2}{1-k_{31}^2} \tag{1.12}$$

$$Q_m = \frac{Y_m^{\max}\ \pi^2}{8\,\omega_0\ C\ K_{31}^2} \tag{1.13}$$

This standard does not include the terminology, "piezoelectric loss", nor discuss the difference of mechanical quality factors Q_A (at resonance) and Q_B (at antiresonance); that is, both are exactly the

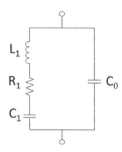

FIGURE 1.10 An equivalent circuit for a k_{31}-type piezoelectric resonator proposed by the IEEE Standard on Piezoelectricity, ANSI/IEEE Std. 176-1987.

same, against many of the practical experimental results and reports.[19,20] In order to solve this IEEE Standard dilemma, this textbook is devoted.

1.3 STRUCTURE OF THIS TEXTBOOK

Being motivated by various engineers' request, Uchino elucidates the comprehensive researches on piezoelectric losses in this book. You have just learned in Chapter 1 "Background of High-Power Piezoelectrics" on the necessity of loss mechanism clarification in order to develop high-energy density piezoelectric components for the portable electronic devices. This book focuses on high-power density "actuator" developments, not on low-power "sensor" applications. Started from the loss phenomenology, we propose typical high-power characterization methods. Then, current high-power piezoelectric materials and drive/operation schemes are introduced. In the latter part, semi-microscopic loss mechanisms are discussed, based on the domain dynamics models, followed by some know-how on commercialization. Note that the "high-power piezoelectrics" in this book does not mean high absolute power devices (kW, MW), but "high-power density", 10–100 MW/m^3 in practice.

For the reader who is not familiar with the fundamentals on piezoelectrics, I put a brief review of piezoelectrics in Chapter 2 "Foundations of Piezoelectrics", and Chapter 3 "Fundamentals of Losses". If you have confidence on these basic knowledge, the reader may skip these two chapters.

The main content starts from Chapter 4 "Phenomenological Approach to Piezoelectric Losses", in which the loss phenomenology in piezoelectrics, including three losses, dielectric, elastic, and piezoelectric losses, is introduced and interrelationships among these losses are discussed. Based on the piezoelectric constitutive and dynamic equations, Chapter 5 "Equivalent Circuits with Piezo Losses" expands the equivalent circuit approach in order to facilitate the experimental analysis easier. Actual heat generation analysis is discussed in piezoelectric materials for pseudo-DC and AC drive conditions in Chapter 6 "Heat Generation in Piezoelectrics". Heat generation at off-resonance is attributed mainly to intensive dielectric loss tan δ', while the heat generation at resonance is mainly originated from the intensive elastic loss tan ϕ'.

Chapter 7 "High-Power Piezo Characterization System" (HiPoCS™) introduces various characterization methodologies of loss factors chronologically, including continuous admittance/impedance spectrum measurement, burst/transient response method, thermal monitoring method, and so on. Mechanical quality factors (Q_A at resonance and Q_B at antiresonance) are primarily measured as a function of vibration velocity. Relating with Chapter 7, Chapter 8 "Drive Schemes of Piezoelectric Transducers" treats the piezo-device driving methods in order to minimize the losses and maximize the transducer efficiency, in particular, at off-resonance capacitive region, at resonance or antiresonance frequency, and at inductive region between the resonance and antiresonance frequencies. If the reader is a materials developer, you may initially skip Chapters 7 and 8.

High-power piezoelectric materials development, based on the loss mechanisms, are introduced in Chapter 9 "Loss Mechanisms in Piezoelectrics". Practical high-power "hard" PZT-based materials are initially described, which exhibit vibration velocities close to 1 m/s (rms), leading to the power density capability ten times that of the commercially available "hard" PZTs. Based on the macroscopic phenomenological study results, we consider microscopic crystallographic and semi-microscopic domain wall dynamics models to understand the experimental results. We propose an internal bias field model to explain the low-loss and high-power origin of these materials. The discussion is expanded to Pb-free piezoelectrics, which have been developed recently. The latter part of this chapter is devoted to the phenomenological handling of the domain wall dynamics, based on Ginzburg–Landau functional.

Chapter 10 "High-Power Piezoelectrics for Practical Applications" is devoted particularly to actual high-power piezo-device developers, since I disclosed a sort of know-how from the company executive viewpoint. You will learn how to suppress the device temperature rise and how to enhance the power density from the viewpoints of ML piezo-device configuration and ML internal electrode

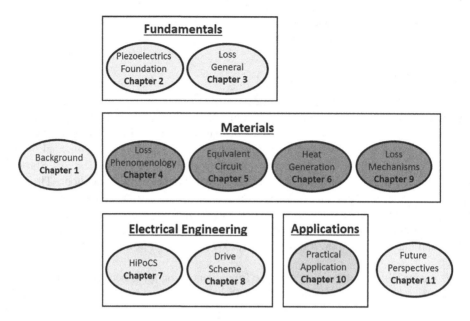

FIGURE 1.11   Textbook construction.

selection. Final Chapter 11 "Concluding Remarks and Future Perspective" aims to encourage the following generation researcher to study on the "imaginary" parameters in order to suppress the heat generation, to enhance the power density and efficiency for the further device miniaturization. The "structure of this textbook" is visualized in Figure 1.11.

The terminologies, "intensive" and "extensive" losses are introduced in my textbook, in the relation with "intensive" and "extensive" parameters in the phenomenology. Suppose a certain volume of material first, and imagine to cut it half. Extensive parameter (material's internal parameter such as displacement/strain x and total dipole moment/electric displacement D) depends on the volume of the material (i.e., becomes a half), while intensive parameter (externally controllable parameter such as force/stress X and voltage/electric field E) is independent on the volume of the material. These are not related with the "intrinsic" and "extrinsic" losses which were introduced to explain the loss contribution from the mono-domain single crystal state and from the others.[21] In this book, our discussion in piezoelectrics is focused primarily on the "extrinsic" losses, in particular, losses originated from domain wall dynamics. However, physical parameters of their performance such as permittivity and elastic compliance still differ in piezoelectric materials, depending on the boundary conditions; mechanically free or clamped and electrically short-circuit or open-circuit. These are distinguished as "intensive" or "extensive" parameters and their associated losses.

CHAPTER ESSENTIALS

1. Advantages of piezoelectric devices over EM types:
   a. More suitable to miniaturization – 1/10 smaller in volume and weight than EM types
   b. No EM noise generation – No magnetic shielding, more compact design
   c. Higher efficiency – Size-insensitive efficiency, effective in the power range lower than 30 W
   d. Non-flammable – Piezo-devices are safer for the overload or the short-circuit at the output terminal
2. Energy density of piezoelectric devices is still limited to 30 W/cm^3. Additional input electrical or mechanical energy is converted to heat generation.

3. The efficiency of the piezoelectric motors is size-insensitive (30% at 30 W motors). Only in low-profile level (<100 W), piezo-motors are superior to counter EM components.
4. Keys to high-power piezoelectric developments:
   a. Low-loss piezoelectric materials
   b. ML technology
   c. High-power supply/drive circuit
5. Three losses in piezoelectrics:
   a. Dielectric loss
   b. Mechanical/elastic loss
   c. Piezoelectric loss – This loss has not been focused previously, but it is the key to understanding frequency dependence of the mechanical quality factor and/or the transducer efficiency.

CHECK POINT

1. There is a copper wire (resistivity = 17 n Ωm) with 50 μm in diameter and 50 cm in length to make an EM motor coil. What is the resistance of this wire, which will generate heat via Joule heat when voltage is increased? 5 mΩ, 0.5 Ω, 5 Ω, or 50 Ω.
2. (T/F) The efficiency of a micro-EM motor in a wrist watch is less than 1%. True or false?
3. (T/F) Displacement vs. electric field hysteresis is one of the origins for heat generation in piezoelectric actuators. True or false?
4. (T/F) The IEEE Standard on Piezoelectricity, ANSI/IEEE Std. 176-1987 suggests that the mechanical quality factors at the resonance Q_A and at the antiresonance Q_B are the same. True or false?

CHAPTER PROBLEMS

We consider Joule heat problem in EM motors and transformers in the miniaturization process. The right-hand side figure illustrates a simple solenoid coil miniaturization by the factor of 1/2. We consider merely 1/2 scaling down all dimensions, solenoid radius R, length L, and coil wire radius r and length l by keeping the number of turn N. The solenoid inductance and the resistance are given by $L = \mu_0 \dfrac{N^2 \pi R^2}{L}$ and $\Omega = \rho \dfrac{l}{r^2}$. In order to keep the solenoid EM energy the same (for the same output work), estimate the Joule heat enhancement by this scale reduction.

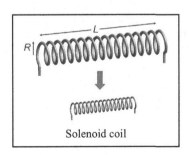

Solenoid coil

Hint: Solenoid electric energy under current I is given by $\dfrac{1}{2}LI^2$. Joule heat will increase by 2^2 times.

Taking into account the coil wire volume reduction by $1/2^3$, the temperature rise enhancement is by a factor of 2^5.

REFERENCES

1. Information Resources: Multiple year market research data from iRAP (Innovative Research & Products); Market Publishers; IDTechEx
2. K.H. Haerdtl, Ceram, *International.* **8**, 121–127 (1982).
3. P. Gerthsen, K.H. Haerdtl and N.A. Schmidt, *Journal of Applied Physics.* **51**, 1131–1134 (1980).
4. T. Ikeda, *Fundamentals of Piezoelectric Materials Science*, (Ohm Publication Co., Tokyo, 1984), p. 83.
5. J.M. Seo, H.W. Joo, H.K. Jung, Optimal design of piezoelectric transformer for high efficiency and high power density, *Sensors and Actuators a-Physical.* **121**, 520–526 (2005).
6. P. de Paco, O. Menendez, E. Corrales, Equivalent circuit modeling of coupled resonator filters, *IEEE Transactions on Ultrasonics Ferroelectrics and Frequency Control.* **55**, 2030–2037(2008).
7. S. Lin, An improved cymbal transducer with combined piezoelectric ceramic ring and metal ring, *Sensors and Actuators a-Physical.* **163**, 266–276 (2010).
8. K. Smyth, S.-G. Kim, Experiment and simulation validated analytical equivalent circuit model for piezoelectric micromachined ultrasonic transducers, *IEEE Transactions on Ultrasonics Ferroelectrics and Frequency Control.* **62**, 744–765 (2015).
9. T. Nakamoto, T. Moriizumi, A Theory of a Quartz Crystal Microbalance Based upon a Mason Equivalent Circuit, *Japanese Journal of Applied Physics Part 1-Regular Papers Short Notes & Review Papers.* **29**, 963–969 (1990).
10. O. Krauss, R. Gerlach, J. Fricke, Experimental and theoretical investigations of SiO2-aerogel matched piezo-transducers, *Ultrasonics.* **32**, 217–222 (1994).
11. Y. C. Chen, L. Wu, K.K. Chang, C.L. Huang, Analysis and simulation of stacked-segment electrome-chanical transducers with partial electrical excitation by PSPICE, *Japanese Journal of Applied Physics Part 1-Regular Papers Short Notes & Review Papers.* 36, 6550–6557 (1997).
12. F.J. Arnold, S.S. Muhlen, The influence of the thickness of non-piezoelectric pieces on pre-stressed piezotransducers, *Ultrasonics.* **41**, 191–196 (2003).
13. Y. Sun, Z. Li, Q. Li, An indirect method to measure the variation of elastic constant c(33) of piezoelec-tric ceramics shunted to circuit under thickness mode, *Sensors and Actuators a-Physical.* **218**, 105–1154 (2014).
14. M. Budinger, J.F. Rouchon, B. Nogarede, A mason type analysis of cylindrical ultrasonic micromotors, *European Physical Journal-Applied Physics.* **25**, 57–65 (2004).
15. H.-M. Zhou, C. Li, L.-M. Xuan, J. Wei, J.-X. Zhao, Equivalent circuit method research of resonant magnetoelectric characteristic in magnetoelectric laminate composites using nonlinear magnetostric-tive constitutive model, *Smart Materials and Structures.* **20**, 35001 (2011).
16. S. Lin, J. Xu, H. Cao, Analysis on the ring-type piezoelectric ceramic transformer in radial vibration, *IEEE Transactions on Power Electronics.* **31**, 5079–5088 (2016).
17. D. Damjanovic, An equivalent electric-circuit of a piezoelectric bar resonator with a large piezoelectric phase-angle, *Ferroelectrics.* **110**, 129–135 (1990).
18. ANSI/IEEE Std 176–1987, *IEEE Standard on Piezoelectricity*, (The Institute of Electrical and Electronics Engineers, New York, 1987), p. 56.
19. S. Hirose, M. Aoyagi, Y. Tomikawa, S. Takahashi, K. Uchino, High power characteristics at antireso-nance frequency of piezoelectric transducers, *Ultrasonics.* **34**, 213–217 (1996).
20. A. V. Mezheritsky, Efficiency of excitation of piezoceramic transducers at antiresonance frequency, *IEEE Transaction of Ultrasonics, Ferroelectrics, and Frequency Control.* **49**(4), 484–494 (2002).
21. N. Setter ed., *Piezoelectric Materials in Devices*, (Ceramics Laboratory, Lausanne, 2002).

2 Foundations of Piezoelectrics

ABSTRACT

This chapter is written to refresh the reader's knowledge on the theoretical background of piezoelectrics, then, practical materials, and typical applications. We cover (1) crystal structure and ferroelectricity, Dielectrics > Piezoelectrics > Pyroelectrics > Ferroelectrics; (2) origin of spontaneous polarization, the dipole coupling with the local field (i.e., Lorentz factor), in conjunction with nonlinear atomic elasticity; (3) origin of field-induced strain, which includes (a) *Inverse Piezoelectric Effect*: $x = d \cdot E$, (b) *Electrostriction*: $x = M \cdot E^2$, (c) *Domain reorientation*: strain hysteresis – this is mostly discussed in this textbook, and (d) *Phase Transition* (antiferroelectric↔ferroelectric): strain jump; (4) piezoelectric constitutive equations, $x = s^E X + d \cdot E$, $D = d \cdot X + \varepsilon_0 \varepsilon^X E$; (5) ferroelectric materials; and (6) applications of ferroelectrics.

2.1 SMART MATERIALS

Ferroelectrics belong to "smart materials". Let us start with the "smartness" of a material. Table 2.1 lists the various effects relating the input (electric field, magnetic field, stress, heat, and light) with the output (charge/current, magnetization, strain, temperature, and light). "Electrical conductor" and "elastic" materials, which generate current and strain outputs, respectively, for the input, voltage or stress (well-known phenomena, as Ohm's and Hooke's laws!), are sometimes called "trivial" materials. On the other hand, *pyroelectric* and *piezoelectric* materials, which generate an electric field with the input of heat and stress (unexpected phenomena!), respectively, are called "smart" materials. These off-diagonal couplings have corresponding converse effects, the *electrocaloric* and *converse-piezoelectric effects*, and both "sensing" and "actuating" functions can be realized in the same materials. Ferroelectric materials exhibit most of these effects with the exception of the magnetic phenomena. Thus, ferroelectrics are said to be very "smart" materials.

Magnetite (iron oxide, Fe_3O_4) and iron (Fe) were discovered 2,000 years ago to acquire attractive force between separated pieces at room temperature. These materials were named "ferromagnets" (because of the Fe inclusion originally) in an area of "magnetism" in the 19th century. Ferroelectricity was discovered in 1920 in Rochelle salt by Valasek.[1] Though Rochelle salt does not include Fe ion, because analogous spontaneous polarization (i.e., large hysteresis in the polarization P vs. electric field E curve) was observed, the terminology "ferroelectricity" was created ("segneto-electricity" was alternatively used when the author was young). In 1975, K. Aizu introduced a new terminology "ferroelasticity", for describing the spontaneous unit cell deformation of a crystal, then introduced the "ferroic" concept to discuss ferromagnetic, ferroelectric, and ferroelastic materials comprehensively.[2] Perovskite ferroelectrics mostly described in this book exhibit "ferroelasticity" simultaneously, owing to the *electrostrictive coupling* introduced in Section 2.3.2, and thus, the "piezoelectricity".

2.2 CRYSTAL STRUCTURE AND FERROELECTRICITY

In the so-called *dielectric* materials, the constituent atoms are considered to be ionized to a certain degree and are either positively or negatively charged. In such ionic crystals, when an electric field is applied, cations are attracted to the cathode and anions to the anode due to electrostatic interaction (i.e., ionic polarization). The electron clouds also deform, causing electronic dipoles. This phenomenon is known as *electric polarization* of the dielectric, and the polarization is expressed quantitatively as the sum of the electric dipoles per unit volume $[C/m^2]$. Figure 2.1 shows schematically

TABLE 2.1

Various Effects in Materials

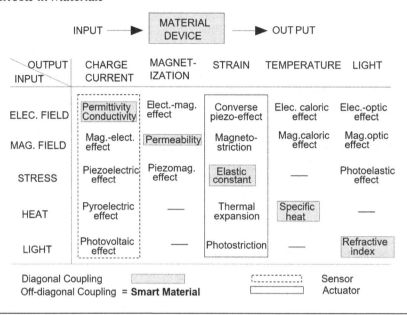

Diagonal Coupling
Off-diagonal Coupling = Smart Material   Sensor   Actuator

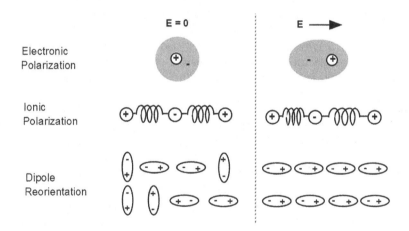

FIGURE 2.1 Microscopic origins of the electric polarization.

the origin of the electric polarization. There are three primary contributions: *electronic, ionic,* and *dipole reorientation-related.* The last one is found in a crystal which originally include spontaneous dipoles such as ice with a water molecule. The degree to which each mechanism contributes to the overall polarization of the material depends on the frequency of the applied field. Electronic polarization can follow alternating fields with frequencies up to THz–PHz (10^{12}–10^{15} cycle/second, higher than visible light wave) and ionic polarization responds up to GHz–THz (10^9–10^{12} cycle/sec, microwave region). Thus, you should understand that a famous relation between the relative permittivity ε and refractive index n:

$$\varepsilon = n^2 \tag{2.1}$$

is valid only when the applied electric field has a frequency on the order of THz or higher. Taking into account the actual material's refractive index < 3 in general, the permittivity at this high temperature should not be higher than 10. If you find higher permittivity, it is from the ionic polarization. Permanent dipole reorientation can follow only up to MHz–GHz (10^6–10^9 cycle/sec). You know a microwave oven in your kitchen is using 300 MHz or higher to resonate primarily water dipole molecule. This is why ferroelectric materials with permanent dipoles cannot be used for microwave dielectric materials; their permittivity is typically high at low frequencies (kHz) but decrease significantly with increasing applied electric field frequency. Frequency dependence of the total polarizability (or permittivity) is depicted in Figure 2.2.

Compared with air-filled capacitors, dielectric capacitors can store more electric charge due to the dielectric polarization P, as shown in Figure 2.3. The physical quantity corresponding to the stored electric charge per unit area is called the *electric displacement* D and is related to the electric field E by the following expression:

$$D = \varepsilon_0 E + P = \varepsilon \varepsilon_0 E. \tag{2.2}$$

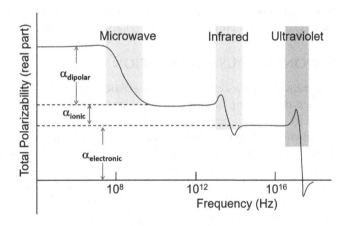

FIGURE 2.2 Frequency dependence of the polarizability (or permittivity).

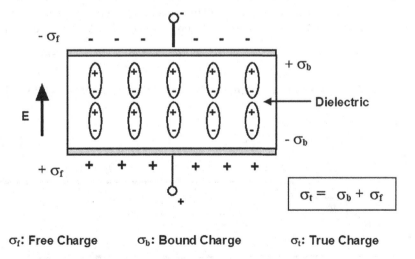

FIGURE 2.3 Charge accumulation in a dielectric capacitor.

Here, ε_0 is the vacuum permittivity (= 8.854×10^{-12} F/m), ε is the material's *relative permittivity* (also simply called permittivity or *dielectric constant*, and in general is a tensor property).

Depending on the crystal structure, the centers of the positive and negative charges may not coincide even without the application of an external electric field. Such crystals are said to possess a *spontaneous polarization* (or *pyroelectric*). When the spontaneous polarization of the dielectric can be reversed by an electric field, it is called *ferroelectric*.

Not every dielectric is a ferroelectric. Crystals can be classified into 32 point groups according to their crystallographic symmetry, and these point groups can be divided into two classes, one with a center of symmetry and the other without as indicated in Table 2.2. There are 21 point groups which do not have a center of symmetry. In crystals belonging to 20 of these point groups [point group (432) being the sole exception], positive and negative charges are generated on the crystal surfaces when appropriate stresses are applied. These materials are known as *piezoelectrics*.

Pyroelectricity is the phenomenon whereby, due to the temperature dependence of the spontaneous polarization, as the temperature of the crystal is changed, electric charges corresponding to the change of the spontaneous polarization appear on the surface of the crystal. Among the pyroelectric crystals, those whose spontaneous polarization can be reversed by an electric field (not exceeding the breakdown limit of the crystal) are called *ferroelectrics*. Thus, there is some experimental ambiguity in this definition; in establishing "ferroelectricity", it is necessary to apply an electric field to a pyroelectric material and experimentally ascertain the polarization reversal.

2.3   ORIGIN OF SPONTANEOUS POLARIZATION

Why is it that crystals which, from a consideration of elastic energy, should be stable by being nonpolar, still experience the shifting of cations and anions and become spontaneously polarized? The reason is briefly explained below. For simplicity, let us assume that dipole moments result from the displacement of one type of ion A (electric charge q) relative to the crystal lattice. Consider the case in which the polarization is caused by all the A ions being displaced equally in a lattice. This kind of ionic displacement can be expected through lattice vibrations at a finite temperature. Figure 2.4 shows some of the possible eigen lattice vibrations in a perovskite-like crystal. (a) shows an initial cubic (symmetrical) structure, (b) is a symmetrically elongated one (i.e., no polarization is generated), (c) has coherently shifted center cations (i.e., the right-ward polarization), and (d) exhibits an anti-polarized shift of the center cations (i.e., no net polarization). If one particular lattice vibration lowers the crystal energy, the ions will shift and stabilize the crystal structure so as to minimize the energy. Starting from the original cubic structure (a), if (b) is stabilized, only oxygen octahedra are distorted without generating dipole moments (acoustic mode). On the other hand, when (c) or (d) is stabilized, dipole moments are generated (optical mode).

TABLE 2.2
Crystallographic Classification According to Crystal Centrosymmetry and Polarity

Polarity	Symmetry	Crystal System										
		Cubic		Hexagonal		Tetragonal		Rhombohedral		Ortho-rhombic	Mono-clinic	Tri-clinic
Nonpolar (22)	Centro (11)	$m3m$	$m3$	$6/mmm$	$6/m$	$4/mmm$	$4/m$	$\overline{3}m$	$\overline{3}$	mmm	$2/m$	$\overline{1}$
	Noncentro (21)	432 $\overline{4}3m$	23	622 $\overline{6}m2$	$\overline{6}$	422 $\overline{4}2m$	$\overline{4}$	32		222		
Polar (Pyroelectric) (10)				$6mm$	6	$4mm$	4	$3m$	3	$mm2$	2 m	1

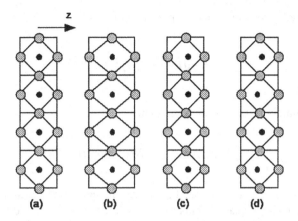

FIGURE 2.4 (a–d) Some possible eigen lattice vibration modes in a perovskite crystal.

The final stabilized states (c) and (d) correspond to ferroelectric and antiferroelectric states, respectively. If this particular mode becomes stabilized, with decreasing temperature, the vibration mode frequency decreases (i.e., *soft phonon mode*), and finally at a certain phase transition temperature, this frequency becomes zero.

It follows that, at any individual A ion site, there exists a local field from the surrounding polarization P, even if there is no external field. The concept of the local field is shown schematically in Figure 2.5. It can be shown that:

$$E^{loc} = E_0 + \sum_i \left[3(p_i \cdot r_i)r_i - r_i^2 p_i \right] / 4\pi\varepsilon_0 \cdot r_i^5. \quad \text{[Refer to Example Problem 2.1]}$$

$$= (\gamma/3\varepsilon_0)P \tag{2.3}$$

This local field is the driving force for the ion shift. Here γ is called the *Lorentz factor*. For an isotropic cubic system, it is known that $\gamma = 1$.[3] ε_0 is the permittivity of vacuum and is equal to 8.854×10^{-12} F/m. If the *ionic polarizability* of ion A is α, then the dipole moment of the unit cell of this crystal is:

$$\mu = (\alpha\gamma/3\varepsilon_0)P \tag{2.4}$$

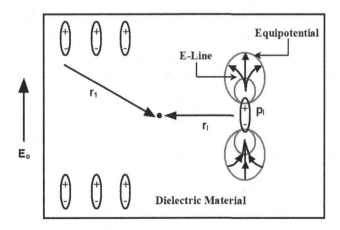

FIGURE 2.5 Concept of the local field. E^{loc} is given by $E^{loc} = E_0 + \sum_i \left[3(p_i \cdot r_i)r_i - r_i^2 p_i \right] / 4\pi\varepsilon_0 \cdot r_i^5$.

The energy of this dipole moment (dipole–dipole coupling) is

$$w_{\text{dip}} = -\mu E^{loc} = -\left(\alpha\gamma^2/9\varepsilon_0^2\right)P^2 \tag{2.5}$$

Defining N to be the number of atoms per unit volume:

$$W_{\text{dip}} = Nw_{\text{dip}} = -\left(N\alpha\gamma^2/9\varepsilon_0^2\right)P^2 \tag{2.6}$$

Furthermore, when the A ions are displaced from their nonpolar equilibrium positions, the elastic energy also increases. If the displacement is u, and the force constants k and k', then the increase of the elastic energy per unit volume can be expressed as:

$$W_{\text{elas}} = N\left[(k/2)u^2 + (k'/4)u^4\right] \tag{2.7}$$

Here, k' (> 0) is the higher-order force constant. It should be noted that in pyroelectrics (i.e., polar status), k' plays an important role in determining the magnitude of the dipole moment. Rewriting Eq. (2.7) with:

$$P = Nqu, \tag{2.8}$$

where q is the electric charge, and combining with Eq. (2.6), the total energy can be expressed as follows (see Figure 2.6):

$$
\begin{aligned}
W_{\text{tot}} &= W_{\text{dip}} + W_{\text{elas}} \\
&= \left[\left(k/2Nq^2\right) - \left(N\alpha\gamma^2/9\varepsilon_0^2\right)\right]P^2 + \left[k'/4N^3q^4\right]P^4
\end{aligned}
\tag{2.9}
$$

Taking the total energy minimum from this, one can see that if the coefficient of the harmonic term of the elastic energy is equal to or greater than the coefficient of the dipole–dipole coupling, then $P = 0$; the A ions are stable and remain at the nonpolar equilibrium positions. Otherwise, a shift from the equilibrium position $P^2 = \left[\left(2N\alpha\gamma^2/9\varepsilon_0^2\right) - \left(k/Nq^2\right)\right]/\left[k'/N^3q^4\right]$ is stable. Spontaneous polarization can occur more easily in perovskite-type crystal structure (e.g., barium titanate) due to a higher value of Lorentz factor $\gamma\,(= 10)^4$ than found for other crystal structures. Note also that the polarizability is changed with temperature, leading to the phase transition. Suppose that the ionic polarizability of ion A, α, increases with decreasing temperature, even if $\left[\left(k/2Nq^2\right) - \left(N\alpha\gamma^2/9\varepsilon_0^2\right)\right] > 0$ (paraelectric!) at a high temperature, this value may become negative with decreasing temperature, leading to a ferroelectric phase transition. Considering a first approximation, a linear relation of the α with temperature, that is, the well-known *Curie–Weiss law*:

$$\left[\left(k/2Nq^2\right) - \left(N\alpha\gamma^2/9\varepsilon_0^2\right)\right] = (T - T_0)/C \tag{2.10}$$

can be derived. Note also that the non-linear higher-order force constant k' is essential, which determines the spontaneous polarization value.

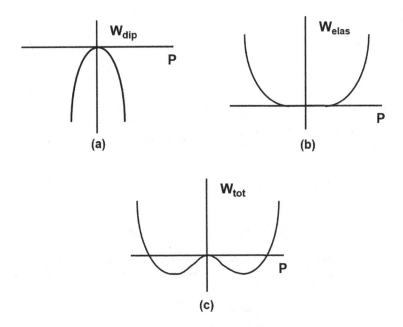

FIGURE 2.6 Energy explanation of the origin of spontaneous polarization. (a) Dipole interaction $W_{dip} = -\left(N\alpha\gamma^2/9\varepsilon_0^2\right)P^2$, (b) elastic energy $W_{elas} = \left(k/2Nq^2\right)P^2 + \left(k'/4N^3q^4\right)P^4$, and (c) total energy $W_{tot} = W_{dip} + W_{elas}$.

Example Problem 2.1

The electric potential arising from a point charge q, at a distance r from the charge is given by $V = \dfrac{1}{4\pi\varepsilon_0}\dfrac{q}{r}$, and the electric field is obtained as $\boldsymbol{E} = -grad(V) = \dfrac{1}{4\pi\varepsilon_0}\dfrac{q}{r^3}\boldsymbol{r}$, where $\boldsymbol{r}$ stands for the position vector along radius direction. Now, calculate the electric field distribution surrounding a dipole $(= q\cdot\boldsymbol{u})$, that is, charges $+q$ and $-q$ are situated in a short distance u, as illustrated in Figure 2.7.

Hint
The distance $+q$ to point P, r_1 and the distance $-q$ to point P, r_2 are expressed as:

$$r_1 = \sqrt{r^2 + \frac{u^2}{4} - ru\cdot\cos\theta}, \quad r_2 = \sqrt{r^2 + \frac{u^2}{4} + ru\cdot\cos\theta}$$

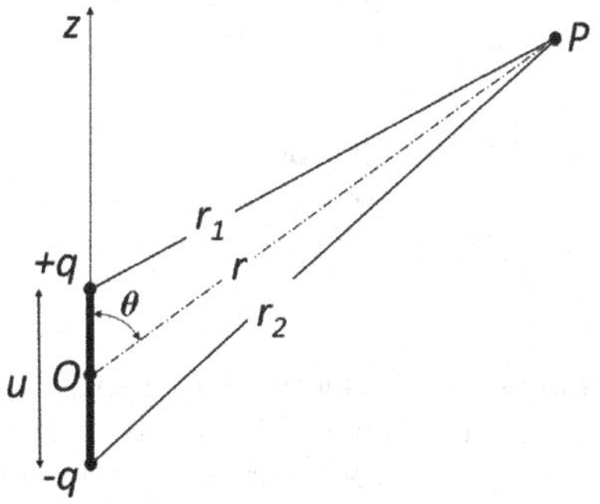

FIGURE 2.7 Electric potential calculation surrounding a dipole.

Solution

The electric potential at the point P in Figure 2.7 is expressed by superposing the potential from $+q$ and $-q$ as

$$V = \frac{1}{4\pi\varepsilon_0}\frac{q}{r_1} - \frac{1}{4\pi\varepsilon_0}\frac{q}{r_2} = \frac{q}{4\pi\varepsilon_0}\left(\frac{1}{\sqrt{r^2 + \dfrac{u^2}{4} - ru\cdot\cos\theta}} - \frac{1}{\sqrt{r^2 + \dfrac{u^2}{4} + ru\cdot\cos\theta}}\right). \tag{P2.1.1}$$

Since we are interested in the case in which $r \gg u$, we can rewrite Eq. (P2.1.1) in the approximate form

$$V = q\cdot u\,\cos\theta/4\pi\varepsilon_0 r^2 = p\cos\theta/4\pi\varepsilon_0 r^2 = \boldsymbol{p}\cdot\boldsymbol{r}/4\pi\varepsilon_0 r^3 \tag{P2.1.2}$$

The quantity $\boldsymbol{p}$ here is the dipole moment (product of the charge q and small distance u) with the vector direction from $-q$ to $+q$. Then, the electric field is obtained as $\boldsymbol{E} = -grad(V)$, and knowing the gradient operator in spherical coordinate as

$$grad(V) = \hat{\boldsymbol{r}}\frac{\partial V}{\partial r} + \hat{\boldsymbol{\theta}}\frac{1}{r}\frac{\partial V}{\partial\theta} + \hat{\boldsymbol{\varphi}}\frac{1}{r\sin\theta}\frac{\partial V}{\partial\varphi}, \tag{P2.1.3}$$

where $\hat{\boldsymbol{r}}, \hat{\boldsymbol{\theta}}, \hat{\boldsymbol{\varphi}}$ are unit vectors along r, θ, and φ directions, we obtain the following:

$$\begin{aligned}\boldsymbol{E} = -grad(V) &= \hat{\boldsymbol{r}}\left(2p\frac{\cos\theta}{4\pi\varepsilon_0 r^3}\right) + \hat{\boldsymbol{\theta}}\frac{1}{r}\left(p\frac{\sin\theta}{4\pi\varepsilon_0 r^2}\right)\\ &= \frac{1}{4\pi\varepsilon_0 r^3}\left[\hat{\boldsymbol{r}}(2p\cdot\cos\theta) + \hat{\boldsymbol{\theta}}(p\cdot\sin\theta)\right]\\ &= \left[3(\boldsymbol{p}\cdot\boldsymbol{r})\boldsymbol{r} - r^2\boldsymbol{p}\right]/4\pi\varepsilon_0 r^5\end{aligned} \tag{P2.1.4}$$

Above is the derivation process of Eq. (2.3). The electric potential and field contour are illustrated in Figure 2.5, which looks like a "dumbbell" shape.

Example Problem 2.2

Perovskite-type barium titanate, $BaTiO_3$, exhibits tetragonal symmetry at room temperature, and the ion shift is illustrated in Figure 2.8. The lattice constants are $c = 4.036$ (Å) and $a = 3.992$ (Å). Calculate the magnitude of the spontaneous polarization for barium titanate.

Hint
Calculate first the dipole moment μ by the product of ionic charge and the ionic displacement, then the polarization $P = N\cdot\mu$ (N: number of the dipole moments included in a unit volume). After calculating dipole moment, sum in a unit cell, then divide it with unit volume.

Solution

The dipole moment is defined to be the product of the magnitude of the ion charge and its displacement. The total dipole moment in a unit cell is calculated by summing the contributions of all the Ba^{2+}, Ti^{4+}, and O^{2-}-related dipoles. Each corner ion Ba contributes 1/8, each face ion O contributes 1/2, and the center Ti contributes 1. Note that four O^{2-} ions of the oxygen octahedron do not shift (this position is taken as the origin), leading to zero contribution to the dipole moment.

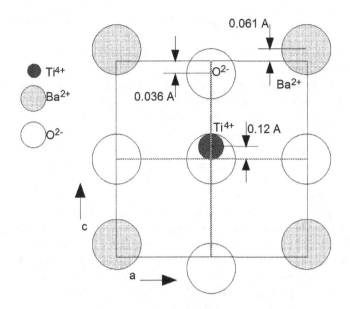

FIGURE 2.8 Ionic shifts in BaTiO$_3$ at room temperature.

$$p = 8[2e/8]\left[0.061\times10^{-10}\,(\text{m})\right] + [4e]\left[0.12\times10^{-10}\,(\text{m})\right] + 2[-2e/2]\left[-0.036\times10^{-10}\,(\text{m})\right]$$

$$= e\left[0.674\times10^{-10}\,(\text{m})\right]$$

$$= 1.08\times10^{-29}\,(\text{Cm}) \tag{P2.2.1}$$

where e is the fundamental charge: 1.602×10^{-19} (C).

Next, the unit cell volume is given by

$$v = a^2 c = (3.992)^2(4.036)\times10^{-30}\,(\text{m}^3)$$

$$= 64.3\times10^{-30}\,(\text{m}^3) \tag{P2.2.2}$$

The spontaneous polarization represents the number of (spontaneous) electric dipoles, p, per unit volume:

$$P_S = p/v = 1.08\times10^{-29}\,(\text{Cm})\big/64.3\times10^{-30}\,(\text{m}^3)$$

$$= 0.17\,(\text{C/m}^2)$$

This theoretical value of P_S is in reasonable agreement with the experimental value of 0.25 (C/m^2).

2.4 ORIGIN OF FIELD-INDUCED STRAIN

Solids, especially ceramics (inorganic materials), are relatively hard mechanically but still expand or contract depending on the change of the state parameters. The *strains* (defined as the *displacement* Δ*L/initial length L*) caused by temperature change and stress are known as thermal expansion and elastic deformation, respectively. In insulating materials, the application of an electric field can also cause deformation. This is called *electric field-induced strain*. We consider the microscopic origin in this section.

Generally speaking, the word "electrostriction" is used in a general sense to describe electric field-induced strain and, hence, frequently also implies the "converse piezoelectric effect". However, in solid-state theory, the converse piezoelectric effect is defined as a primary electromechanical coupling effect, that is, the strain is proportional to the electric field, while electrostriction is a secondary coupling in which the strain is proportional to the square of the electric field. Thus, strictly speaking, they should be distinguished. However, the piezoelectricity of a ferroelectric which has a centrosymmetric prototype (high temperature) phase such as in barium titanate is considered to originate from the electrostrictive interaction, and hence, the two effects are related. Section 2.5 handles the phenomenological approach on this issue. The above phenomena hold strictly under the assumptions that the object material is a mono-domain single crystal and that its state does not change under the application of an electric field. In a practical piezoelectric ceramic, additional strains accompanied by the reorientation of ferroelectric domains are also important.

Why a strain is induced by an electric field is explained herewith.[5] For simplicity, let us consider an ionic crystal such as NaCl. Figure 2.9a and b shows a 1D rigid-ion spring model of the crystal lattice. The springs represent equivalently the cohesive force resulting from the electrostatic Coulomb energy and the quantum-mechanical repulsive energy. Figure 2.9b shows the centrosymmetric case, whereas Figure 2.9a shows the more general non-centrosymmetric case. In Figure 2.9b, the springs joining the ions are all the same, whereas in Figure 2.9a, the springs joining the ions are different for the longer and shorter ionic distances, in other words, hard and soft springs existing alternately are important. Next, consider the state of the crystal lattice (a) under an applied electric field. The cations are drawn in the direction of the electric field and the anions in the opposite direction, leading to the relative change in the inter-ionic distance. Depending on the direction of the electric field, the soft spring expands or contracts more than the contraction or expansion of the hard spring, causing a strain x (a unit cell length change) in proportion to the electric field E. This is the *converse piezoelectric effect*. When expressed as

$$x = d \cdot E, \tag{2.11}$$

the proportionality constant d is called the *piezoelectric constant*.

On the other hand, in Figure 2.9b, the amounts of extension and contraction of the spring are nearly the same, and the distance between the two cations (lattice parameter) remains almost the same; hence, there is no strain. However, more precisely, ions are not connected by such idealized

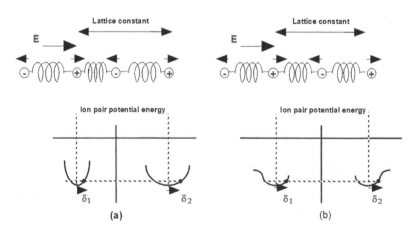

FIGURE 2.9   Microscopic explanation of the (a) piezostriction and (b) electrostriction.

springs (those are called *harmonic springs*, in which force (F) = spring constant (k) × displacement (Δ) holds). In most cases, the springs possess *anharmonicity* ($F = k_1\Delta - k_2\Delta^2$), that is, they are somewhat easy to extend but hard to contract. Such subtle differences in the displacement cause a change in the lattice parameter, producing a strain which is irrelevant to the direction of the applied electric field (+E or −E) and, hence, is an even function of the electric field. This is called the *electrostrictive effect* and can be expressed as

$$x = M \cdot E^2, \tag{2.12}$$

where M is the *electrostrictive constant*.

The 1D crystal pictured in Figure 2.9a also possesses a spontaneous bias of electrical charge or a spontaneous dipole moment. The total dipole moment per unit volume is called the *spontaneous polarization*. When a large reverse-bias electric field is applied to a crystal that has a spontaneous polarization in a particular polar direction, a transition "phase" is formed which is another stable crystal state in which the relative positions of the ions are reversed. In terms of an un-twinned single crystal, this is equivalent to rotating the crystal 180° about an axis perpendicular to its polar axis. This is also understood from the potential double minima in Figure 2.6. This transition, referred to as *polarization reversal*, also causes a remarkable change in strain. This particular class of substances is referred to as *ferroelectrics*, as mentioned in Section 2.2. Generally, what is actually observed as a field-induced strain is a complicated combination of the three basic effects just described.

Figure 2.10 shows typical strain curves for a piezoelectric lead zirconate titanate (PZT) based and an electrostrictive lead magnesium niobate (PMN)-based ceramic.[6] An almost linear strain curve in PZT becomes distorted and shows large hysteresis with increasing applied electric field level, which is due to the polarization reorientation. On the other hand, PMN does not exhibit hysteresis under an electric field cycle. However, the strain curve deviates from the quadratic relation (E^2) at a high electric field level.

We described the converse piezoelectric effect above. Then, what is the normal or *direct piezoelectric effect*? This is the phenomenon whereby charge (polarization = Coulomb per unit area) is generated by applying an external stress (force per unit area). Note that the same piezoelectric coefficient d is used as used in Eq. (2.11), in the relation

$$P = d \cdot X. \tag{2.13}$$

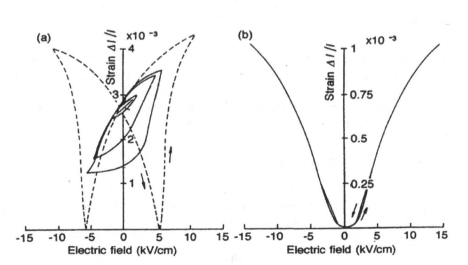

FIGURE 2.10 Typical strain curves for a piezoelectric PZT- based (a) and an electrostrictive PMN-based ceramic (b).

2.5 THEORY OF FERROELECTRIC PHENOMENOLOGY

We explained microscopic origins of spontaneous polarization and electric field-induced strain in ferroelectrics in Sections 2.3 and 2.4. We construct now macroscopic phenomenological treatments of these phenomena.

2.5.1 BACKGROUND OF PHENOMENOLOGY[7]

A thermodynamic phenomenological theory is discussed basically in the form of expansion series of the free energy as a function of the physical properties; polarization P and electric field E, temperature T and entropy S, and stress X and strain x [and if applicable, magnetic field H, and magnetization M, in a magnetic or magnetostrictive material]. In our ferroelectric discussion, the last parameters will be neglected. This derivation process is the "most frequently asked question" from the readers.

2.5.1.1 Polarization Expansion

The free energy can be expanded in general as follows, using the so-called *order parameter*, P:

$$F(P) = a_1 P + a_2 P^2 + a_3 P^3 + a_4 P^4 + a_5 P^5 + a_6 P^6 + \cdots$$

We assume that the free energy of the crystal should not change with polarization reversal ($P \rightarrow -P$). Otherwise, the charge or permittivity in the capacitance would be changed according to the capacitor orientation/upside down, which may cause serious practical problems in electronic equipment. From $F(P) = F(-P)$, the energy expansion series should not contain terms in *odd powers* of a vector component, P. Thus, the expansion series include only even powers of P:

$$F(P) = a_2 P^2 + a_4 P^4 + a_6 P^6 + \cdots$$

2.5.1.2 Temperature Expansion

Next, we take into account the expansion series in terms of P and temperature:

$$F(P,T) = a_2 P^2 + a_4 P^4 + a_6 P^6 + \cdots$$

$$+ b_1 T + b_2 T^2 + \cdots + c_1 T P^2 + \cdots$$

From $S = -\left(\dfrac{\partial F}{\partial T}\right) = -b_1$, and that a constant entropy is meaningless, we may take $b_1 = 0$. The term $b_2 T^2$ is a higher-order term to be neglected. Thus, we adopt only $c_1 T P^2$. Note that a possible term TP is omitted from the same reason $F(P) = F(-P)$ again. It is important to understand that the product TP^2 of the two parameters (P^2 and T) explains the *coupling effect*; that is, T change causes change in P to keep the same free energy (this effect is called "pyroelectricity") or E application causes T change (this is called "electrocaloric" *effect*). By combining $a_2 P^2$ and $c_1 T P^2$, we introduce

$$(1/2)\alpha = a_2 P^2 + c_1 T P^2 = (1/2)\left(\frac{T - T_0}{\varepsilon_0 C}\right) P^2$$

The above is a phenomenological expression in relation with Eq. (2.10). We also introduce the following notations by neglecting temperature dependence of the higher-order coefficients:

$$(1/4)\beta = a_4$$

$$(1/6)\gamma = a_6$$

2.5.1.3 Stress Expansion

Now we introduce the stress expansion series:

$$F(P,T,X) = (1/2)\alpha(T)P^2 + (1/4)\beta P^4 + (1/6)\gamma P^6 + \cdots$$

$$+ d_1 X + d_2 X^2 + \cdots + e_1 P^2 X + \cdots \quad \left[\alpha(T) = (T - T_0)/\varepsilon_0 C\right]$$

From $x = -\left(\dfrac{\partial F}{\partial X}\right) = -d_1$, and that constant strain is meaningless, we take $d_1 = 0$ (as the strain origin). $P^2 X$ is the fundamental electromechanical coupling (i.e., *electrostrictive coupling*), which explains the polarization generation under stress or strain generation under electric field. This argument is valid only when the spontaneous polarization exists in piezoelectric phase, and the polarization disappears in paraelectric phase. It is not valid in quartz, where the piezoelectric phase does not have the spontaneous polarization. (Actually, quartz is not a ferroelectric.) Odd power of stress X can remain in the free energy, because the crystal orientation/upside down does not change the sign of X (X is a tensor, not a vector; negative X means compressive stress). Introducing new notations $d_2 = -(1/2)\,s$ (elastic compliance) and $e_1 = -Q$ (*electrostrictive coefficient*), we finally obtain

$$G_1 = (1/2)\left[(T - T_0)/\varepsilon_0 C\right]P^2 + (1/4)\beta P^4 + (1/6)\gamma P^6 - (1/2)sX^2 - QP^2 X$$

The above free energy, G_1, is particularly called *elastic Gibbs energy*, which is the starting formula of most of the conventional textbooks.

2.5.2 LANDAU THEORY OF THE FERROELECTRIC PHASE TRANSITION

We assume that the Landau free energy F in 1D is represented in terms of polarization P (excluding stress terms) as:

$$F(P,T) = (1/2)\alpha P^2 + (1/4)\beta P^4 + (1/6)\gamma P^6 \tag{2.14}$$

Only the coefficient α is temperature dependent, and β and γ are supposed to be constants. The phenomenological formulation should be applied for the entire temperature range over which the crystal is in its paraelectric and ferroelectric states. It is noteworthy that the temperature dependence of above α is originated from the microscopic origin and temperature dependence of polarizability "α" in Eq. (2.10).

Because the spontaneous polarization should be zero in the paraelectric state, the free energy should be zero in the paraelectric phase at any temperatures above its phase transition temperature (i.e., *Curie temperature*). To stabilize the ferroelectric state, the free energy for a certain polarization P should be lower than "zero". Otherwise, the paraelectric state should be realized. Thus, at least, the coefficient α of the P^2 term must be negative for the polarized state to be stable; while in the paraelectric state, it must be positive passing through zero at some temperature T_0 (*Curie–Weiss temperature*). From this concept, as a linear relation:

$$\alpha = (T - T_0)/\varepsilon_0 C \tag{2.15}$$

where C is taken as a positive constant called the *Curie–Weiss constant* and T_0 is equal to or lower than the actual transition temperature T_C (*Curie temperature*). The temperature dependence of α is related (on a microscopic level) to the temperature dependence of the ionic polarizability "α" in Eq. (2.10), coupled with thermal expansion and other effects of anharmonic lattice interactions.

The equilibrium polarization, P, established with the application of an electric field, E, satisfies the condition:

$$(\partial F/\partial P) = E = \alpha P + \beta P^3 + \gamma P^5 \qquad (2.16)$$

With no electric field applied, Eq. (2.16) provides two cases:

$$P(\alpha + \beta P^2 + \gamma P^4) = 0 \qquad (2.17)$$

  i. $P = 0 \rightarrow$ This trivial solution corresponds to a paraelectric state.
  ii. $\alpha + \beta P^2 + \gamma P^4 = 0 \rightarrow$ This finite polarization solution corresponds to a ferroelectric state.

2.5.2.1   The Second-Order Transition

When β is positive, γ is often neglected because nothing special is added by this term. There are not many examples which show this "second-order" transition, but this description provides intuitive ideas on the phase transition because of its mathematical simplicity. Triglycine sulfate (TGS) is a close example of a ferroelectric exhibiting the second-order transition.

The polarization for zero applied field is obtained from Eq. (2.17) as

$$\alpha P_S + \beta P_S^3 = 0 \quad \left[\alpha = (T - T_0)/\varepsilon_0 C\right]$$

so that either $P_S = 0$ or

$$P_S^2 = -\alpha/\beta = (T_0 - T)/\beta\varepsilon_0 C. \qquad (2.18a)$$

For $T > T_0$, the unique solution $P_S = 0$ is obtained. For $T < T_0$, the minimum of the Landau free energy is obtained at:

$$P_S = \pm\sqrt{(T_0 - T)/(\beta\ \varepsilon_0\ C)} \qquad (2.18b)$$

The phase transition occurs at $T_C = T_0$, and the spontaneous polarization goes continuously to zero at this temperature; this is called a *second-order transition*.

The relative permittivity ε is calculated as:

$$1/\varepsilon = \varepsilon_0/(\partial P/\partial E) = \varepsilon_0\left(\alpha + 3\beta P^2\right) \qquad (2.19)$$

Though $1/(\partial P/\partial E) = (\partial E/\partial P)$ is not precise mathematically, it is an occasionally used "assumption" in corporation with Eq. (2.16).

Then, for $T > T_0$, $P = 0$:

$$\varepsilon = 1/\varepsilon_0\alpha = C/(T - T_0) \quad (T > T_0) \qquad (2.20)$$

Inverse permittivity is proportional to the temperature with the slope of $1/C$, which is well-known "Curie-Weiss Law".

For $T < T_0$, $P_S^2 = -\alpha/\beta = (T_0 - T)/\beta\varepsilon_0 C$:

$$1/\varepsilon = \varepsilon_0\left(\alpha + 3\beta P^2\right) = \varepsilon_0\left[\alpha + 3\beta(-\alpha/\beta)\right] = -2\varepsilon_0\alpha$$

$$\varepsilon = -1/2\varepsilon_0\alpha = C/\left[2(T_0 - T)\right] \quad (T < T_0) \qquad (2.21)$$

Note that the slope of the inverse permittivity is now $2/C$ in the ferroelectric phase, which is a half of the paraelectric phase slope.

2.5.2.2 The First-Order Transition

When β is negative in Eq. (2.14) and γ is taken to be positive (which is essential!), a first-order transition is described. The equilibrium condition for $E = 0$ in this case is expressed by:

$$\left[(T - T_0)/C\right]P_S + bP_S^3 + \gamma P_S^5 = 0 \tag{2.22}$$

and leads to either $P_S = 0$ or to a spontaneous polarization that is a root for:

$$\left[(T - T_0)/C\right] + \beta P_S^2 + \gamma P_S^4 = 0 \tag{2.23}$$

$$P_S^2 = \left[-\beta + \sqrt{\beta^2 - 4\gamma(T - T_0)/\varepsilon_0 C}\right]/2\gamma \tag{2.24}$$

The transition temperature T_C is obtained when the condition that the free energies of the paraelectric and ferroelectric phases are equal is applied such that $F = 0$ and:

$$\left[(T - T_0)/C\right] + (1/2)\beta P_S^2 + (1/3)\gamma P_S^4 = 0 \tag{2.25}$$

which allows us to write [refer to Example Problem 2.3 for the derivation process]:

$$T_C = T_0 + (3/16)\left(\beta^2 C/\gamma\right) \tag{2.26}$$

According to this equation, the Curie temperature T_C is slightly higher than the Curie–Weiss temperature, T_0, and that a discontinuity in the spontaneous polarization, P_S, as a function of temperature occurs at T_C. This is represented in Figure 2.11b where we also see the dielectric constant as a function of temperature peak at T_C. The inverse dielectric constant is also discontinuous at T_C, and when one extends the linear paraelectric portion of this curve back across the temperature axis, it intersects at the Curie–Weiss temperature, T_0. These are characteristic behaviors for the *first-order transition*. BaTiO$_3$ is an example of a ferroelectric that shows the first-order transition.

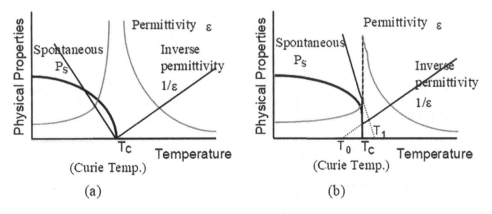

FIGURE 2.11 Phase transitions in a ferroelectric: (a) second-order and (b) first-order phase transitions.

Example Problem 2.3

Landau free energy for a *first-order phase transition* is expanded in terms of the order parameter P as

$$F(P,T) = (1/2)\alpha P^2 + (1/4)\beta P^4 + (1/6)\gamma P^6 \quad \left[\alpha = (T - T_0)/\varepsilon_0 C\right].$$

a. Is $P_S^2 = \left[-\beta - \sqrt{\beta^2 - 4\gamma(T - T_0)/\varepsilon_0 C}\right]/2\gamma$ not another root?

b. Verify the difference between the Curie T_C and Curie–Weiss T_0 temperatures as expressed by:

$$T_C = T_0 + (3/16)\left(\beta^2 \varepsilon_0 C/\gamma\right).$$

Hint

The potential minima are obtained from

$$(\partial F/\partial P) = E = \alpha P + \beta P^3 + \gamma P^5 = 0. \tag{P2.3.1}$$

This equation is valid for any temperature below and above Curie temperature with zero external field. There are generally three minima including $P = 0$ ($F = 0$). Other minima can be solved from $\alpha + \beta P^2 + \gamma P^4 = 0$. At the Curie temperature, the free energy at the non-zero polarization must be equal to the free energy of the paraelectric state; that is, zero ($F = 0$). Thus, we obtain another condition:

$$F = (1/2)\alpha P^2 + (1/4)\beta P^4 + (1/6)\gamma P^6 = 0. \tag{P2.3.2}$$

This equation is only valid at the phase transition temperature T_C.

Solution

a. Equation (P2.3.1) provides two roots, $P_S^2 = \left[-\beta \pm \sqrt{\beta^2 - 4\gamma(T - T_0)/\varepsilon_0 C}\right]/2\gamma$, in general. Since $(T - T_0) < 0$ and $\gamma > 0$, $\sqrt{\beta^2 - 4\gamma(T - T_0)/\varepsilon_0 C} > \sqrt{\beta^2} = -\beta$ (or $|\beta|$). Note that $\beta < 0$. Thus, $\left[-\beta - \sqrt{\beta^2 - 4\gamma(T - T_0)/\varepsilon_0 C}\right]/2\gamma < 0$, which is contradictory with a root of P_S^2. The spontaneous polarization should be positive or negative, but still be a real number, never be an imaginary.

b. Equations (P2.3.1) and (P2.3.2) are reduced for non-zero polarizations to

$$\alpha + \beta P^2 + \gamma P^4 = 0, \tag{P2.3.3}$$

$$\alpha + (1/2)\beta P^2 + (1/3)\gamma P^4 = 0. \tag{P2.3.4}$$

Note that Eq. (P2.3.3) is valid for all temperatures below T_C, while Eq. (P2.3.4) is only valid at $T = T_C$. Eliminating the higher-order P^4 term from these two equations [$3 \times$ (P2.3.4) − (P2.3.3)],

$$\left[(3/2) - 1\right]\beta P^2 + [3 - 1]\alpha = 0 \rightarrow P^2 = -4\alpha/\beta. \tag{P2.3.5}$$

Then, we obtain the following equation from Eq. (P2.3.3):

$$\alpha + \beta(-4\alpha/\beta) + \gamma(-4\alpha/\beta)^2 = 0.$$

or,

$$-3\alpha + \gamma \times 16\alpha^2/\beta^2 = 0.$$

Taking into account $T = T_C$, $\alpha_{T=T_c} = (T_C - T_0)/\varepsilon_0 C$ and $16\gamma\alpha/\beta^2 = 3$, the Curie temperature is calculated as

$$T_C = T_0 + (3/16)(\beta^2 \varepsilon_0 C/\gamma). \tag{P2.3.6}$$

Since $\beta^2 > 0$, $\gamma > 0$, the Curie temperature T_C should be higher than the Curie–Weiss temperature, T_0.

Example Problem 2.4

Draw the polarization vs. electric field hysteresis curve at a certain temperature for a ferroelectric with the second-order phase transition as

$$F(P,T) = (1/2)\alpha P^2 + (1/4)\beta P^4 \quad [\alpha = (T - T_0)/\varepsilon_0 C].$$

a. Obtain the spontaneous polarization P_S at a temperature T ($< T_C$) at $E = 0$.
b. Obtain the permittivity ε at a temperature T ($< T_C$) at $E = 0$.
c. Supposing that all positive polarization P reverse the direction at once (without making any small domain structures), obtain the macroscopic coercive field at a temperature T ($< T_C$).
d. Draw the polarization vs. electric field hysteresis curve at a temperature T ($< T_C$).

Solution

The potential minima are obtained from

$$(\partial F/\partial P) = E = \alpha P + \beta P^3. \tag{P2.4.1}$$

By putting $y_1 = \alpha P + \beta P^3$ and $y_2 = E$, visual geometrical solution technique can be used, as illustrated in Figure 2.12 as a function of polarization P; that is, the intersects of these two curves ($y_1 = y_2$) provide the solution points: only one intersect exists for $T > T_C$, while for $T < T_C$, there are three intersects.

a. Spontaneous polarization is obtained from Eq. (P2.4.1):

$$\alpha P + \beta P^3 = E = 0$$

Thus,

$$P_S^2 = -\alpha/\beta,$$

or

$$P_S = \sqrt{-\alpha/\beta}. \tag{P2.4.2}$$

b. Relative permittivity ε is obtained from $(1/\varepsilon_0)\dfrac{\partial P}{\partial E}$, and from Eq. (P2.4.1),

$$\varepsilon = 1/\varepsilon_0 \left(\frac{\partial E}{\partial P}\right) = 1/\varepsilon_0 (\alpha + 3\beta P^2)$$

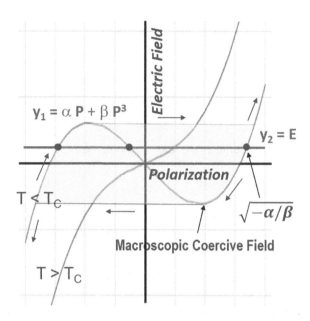

FIGURE 2.12 Polarization vs. electric field hysteresis curve, obtained from a graphic technique.

From Eq. (P2.4.2), relative permittivity is given by

$$\varepsilon = 1/\varepsilon_0 \, (-2\alpha) = C/2(T_0 - T) \tag{P2.4.3}$$

c. The macroscopic coercive field is obtained from the maximum/minimum point of y_1 curve:

$$\frac{\partial y_1}{\partial P} = 0 \rightarrow \alpha + 3\beta P^2 = 0 \rightarrow P = \sqrt{-\alpha/3\beta}$$

Since the coercive field is obtained from y_1 (max point),

$$y_1 = \alpha P + \beta P^3 = \sqrt{-\alpha/3\beta} \left[\alpha + \beta(-\alpha/3\beta) \right]$$
$$= \sqrt{-4\alpha^3/27\beta} \tag{P2.4.4}$$

d. The shadowed area in Figure 2.12 shows the polarization vs. electric field hysteresis loop, when we assume the macroscopic all polarization one-time reversal. Note that the above macroscopic phenomenology is based on the mono-domain uniform single crystal status. In an actual specimen with a multiple-domain configuration, the coercive field is much smaller than Eq. (P2.4.4). The modified phenomenology will be discussed in Chapter 9.

2.5.3 Phenomenological Description of Electrostriction

In a ferroelectric state whose prototype phase (high-temperature paraelectric phase) is centrosymmetric and non-piezoelectric, the piezoelectric coupling term PX is omitted and only the electrostrictive coupling term $P^2 \cdot X$ is introduced. The theories for electrostriction in ferroelectrics were formulated in the 1950s by Devonshire[8,9] and Kay.[10] Let us assume that the elastic Gibbs energy should be expanded in a 1D form:

$$G_1(P, X, T) = (1/2)\alpha P^2 + (1/4)\beta P^4 + (1/6)\gamma P^6$$

$$-(1/2)sX^2 - QP^2X, \quad \left(\alpha = (T - T_0)/\varepsilon_0 C\right) \tag{2.27}$$

where P, X, and T are the polarization, stress, and temperature, respectively, and s and Q are called the elastic compliance and the electrostrictive coefficient, respectively. This leads to Eqs. (2.28) and (2.29).

$$E = \left(\partial G_1/\partial P\right) = \alpha P + \beta P^3 + \gamma P^5 - 2Q PX \tag{2.28}$$

$$x = -\left(\partial G_1/\partial X\right) = sX + QP^2 \tag{2.29}$$

2.5.3.1   Case I: $X = 0$

When the external stress is zero, the following equations are derived:

$$E = \alpha P + \beta P^3 + \gamma P^5 \tag{2.30}$$

$$x = QP^2 \tag{2.31}$$

$$1/\varepsilon_0\varepsilon = \alpha + 3\beta P^2 + 5\gamma P^4 \tag{2.32}$$

If the external electric field is equal to zero ($E = 0$), two different states are derived; $P = 0$ and $P^2 = \left(-\beta + \sqrt{\beta^2 - 4\alpha\gamma}\right)/2\gamma$.

i. Paraelectric phase: $P_S = 0$ or $P = \varepsilon_0 \varepsilon E$ (under small E)

$$\text{Permittivity}: \varepsilon = C/(T - T_0) \quad \text{(Curie-Weiss law)} \tag{2.33}$$

$$\text{Electrostriction}: x = Q\varepsilon_0^2\varepsilon^2 E^2 \tag{2.34}$$

The previously mentioned electrostrictive coefficient M in Eq. (2.12) is related to the electrostrictive Q coefficient through

$$M = Q\varepsilon_0^2\varepsilon^2 \tag{2.35}$$

ii. Ferroelectric phase: $P_S^2 = \left(-\beta + \sqrt{\beta^2 - 4\alpha\gamma}/2\gamma\right)$ or $P = P_S + \varepsilon_0\varepsilon E$ (under small E)

$$x = Q(P_S + \varepsilon_0\varepsilon E)^2 = Q P_S^2 + 2\varepsilon_0\varepsilon Q P_S E + Q\varepsilon_0^2\varepsilon^2 E^2 \tag{2.36}$$

where we define the spontaneous strain x_S and the piezoelectric constant d as:

$$\text{Spontaneous strain}: x_S = QP_S^2 \tag{2.37}$$

$$\text{Piezoelectric constant}: d = 2\varepsilon_0\varepsilon Q P_S \tag{2.38}$$

We neglect the third electrostriction term here because of small E.

We see from Eq. (2.38) that piezoelectricity is equivalent to the *electrostrictive phenomenon biased by the spontaneous polarization*. The temperature dependences of the spontaneous strain and the piezoelectric constant are plotted in Figure 2.13. [Refer to Example Problem 2.6.]

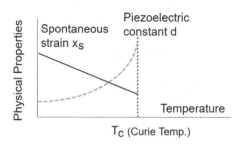

FIGURE 2.13   Temperature dependence of the spontaneous strain and the piezoelectric constant.

2.5.3.2   Case II: $X \neq 0$

When a hydrostatic pressure p ($X = -p$) is applied, the inverse permittivity is changed in proportion to p:

$$1/\varepsilon_0\varepsilon = \alpha + 3\beta P_S^2 + 5\gamma P_S^4 + 2Q\,p \text{ (Ferroelectric state)}$$

$$\alpha + 2Q\,p = \left(T - T_0 + 2Q\varepsilon_0 C\,p\right)/\left(\varepsilon_0 C\right) \text{ (Paraelectric state)}$$

(2.39)

Here, electrostrictive constant Q should be Q_h, which is defined by $(Q_{33} + 2Q_{13})$. Therefore, the pressure dependence of the Curie–Weiss temperature T_0 or the transition temperature T_C is derived as follows:

$$\left(\partial T_0/\partial p\right) = \left(\partial T_C/\partial p\right) = -2Q\varepsilon_0 C$$

(2.40)

In general, the ferroelectric Curie temperature is decreased with increasing hydrostatic pressure (i.e., $Q_h > 0$). More precisely, ~50°C temperature decrease per 1 GPa hydrostatic pressure increase in perovskite ferroelectrics because the product QC is constant in most of the perovskites.[5]

Example Problem 2.5

Barium titanate has $d_{33} = 320 \times 10^{-12}$ C/N, $\varepsilon_c\,(= \varepsilon_3) = 800$ and $Q_{33} = 0.11\,\mathrm{m^4/C^{-2}}$ at room temperature. Estimate the spontaneous polarization P_S.

Solution

Let us use the relationship:

$$d_{33} = 2\,\varepsilon_0\varepsilon_3 Q_{33} P_S.$$

(P2.5.1)

$$P_S = d_{33}/2\,\varepsilon_0\varepsilon_3 Q_{33}$$

$$= 320 \times 10^{-12}[\text{C/N}]/\left\{2 \times 8.854 \times 10^{-12}[\text{F/m}] \times 800 \times 0.11\left[\mathrm{m^4/C^2}\right]\right\}$$

$$= 0.21\left[\text{C/m}^2\right]$$

(P2.5.2)

Example Problem 2.6

In the case of a second-order phase transition, the elastic Gibbs energy is expanded in a 1D form as follows:

$$G_1(P,X,T) = (1/2)\alpha P^2 + (1/4)\beta P^4 - (1/2)sX^2 - QP^2 X,$$

(P2.6.1)

where only the coefficient α is dependent on temperature, $\alpha = (T - T_0)/\varepsilon_0 C$. Obtain the dielectric constant, spontaneous polarization, spontaneous strain, and piezoelectric constant as a function of temperature T (under stress-free condition).

Solution

$$E = \left(\partial G_1/\partial P\right) = \alpha P + \beta P^3 - 2QPX \tag{P2.6.2}$$

$$x = -\left(\partial G_1/\partial X\right) = sX + QP^2 \tag{P2.6.3}$$

When an external stress is zero, we can deduce the three characteristic equations:

$$E = \alpha P + \beta P^3 \tag{P2.6.4}$$

$$x = QP^2 \tag{P2.6.5}$$

$$1/\varepsilon_0\varepsilon = (\partial E/\partial P) = \alpha + 3\beta P^2 \tag{P2.6.6}$$

By setting $E = 0$ initially, we obtain the following two stable states: $P_S^2 = 0$ or $-\alpha/\beta$

i. Paraelectric phase $-T > T_0 - P_S = 0$

$$1/\varepsilon_0\varepsilon = \alpha, \text{ then } \varepsilon = C/(T - T_0) \text{ (Curie-Weiss Law)} \tag{P2.6.7}$$

ii. Ferroelectric phase $-T < T_0 - P_S = \pm\sqrt{(T_0-T)/\varepsilon_0 C\beta}$ $\qquad$ (P2.6.8)

$$1/\varepsilon_0\varepsilon = \alpha + 3\beta P^2 = -2\alpha, \text{ then } \varepsilon = C/2(T_0 - T) \tag{P2.6.9}$$

$$x_S = QP_S^2 = Q(T_0 - T)/\varepsilon_0 C\beta \tag{P2.6.10}$$

From Eqs. (P2.6.8) and (P2.6.9), the piezoelectric constant is obtained as

$$d = 2\varepsilon_0\varepsilon Q P_S$$
$$= Q\sqrt{\varepsilon_0 C/\beta}(T_0 - T)^{-1/2} \tag{P2.6.11}$$

Referring to Figures 2.11b and 2.13, which shows the case of the first-order phase transition, draw by yourself the temperature dependence of inverse permittivity, spontaneous polarization, spontaneous strain and piezoelectric constant.

2.5.4 Converse Effects of Electrostriction

So far we have discussed the electric field-induced strains, that is, piezoelectric strain (*converse piezoelectric effect*, $x = d\,E$) and electrostriction (*electrostrictive effect*, $x = M\,E^2$). Let us consider here the converse effect, that is, the material's response to an external stress, which is applicable to sensors. Actually, for the piezoelectricity, this is the "direct piezoelectric effect", that is, the *increase of the spontaneous polarization by an external stress*, and expressed as

$$\Delta P = d \cdot X. \tag{2.41}$$

To the contrary, since an electrostrictive material does not have a spontaneous polarization, it does not generate any charge under stress but does exhibit a change in permittivity (first derivative of polarization P in terms of field E) [see Eq. (2.39)]:

$$\Delta\left(1/\varepsilon_0\varepsilon\right) = -2QX \qquad (2.42)$$

This is the *converse electrostrictive effect*. The converse electrostrictive effect, the stress dependence of the permittivity, is used in stress sensors.[11] A bimorph structure which subtracts the static capacitances of two dielectric ceramic plates can provide superior stress sensitivity and temperature stability. The capacitance changes of the top and bottom plates have opposite signs for uniaxial transversal stress and the same sign for temperature change. The response speed is limited by the capacitance measuring frequency up to about 1 kHz. Unlike piezoelectric sensors, electrostrictive sensors are effective in the low-frequency range, especially pseudo-DC.

2.5.5 TEMPERATURE DEPENDENCE OF ELECTROSTRICTION

We have treated the electrostrictive coefficient as a temperature-independent constant in the last section. How is the actual situation? Several expressions for the electrostrictive coefficient Q have been proposed so far. From the data obtained by independent experimental methods such as

1. Electric field-induced strain in the paraelectric phase,
2. Spontaneous polarization and spontaneous strain (x-ray diffraction) in the ferroelectric phase,
3. d constants from the field-induced strain in the ferroelectric phase or from piezoelectric resonance, with permittivity, spontaneous polarization,
4. Pressure dependence of permittivity in the paraelectric phase,

nearly equal values of Q were obtained. Figure 2.14 shows the temperature dependence of the electrostrictive coefficients Q_{33} and Q_{31} observed for a complex perovskite $Pb(Mg_{1/3}Nb_{2/3})O_3$ single-crystal sample, whose Curie temperature is near 0°C.[12] It is seen that there is no significant anomaly in the electrostrictive coefficient Q through the temperature range in which the paraelectric to ferroelectric phase transition occurs and piezoelectricity appears. Q is verified to be almost temperature independent. Note, however, that since the electrostriction under an applied field E is obtained by $Q\,\varepsilon_0^2\varepsilon^2E^2$, rather significant temperature dependence is observed from the permittivity change with temperature.

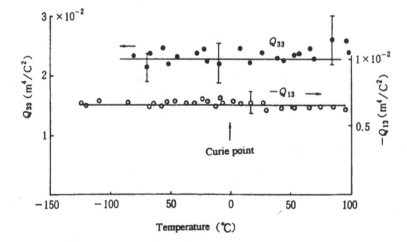

FIGURE 2.14 Temperature dependence of the electrostrictive constants Q_{33} and Q_{31} measured in a single-crystal $Pb(Mg_{1/3}Nb_{2/3})O_3$.

2.5.6 Electromechanical Coupling Factor

2.5.6.1 Piezoelectric Constitutive Equations

We introduce the concept of electromechanical coupling factor k in this section. We start from the fundamental piezoelectric constitutive equations (1D expression) without including losses:

$$D = \varepsilon_0 \varepsilon^X E + d\,X \qquad (2.43)$$

$$x = d\,E + s^E X, \qquad (2.44)$$

where D is electric displacement (which is almost equal to polarization P, as long as the relative permittivity ε is large), x strain, and these *extensive* parameters (material's properties) are controlled by the *intensive* parameters (externally controllable parameters), electric field E and stress X. Note that the spontaneous polarization and strain do not appear explicitly in the constitutive equations, but implicitly included in piezoelectric constant; that is, the piezoelectric constitutive equations are merely to discuss the deviation of D and x under the external parameter E and X in the linear relations. The proportional constant ε^X is the relative permittivity under *stress-free* condition (ε_0 is the vacuum permittivity 8.854×10^{-12} F/m), s^E elastic compliance under *electric field-free* (or *short-circuit*) condition, and d is the piezoelectric constant. It is worth to note that the d's appearing in Eqs. (2.43) and (2.44) should be thermodynamically the same. The verification process is beyond this textbook level.

2.5.6.2 Electromechanical Coupling Factor

The term, *electromechanical coupling factor k*, is defined as the square value k^2 be the ratio of the converted energy over the input energy: when electric to mechanical

$$k^2 = \left(\text{Stored mechanical energy/Input electrical energy}\right), \qquad (2.45)$$

and when mechanical to electric,

$$k^2 = \left(\text{Stored electrical energy/Input mechanical energy}\right) \qquad (2.46)$$

Let us calculate Eq. (2.45) first in a static (pseudo-DC) condition, when an external electric field E_3 is applied to a piezoelectric material under pseudo static condition. See Figure 2.15a first, when we apply electric field on the top and bottom electrodes under stress-free condition ($X = 0$). Input electric energy must be equal to $(1/2)\,\varepsilon_0 \varepsilon_3^X E_3^2$ from Eq. (2.43), and the strain generated by E_3 should be $d_{33}E_3$ from Eq. (2.44). Since the converted and stored mechanical energy is obtained as $(1/2\,s_{33}^E)\,x_3^2$ (note that the superscript E of the elastic compliance is E-constant condition, because the output impedance of the voltage supply is very small), we obtain:

$$k_{33}^2 = \left[(1/2)(d_{33}E_3)^2 / s_{33}^E\right]\left[(1/2)\varepsilon_0 \varepsilon_3^X E_3^2\right]$$

$$= d_{33}^2 / \varepsilon_0 \varepsilon_3^X \cdot s_{33}^E. \qquad (2.47a)$$

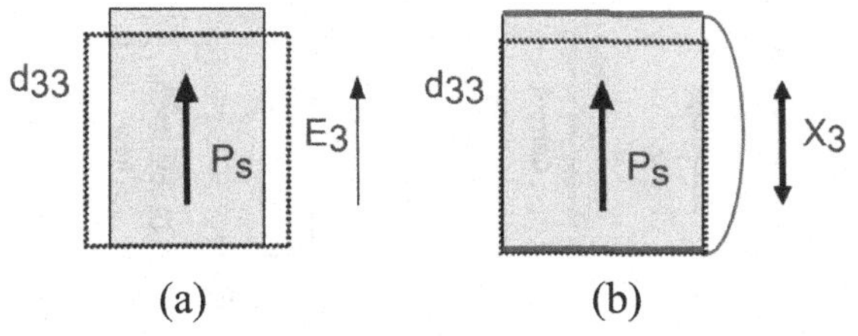

FIGURE 2.15 Calculation models of electromechanical coupling factor k for (a) electric input and (b) stress input.

Let us now consider Eq. (2.46), when an external stress X_3 is applied to a piezoelectric material. Refer to Figure 2.15b. Under short-circuit condition ($E_3 = 0$), the input mechanical energy must be equal to $(1/2)\, s_{33}{}^E X_3{}^2$ from Eq. (2.44), and the electric displacement D_3 (or polarization P_3) generated by X_3 should be equal to $d_{33}X_3$ from Eq. (2.57). This D_3 can be obtained by integrating the short-circuit current (E-constant condition) in terms of time through the electric lead. Since the converted and stored electric energy is obtained as $(1/2\, \varepsilon_0 \varepsilon_3^X)\, D_3^2$, we obtain:

$$k_{33}^2 = \left[\left(1/2\, \varepsilon_0 \varepsilon_3^X \right) \left(d_{33} X_3 \right)^2 \right] \Big/ \left[(1/2)\, s_3^E X_3^2 \right]$$

$$= d_{33}^2 \big/ \varepsilon_0 \varepsilon_3^X \cdot s_{33}^E . \tag{2.47b}$$

It is essential to understand that the electromechanical coupling factor k (or k^2, which has a physical meaning of energy transduction/conversion rate) can be exactly the same for both converse (2.47(a)) and direct (2.47(b)) piezoelectric effects. The conditions under constant X (free stress) or constant E (short-circuit) are considered to be non-constrained. This k_{33} is called as "static" or "material's" electromechanical coupling factor.

Figure 2.16 illustrates further discussion on two models of electromechanical coupling factor k for (a) short-circuit and (b) open-circuit conditions, by including the concept of "depolarization field". The former is the same as Figure 2.15b, and the result was calculated in Eq. 2.47(b). Under the open-circuit condition, because of the electric constraint (or boundary) condition difference, the elastic compliance may be different. Thus, let us denote it as s^D (electric displacement constant condition or $D =$ constant and $E \neq 0$). Under the stress X_3, the input mechanical energy is now given by the sum of the elastic energy $(1/2)\, s_{33}{}^D X_3{}^2$ per unit volume and the converted (transduced or stored) electrical energy per unit volume under open-circuit is given by $(1/2)\, \varepsilon_0 \varepsilon_3^X E_3^2$ (D_3 is supposed to be zero). When the top and bottom electrodes are open-circuited in Figure 2.16b, the *depolarization field E* is induced because of the piezoelectrically induced charge $P = d_{33}X_3$ (i.e., bound charge) in order to satisfy $D_3 =$ constant (from Gauss law div $(D) = 0$ (no free charge) in a highly resistive material) in Eq. (2.43):

$$E_3 = -\left(d_{33} \big/ \varepsilon_0 \varepsilon_3^X \right) X_3 . \tag{2.48}$$

Note that this "depolarization field" is valid only for the short time period, by neglecting the charge drift in the sample or air-floating free charge in a long time scale (typically much longer than 1 minute). When quickly movable free charge exists, a sort of "field screening" occurs. Thus, from Eq. (2.44), the strain induced under the open-circuit condition should be smaller than that under the short-circuit, as calculated:

$$x_3 = s_{33}^E X_3 + d_{33} E_3 = s_{33}^E X_3 - \left(d_{33}^2 \big/ \varepsilon_0 \varepsilon_3^X \right) X_3 = s_{33}^E \left[1 - \left(d_{33}^2 \big/ \varepsilon_0 \varepsilon_3^X s_{33}^E \right) \right] X_3 . \tag{2.49}$$

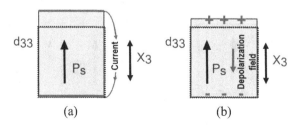

(a)　　　　　　　　　　　　　　(b)

FIGURE 2.16 Calculation models of electromechanical coupling factor k for (a) short-circuit and (b) open-circuit conditions.

If we denote $x_3 = s_{33}^D X_3$ under the open-circuit, we can obtain the following important relation between E-constant and D-constant elastic compliances through k:

$$s_{33}^D = s_{33}^E \left(1 - k_{33}^2\right). \ \left(k : \text{electromechanical coupling factor}\right) \tag{2.50}$$

The elasticity under open-circuit condition is stiffer than that under short-circuit condition. Finally, let us reconfirm the electromechanical coupling factor definition Eq. (2.47b) under the open-circuit condition: input mechanical energy is given by sum of $(1/2) s_{33}^D X_3^2$ and the stored electrical energy as charges given by $(1/2) \varepsilon_0 \varepsilon_3^X E_3^2$ (which coincides with the converted electric energy). Taking into account Eqs. (2.48) and (2.50),

$$k_{33}^2 = (1/2)\varepsilon_0\varepsilon_3^X E_3^2 \Big/ \left[(1/2)s_{33}^D X_3^2 + (1/2)\varepsilon_0\varepsilon_3^X E_3^2\right]$$

$$= (1/2)\varepsilon_0\varepsilon_3^X \left[-\left(d_{33}/\varepsilon_0\varepsilon_3^X\right)X_3\right]^2 \Big/ \left[(1/2)\left[s_{33}^E\left(1-k_{33}^2\right)\right]X_3^2 + (1/2)\varepsilon_0\varepsilon_3^X\left[-\left(d_{33}/\varepsilon_0\varepsilon_3^X\right)X_3\right]^2\right]$$

$$= d_{33}^2 \Big/ \varepsilon_0\varepsilon_3^X \cdot s_{33}^E.$$

The electromechanical coupling factor k can be calculated exactly the same, irrelevant to the short-circuit or open-circuit condition virtual models.

Example Problem 2.7

One of the PZT ceramics has a piezoelectric coefficient, $d_{33} = 590 \times 10^{-12}$ (C/N), a dielectric constant, $\varepsilon_3 = 3,400$, and an elastic compliance, $s_{33}^E = 20 \times 10^{-12}$ (m²/N).

a. Calculate the induced strain under an applied electric field of $E_3 = 10 \times 10^5$ (V/m). Then, calculate the generative stress under a completely clamped (mechanically constrained) condition.
b. Calculate the induced electric field under an applied stress of $X_3 = 3 \times 10^7$ (N/m²), which corresponds to the generative stress in (a). The magnitude of the induced electric field does not correspond to the magnitude of the applied electric field in (a) [i.e., 10×10^5(V/m)]. Explain why.

Solution

Part (a):

$$x_3 = d_{33}E_3 = \left[590 \times 10^{-12}(\text{C/N})\right]\left[10 \times 10^5(\text{V/m})\right]$$

$$x_3 = 5.9 \times 10^{-4} \tag{P2.7.1}$$

Under completely clamped (mechanically constrained) conditions,

$$X_3 = x_3/s_{33} = 5.9 \times 10^{-4}/20 \times 10^{-12}\left(\text{m}^2/\text{N}\right)$$

$$X_3 = 3.0 \times 10^7\left(\text{N/m}^2\right) \tag{P2.7.2}$$

Part (b):

$$P_3 = d_{33}X_3 = \left[590 \times 10^{-12}\ (\text{C/N})\right]\left[3 \times 10^7\left(\text{N/m}^2\right)\right]$$

$$P_3 = 1.77 \times 10^{-2}\left(\text{C/m}^2\right) \tag{P2.7.3}$$

$$E_3 = P_3/\varepsilon_0\varepsilon_3 = \left[1.77\times10^{-2}\left(\text{C/m}^2\right)\right]/[3,400]\left[8.854\times10^{-12}\ (\text{F/m})\right]$$

$$E_3 = 5.9\times10^5\ (\text{V/m}) \tag{P2.7.4}$$

The induced field is only 59% of the electric field applied for the case given in **Part (a)**. This is most readily explained in terms of the electromechanical coupling factor. Considering the mechanical energy produced through the electromechanical response of the piezoelectric with respect to the electrical energy supplied to the material, the so-called *electromechanical coupling factor, k,* is defined as follows:

$$k^2 = \left(\text{Stored mechanical energy}/\text{Input electrical energy}\right).$$

An alternative way of defining the electromechanical coupling factor is by considering the electrical energy produced through the electromechanical response of the piezoelectric with respect to the mechanical energy supplied to the material, which leads to the same outcome:

$$k^2 = \left(\text{Stored electrical energy}/\text{Input mechanical energy}\right)$$

In this case, we may evaluate the electromechanical coupling factor associated with the piezoelectric response, taking place through d_{33}, and find that:

$$k_{33}^2 = \frac{d_{33}^2}{s_{33}\varepsilon_0\varepsilon_3} = \frac{\left[590\times10^{-12}\left(\dfrac{\text{C}}{\text{N}}\right)\right]^2}{\left[20\times10^{-12}\left(\dfrac{\text{m}^2}{\text{N}}\right)\right]\left[8.854\times10^{-12}\left(\dfrac{\text{F}}{\text{m}}\right)\right][3,400]}$$

$$k_{33}^2 = 0.58 \tag{P2.7.5}$$

So, we see why the induced electric field determined in **Part (b)** has a magnitude that is only a fraction of the field applied in **Part (a)** of 10×10^5 (V/m). Here we see that this fraction corresponds to about k^2; that is, each energy transduction ratio accompanying the $E\to M$ and $M\to E$ processes is k^2. Thus (last electrical energy)/(initial electrical energy) = k^4, leading to (last induced field)/(initial applied field) = k^2. In this sense, we may regard the quantity k^2 to be the transduction ratio associated with a particular electrical-to-mechanical or mechanical-to-electrical piezoelectric transduction event. In the case of $k = 100\%$, the magnitude of the induced electric field (b) should correspond to the same field in (a) [i.e., 10×10^5(V/m)].

Example Problem 2.8

Derive the electromechanical coupling factor k_{ij} of piezoelectric ceramic vibrators for the following vibration modes (refer to Figure 2.17):

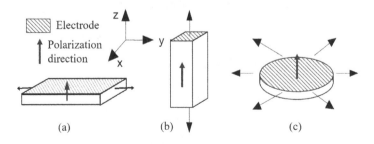

(a)   (b)   (c)

FIGURE 2.17 (a) Longitudinal plate extension via d_{31}, (b) longitudinal rod extension via d_{33}, and (c) planar extensional vibration modes of piezoelectric devices.

a. Longitudinal plate extension mode ($\perp E$): k_{31}
b. Longitudinal length extension mode ($/\!/ E$): k_{33}
c. Planar extension mode of the circular plate: k_p

Hint
Using the expression:

$$U = U_{MM} + 2U_{ME} + U_{EE}$$

$$= (1/2)\sum_{i,j} s_{ij}^E X_j X_i \sum_{m,i} d_{mi} E_m X_i + (1/2)\sum_{k,m} \varepsilon_0 \varepsilon_{mk}^X E_k E_m \qquad \text{(P2.8.1)}$$

the electromechanical coupling factor is given by

$$k = U_{ME}\Big/\sqrt{U_{MM} \cdot U_{EE}} \qquad \text{(P2.8.2)}$$

Solution

a. The relating equations for the k_{31} mode are:

$$x_1 = s_{11}^E X_1 + d_{31} E_3$$

$$D_3 = d_{31} X_1 + \varepsilon_0 \varepsilon_{33}^X E_3$$

Note that when L (length)$>>w$ (width), b (thickness), we can assume X_2 and X_3 are almost zero, and the above two equations are only essential.

U_{ME} comes from the d_{31} term as $(1/2)d_{31}E_3 X_1$, U_{MM} comes from the $s_{11}{}^E$ term as $(1/2)s_{11}^E X_1^2$, and U_{EE} comes from the $\varepsilon_{33}{}^X$ term as $(1/2)\varepsilon_{33}^X E_3^2$.

Thus,

$$k_{31} = (1/2)d_{31}E_3 X_1\Big/\sqrt{(1/2)s_{11}^E X_1^2 \cdot (1/2)\varepsilon_0 \varepsilon_{33}^X E_3^2}$$
$$= d_{31}\Big/\sqrt{s_{11}^E \cdot \varepsilon_0 \varepsilon_{33}^X} \qquad \text{(P2.8.3)}$$

Since $d_{31} < 0$, k_{31} should be negative value. However, the spec data usually show the absolute value.

b. Similarly, the relating equations for the k_{33} mode are:

$$x_3 = s_{33}^E X_3 + d_{33} E_3$$

$$D_3 = d_{33} X_3 + \varepsilon_{33}^X E_3$$

U_{ME} comes from the d_{33} term as $(1/2)d_{33}E_3 X_3$, U_{MM} comes from the $s_{33}{}^E$ term as $(1/2)s_{33}^E X_3^2$, and U_{EE} comes from the $\varepsilon_{33}{}^X$ term as $(1/2)\varepsilon_{33}^X E_3^2$.

Thus,

$$k_{33} = (1/2)d_{33}E_3 X_3\Big/\sqrt{(1/2)s_{33}^E X_3^2 \cdot (1/2)\varepsilon_{33}^X E_3^2}$$
$$= d_{33}\Big/\sqrt{s_{33}^E \cdot \varepsilon_{33}^X} \qquad \text{(P2.8.4)}$$

c. The relating equations for the k_p mode are following three, including 2D x_1 and x_2 equations:

$$x_1 = s_{11}^E X_1 + s_{12}^E X_2 + d_{31} E_3$$

$$x_2 = s_{12}^E X_1 + s_{22}^E X_2 + d_{32} E_3$$

$$D_3 = d_{31} X_1 + d_{32} X_2 + \varepsilon_{33}^X E_3$$

Assuming axial symmetry, $s_{11}^E = s_{22}^E$, $d_{31} = d_{32}$ and $X_1 = X_2 (= X_p)$ in a random ceramic specimen, the above equations are transformed to the following two equations:

$$x_1 + x_2 = 2\left(s_{11}^E + s_{12}^E\right) X_p + 2 d_{31} E_3$$

$$D_3 = 2 d_{31} X_p + \varepsilon_{33}^X E_3$$

U_{ME} comes from the d_{33} term as $(1/2) \cdot 2 d_{31} E_3 X_p$, U_{MM} comes from the s_{33}^E term as $(1/2) \cdot 2\left(s_{11}^E + s_{12}^E\right) X_p^2$, and U_{EE} comes from the ε_{33}^X term as $(1/2)\varepsilon_{33}^X E_3^2$. Thus,

$$
\begin{aligned}
k_p &= 2 d_{31} \Big/ \sqrt{2\left(s_{11}^E + s_{12}^E\right) \cdot \varepsilon_{33}^X} \\
&= \left[d_{31} \Big/ \sqrt{s_{11}^E \cdot \varepsilon_{33}^X} \right] \cdot \sqrt{2/(1-\sigma)} = k_{31} \cdot \sqrt{2/(1-\sigma)}
\end{aligned}
\tag{P2.8.5}
$$

where σ is Poisson's ratio given by

$$\sigma = -s_{12}^E \big/ s_{11}^E . \tag{P2.18.6}$$

When we compare the values k_{31} and k_{33}, since $|d_{31}|/d_{33} \approx 1/3$, similar difference is expected for k_{31}/k_{33}, under a supposition that the elastic compliances s_{11}^E and s_{33}^E are not different more than 20%. On the other hand, since $\sigma \approx 1/3$, $k_p \approx \sqrt{3} k_{31}$ is expected. The reader needs to understand that the electromechanical coupling factor or energy transduction rate is significantly different according to the device structure/mode. Example data for PZT 4 are: $k_{31} = 33\%$, $k_{33} = 70\%$, and $k_p = 58\%$.

2.5.6.3   Intensive and Extensive Parameters

According to IUPAC (the International Union of Pure and Applied Chemistry), an extensive parameter depends on the volume of the material, while an intensive parameter is the ratio of two extensive ones and, therefore, is independent on the volume of the material.[13] Consequently, stress (X) and electric field (E) are intensive parameters, which are externally controllable, while strain (x) and dielectric displacement (D) (or polarization (P)) are extensive parameters, which are internally determined in the material. We start with the Gibbs free energy, G, which in this case is expressed as:

$$G = -(1/2)s^E X^2 - d\,X\,E - (1/2)\varepsilon_0 \varepsilon^X E^2 \tag{2.51}$$

and in general differential form as:

$$dG = -x\,dX - D\,dE - S\,dT \tag{2.52}$$

where x is the strain, X is the stress, D is the electric displacement, E is the electric field, S is the entropy, and T is the temperature. Equation (2.52) is the energy expression in terms of the externally controllable (which is denoted as "intensive") physical parameters X and E. The temperature dependence of the function is associated with the elastic compliance, s^E, the dielectric constant, ε^X, and the piezoelectric charge coefficient, d. Different from the discussion on Eq. (2.14), where the terms with odd number power of the extensive parameter P (spontaneous polarization) need to be eliminated, Eq. (2.51) can include the E proportional term, which reflects to the extension or shrinkage of the sample depending on the E direction to keep the Gibbs energy constant. We obtain from the Gibbs energy function the following two piezoelectric equations:

$$x = -\frac{\partial G}{\partial X} = s^E X + dE \tag{2.53}$$

$$D = -\frac{\partial G}{\partial E} = dX + \varepsilon_0 \varepsilon^X E \tag{2.54}$$

Note that the Gibbs energy function provides "intensive" physical parameters: E-constant elastic compliance, s^E, and X-constant permittivity, ε^X.

When we consider the free energy in terms of the "extensive" (i.e., material-related) parameters of strain, x, and electric displacement, D, we start from the differential form of the Helmholtz free energy designated by A, such that:

$$dA = X\,dx + E\,dD - S\,dT, \tag{2.55}$$

We obtain from this energy function the following two piezoelectric equations:

$$X = \frac{\partial A}{\partial x} = c^D x - hD \tag{2.56}$$

$$E = \frac{\partial A}{\partial D} = -hx + \left(\frac{1}{\varepsilon_0}\right)\kappa^x D \tag{2.57}$$

where c^D is the elastic stiffness at constant electric displacement (open-circuit conditions), h is the inverse piezoelectric charge coefficient, and κ^x is the inverse dielectric constant at constant strain (mechanically clamped conditions). We sometimes use κ_0 for ($1/\varepsilon_0$).

Example Problem 2.9

Verify the relationship:

$$\frac{d^2}{s^E \varepsilon^X \varepsilon_0} = \frac{h^2}{c^D \left(\kappa^x/\varepsilon_0\right)} \tag{P2.9.1}$$

This value is defined as the square of an electromechanical coupling factor (k^2), which should be the same even for different energy description systems.

Solution

When Eqs. (2.53) and (2.54) are combined with Eqs. (2.56) and (2.57), we obtain:

$$X = c^D \left(s^E X + d E\right) - h\left(d X + \varepsilon_0 \varepsilon^X E\right) \tag{P2.9.2}$$

$$E = -h\left(s^E X + d E\right) + \left(\kappa^x/\varepsilon_0\right)\left(d X + \varepsilon_0 \varepsilon^X E\right) \tag{P2.9.3}$$

or upon rearranging:

$$\left(1 - c^D s^E + h d\right) X + \left(h\varepsilon_0 \varepsilon^X - c^D d\right) E = 0 \tag{P2.9.4}$$

$$\left[h s^E - \left(\kappa^x/\varepsilon_0\right)d\right] X + \left[1 - \left(\kappa^x/\varepsilon_0\right)\varepsilon_0 \varepsilon^X + h d\right] E = 0 \tag{P2.9.5}$$

Combining the latter two equations yields:

$$\left(1 - c^D s^E + hd\right)\left[1 - \left(\kappa^x/\varepsilon_0\right)\varepsilon_0\varepsilon^X + hd\right] - \left(h\varepsilon_0\varepsilon^X - c^D d\right)\left[hs^E - \left(\kappa^x/\varepsilon_0\right)d\right] = 0 \qquad \text{(P2.9.6)}$$

which, when simplified, produces the desired relationship:

$$\frac{d^2}{s^E \varepsilon^X \varepsilon_0} = \frac{h^2}{c^D \left(\kappa^x/\varepsilon_0\right)} \qquad \text{(P2.9.7)}$$

It is important to consider the conditions under which a material will be operated when characterizing the dielectric constant and elastic compliance of that material. When a constant electric field is applied to a piezoelectric sample as illustrated in Figure 2.18a, the total input electric energy (*left*) should be equal to a combination of the energies associated with two distinct mechanical conditions that may be applied to the material: (1) stored electric energy under the *mechanically clamped state*, where a constant strain (*zero strain*) is maintained and the specimen cannot deform, and (2) converted mechanical energy under the *mechanically free state*, in which the material is not constrained and is free to deform. This situation can be expressed by:

$$\left(\frac{1}{2}\right)\varepsilon^X\varepsilon_0 E_0^2 = \left(\frac{1}{2}\right)\varepsilon^x\varepsilon_0 E_0^2 + \left(\frac{1}{2s^E}\right)x^2 = \left(\frac{1}{2}\right)\varepsilon^x\varepsilon_0 E_0^2 + \left(\frac{1}{2s^E}\right)(dE_0)^2$$

such that

$$\varepsilon^X\varepsilon_0 = \varepsilon^x\varepsilon_0 + \left(\frac{d^2}{s^E}\right) \qquad \text{(2.58a)}$$

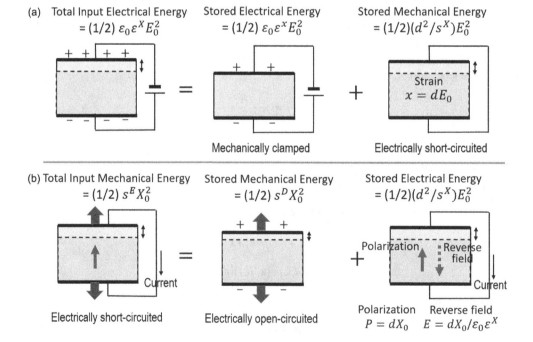

FIGURE 2.18 Schematic representation of the response of a piezoelectric material under: (a) constant applied electric field and (b) constant applied stress conditions.

$$\varepsilon^x = \varepsilon^X \left(1 - k^2\right) \left[k^2 = \frac{d^2}{\varepsilon^X \varepsilon_0 \, s^E} \right] \qquad (2.58b)$$

When a constant stress is applied to the piezoelectric as illustrated in Figure 2.18b, the total input mechanical energy will be a combination of the energies associated with two distinct electrical conditions that may be applied to the material: (1) stored mechanical energy under the *open-circuit state*, where a constant electric displacement is maintained, and (2) converted electric energy (i.e., "depolarization" field) under the *short-circuit condition*, in which the material is subject to a constant electric field. This can be expressed as:

$$\left(\frac{1}{2}\right) s^E X_0^2 = \left(\frac{1}{2}\right) s^D X_0^2 + \left(\frac{1}{2}\right) \varepsilon^X \varepsilon_0 E^2 = \left(\frac{1}{2}\right) s^D X_0^2 + \left(\frac{1}{2}\right) \varepsilon^X \varepsilon_0 \left(d/\varepsilon_0 \varepsilon^X\right)^2 X_0^2$$

which leads to:

$$s^E = s^D + \left(\frac{d^2}{\varepsilon^X \varepsilon_0}\right) \qquad (2.59a)$$

$$s^D = s^E \left(1 - k^2\right) \left[k^2 = \frac{d^2}{\varepsilon^X \varepsilon_0 \, s^E} \right] \qquad (2.59b)$$

Hence, we obtain the following equations:

$$\varepsilon^x / \varepsilon^X = \left(1 - k^2\right) \qquad (2.60)$$

$$s^D / s^E = \left(1 - k^2\right) \qquad (2.61)$$

where

$$k^2 = \frac{d^2}{s^E \varepsilon^X \varepsilon_0} \qquad (2.62)$$

We may also write equations of similar form for the corresponding reciprocal quantities:

$$\kappa^X / \kappa^x = \left(1 - k^2\right) \qquad (2.63)$$

$$c^E / c^D = \left(1 - k^2\right) \qquad (2.64)$$

where, in this context,

$$k^2 = \frac{h^2}{c^D \left(\kappa^x / \varepsilon_0\right)} \qquad (2.65)$$

This new parameter k is also the *electromechanical coupling factor* in the extensive parameter description and identical to the k in Eq. (2.62). It will be regarded mostly as a real quantity for the cases we examine in this text (see Example Problem 2.9). There are following three relationships between the intensive and extensive parameters: permittivity under constant stress ($\varepsilon^X \varepsilon_0$), the elastic compliance under constant electric field (s^E), and the piezoelectric charge coefficient (d)

in terms of their corresponding reciprocal quantities: inverse permittivity under constant strain (κ/ε_0), the elastic stiffness under constant electric displacement (c^D), and the inverse piezoelectric coefficient (h). Note also $k^2 = \dfrac{d^2}{s^E \varepsilon^X \varepsilon_0} = \dfrac{h^2}{c^D \left(\kappa^x / \varepsilon_0 \right)}$

$$\varepsilon^X \varepsilon_0 = \frac{1}{\left(\dfrac{\kappa^x}{\varepsilon_0} \right) \left[1 - \dfrac{h^2}{c^D \left(\kappa^x / \varepsilon_0 \right)} \right]} = \frac{1}{\left(\kappa^x / \varepsilon_0 \right) \left(1 - k^2 \right)} \tag{2.66}$$

$$s^E = \frac{1}{c^D \left[1 - \dfrac{h^2}{c^D \left(\kappa^x / \varepsilon_0 \right)} \right]} = \frac{1}{c^D \left(1 - k^2 \right)} \tag{2.67}$$

$$d = \frac{\dfrac{h^2}{c^D \left(\kappa^x / \varepsilon_0 \right)}}{h \left[1 - \dfrac{h^2}{c^D \left(\kappa^x / \varepsilon_0 \right)} \right]} = \frac{k^2}{h \left(1 - k^2 \right)} \tag{2.68}$$

It is important to note the "extensive" parameters, that is, x-constant or D-constant conditions. Under the above discussed hypothetical 1D treatment, all equations above stand correct. However, when we discuss on a practical 3D piezoelectric specimen, true x-constant status is realized only when the specimen is three-dimensionally clamped, while D-constant status is obtained only when no electrode is prepared on a highly resistive specimen. We will discuss this issue again in Chapter 4.

2.6 TENSOR/MATRIX DESCRIPTION OF PIEZOELECTRICITY

The 1D phenomenological analysis was introduced to explain the phase transition from the high crystallographically symmetric mother phase to the ferroelectric state with a change in temperature, stress, or electric field by introducing higher-order nonlinear terms. To the contrary, the 3D tensor analysis in this section is basically merely linear treatment (except for a tricky treatment of the "electrostrictive" effect) at a certain temperature with a stable phase such as a polar and piezoelectric state in order to discuss mainly the coupling effect between "order parameters" (here polarization P).

2.6.1 TENSOR REPRESENTATION

In the solid-state theoretical treatment of the phenomenon of piezoelectricity or electrostriction, the strain x_{kl} is expressed in terms of the electric field, E_i, or electric polarization, P_i, as follows:

$$x_{kl} = \sum_i d_{ikl} E_i + \sum_{i,j} M_{ijkl} E_i E_j$$

$$= \sum_i g_{ikl} P_i + \sum_{i,j} Q_{ijkl} P_i P_j \tag{2.69}$$

where d_{ikl} and g_{ikl} are the piezoelectric coefficients and M_{ijkl} and Q_{ijkl} are the electrostrictive coefficients. Represented in this way, we regard the quantities E_i and x_{kl} as first-rank and second-rank tensors, respectively, while d_{ikl} and M_{ijkl} are considered third-rank and fourth-rank tensors, respectively. Generally speaking, if two physical properties are represented by tensors of p-rank and q-rank, *the quantity that combines the two properties in a linear relation is represented by a tensor of $(p + q)$-rank.*

The subscript "*i*" of d_{ijk} is related with a vector E_i and "*jk*" corresponds to a tensor x_{jk}, and d_{ijk} tensor can be viewed as a 3D array of coefficients comprised of three "layers" of the following form:

$$\text{1st layer } (i = 1) \quad \begin{pmatrix} d_{111} & d_{112} & d_{113} \\ d_{121} & d_{122} & d_{123} \\ d_{131} & d_{132} & d_{133} \end{pmatrix}$$

$$\text{2nd layer } (i = 2) \quad \begin{pmatrix} d_{211} & d_{212} & d_{213} \\ d_{221} & d_{222} & d_{223} \\ d_{231} & d_{232} & d_{233} \end{pmatrix} \tag{2.70}$$

$$\text{3rd layer } (i = 3) \quad \begin{pmatrix} d_{311} & d_{312} & d_{313} \\ d_{321} & d_{322} & d_{323} \\ d_{331} & d_{332} & d_{333} \end{pmatrix}$$

2.6.2 CRYSTAL SYMMETRY AND TENSOR FORM

A physical property measured along two different directions must have the same value, if these two directions are crystallographically equivalent. This consideration sometimes reduces the number of independent tensor components representing a given physical property. Let us consider the third-rank piezoelectricity tensor as an example. The converse piezoelectric effect is expressed in tensor notation as:

$$x_{kl} = \sum_i d_{ikl} E_i = d_{ikl} E_i \tag{2.71}$$

Just from the description simplicity, the symbol Σ is omitted from now on, and long as the tensor components keep the same suffices (in the above case, "*i*" in d_{ikl} and E_i.

An electric field, E, initially defined in terms of an (x, y, z) coordinate system, is redefined as E' in terms of another system rotated with respect to the first with coordinates (x', y', z') by means of a transformation matrix (a_{ij}) such that:

$$E'_i = a_{ij} E_j \tag{2.72}$$

or

$$\begin{bmatrix} E'_1 \\ E'_2 \\ E'_3 \end{bmatrix} = \begin{pmatrix} a_{11} & a_{12} & a_{13} \\ a_{21} & a_{22} & a_{23} \\ a_{31} & a_{32} & a_{33} \end{pmatrix} \begin{bmatrix} E_1 \\ E_2 \\ E_3 \end{bmatrix} \tag{2.73}$$

The matrix (a_{ij}) is thus seen to be simply the array of direction cosines that allows us to transform the components of vector E referred to the original coordinate axes to components referred to the axes of the new coordinate system. The second-rank strain tensor is, thus, transformed in the following manner:

$$x'_{ij} = a_{ik} a_{jl} x_{kl} \tag{2.74}$$

or

$$
\begin{bmatrix} x'_{11} & x'_{12} & x'_{13} \\ x'_{21} & x'_{22} & x'_{23} \\ x'_{31} & x'_{32} & x'_{33} \end{bmatrix} = \begin{pmatrix} a_{11} & a_{12} & a_{13} \\ a_{21} & a_{22} & a_{23} \\ a_{31} & a_{32} & a_{33} \end{pmatrix} \begin{bmatrix} x_{11} & x_{12} & x_{13} \\ x_{21} & x_{22} & x_{23} \\ x_{31} & x_{32} & x_{33} \end{bmatrix} \begin{pmatrix} a_{11} & a_{21} & a_{31} \\ a_{12} & a_{22} & a_{32} \\ a_{13} & a_{23} & a_{33} \end{pmatrix}
$$

$$(2.75)$$

while the transformation of the third-rank piezoelectric tensor is expressed as:

$$d'_{ijk} = a_{il}a_{jm}a_{kn}d_{lmn} \tag{2.76}$$

Because of the strain symmetric relation: $x_{ij} = x_{ji}$, from Eq. (2.70), we can derive that the d_{lmn} tensor components are symmetric with respect to m and n such that $d_{lmn} = d_{lnm}$, and the following equivalences can be established:

$$
\begin{array}{ccc}
d_{112} = d_{121} & d_{113} = d_{131} & d_{123} = d_{132} \\
d_{221} = d_{212} & d_{213} = d_{231} & d_{223} = d_{232} \\
d_{321} = d_{312} & d_{313} = d_{331} & d_{323} = d_{332}
\end{array}
$$

The number of independent coefficients is thus reduced from an original 27 (=3^3) to only 18, and the d_{lmn} tensor may then be represented by layers of the following form. It is noteworthy that the matrix for each layer is a symmetric matrix, leading to 6 independent components for each layer (6 components × 3 layers = 18 components).

$$
\text{1st layer} \quad \begin{pmatrix} d_{111} & d_{112} & d_{131} \\ d_{112} & d_{122} & d_{123} \\ d_{131} & d_{123} & d_{133} \end{pmatrix}
$$

$$
\text{2nd layer} \quad \begin{pmatrix} d_{211} & d_{212} & d_{231} \\ d_{212} & d_{222} & d_{223} \\ d_{231} & d_{123} & d_{233} \end{pmatrix} \tag{2.77}
$$

$$
\text{3rd layer} \quad \begin{pmatrix} d_{311} & d_{312} & d_{331} \\ d_{312} & d_{322} & d_{323} \\ d_{331} & d_{323} & d_{333} \end{pmatrix}
$$

2.6.3 MATRIX NOTATION

The reduction in number of tensor components just carried out for the tensor quantity d_{ijk} makes it possible to render the 3D array of coefficients in a more tractable 2D matrix form. This reduced notation is also practically easy to print out on a 2D sheet. The process is accomplished by abbreviating the suffix notation used to designate the tensor components according to the following scheme:

Tensor notation	11	22	33	23,32	31,13	12,21
Matrix notation	1	2	3	4	5	6

The layers of tensor components represented by Eq. (2.77) may now be rewritten as:

$$\begin{array}{ll}
\text{1st layer} & \left\{ \begin{array}{ccc} d_{11} & (1/2)d_{16} & (1/2)d_{15} \\ (1/2)d_{16} & d_{12} & (1/2)d_{14} \\ (1/2)d_{15} & (1/2)d_{14} & d_{13} \end{array} \right\} \\[3em]
\text{2nd layer} & \left\{ \begin{array}{ccc} d_{21} & (1/2)d_{26} & (1/2)d_{25} \\ (1/2)d_{26} & d_{22} & (1/2)d_{24} \\ (1/2)d_{25} & (1/2)d_{24} & (1/2)d_{23} \end{array} \right\} \\[3em]
\text{3rd layer} & \left\{ \begin{array}{ccc} d_{31} & (1/2)d_{36} & (1/2)d_{35} \\ (1/2)d_{36} & d_{32} & (1/2)d_{34} \\ (1/2)d_{35} & (1/2)d_{34} & d_{33} \end{array} \right\}
\end{array} \tag{2.78}$$

Since we define "4, 5, and 6" as $d_{16} = d_{112} + d_{121}$, the matrix components above include (1/2) for the "4, 5, and 6" components. The last two suffixes in the tensor notation correspond to those of the strain components; therefore, for the sake of consistency, we will also make similar substitutions in the notation for the strain components.

$$\begin{bmatrix} x_{11} & x_{12} & x_{13} \\ x_{21} & x_{22} & x_{23} \\ x_{31} & x_{32} & x_{33} \end{bmatrix} \rightarrow \begin{bmatrix} x_1 & (1/2)x_6 & (1/2)x_5 \\ (1/2)x_6 & x_2 & (1/2)x_4 \\ (1/2)x_5 & (1/2)x_4 & x_3 \end{bmatrix} \tag{2.79}$$

Note here that again, the number of independent components for this second-rank tensor may also be reduced from 9 (=3^2) to 6, because it is a symmetric tensor and $x_{ij} = x_{ji}$. The factors of (1/2) in this and the piezoelectric layers of Eq. (2.78) are included in order to retain the general form that, like the corresponding tensor equation expressed by Eq. (2.71), includes no factors of 2 so that we may write:

$$x_j = d_{ij}E_i \quad \left(i = 1,2,3; \ j = 1,2,\dots,6\right) \tag{2.80a}$$

or

$$\begin{bmatrix} x_1 \\ x_2 \\ x_3 \\ x_4 \\ x_5 \\ x_6 \end{bmatrix} = \begin{pmatrix} d_{11} & d_{21} & d_{31} \\ d_{12} & d_{22} & d_{32} \\ d_{13} & d_{23} & d_{33} \\ d_{14} & d_{24} & d_{34} \\ d_{15} & d_{25} & d_{35} \\ d_{16} & d_{26} & d_{36} \end{pmatrix} \begin{bmatrix} E_1 \\ E_2 \\ E_3 \end{bmatrix} \tag{2.80b}$$

When deriving a matrix expression for the direct piezoelectric effect in terms of the matrix form of the stress X_{ij}, the factors of $(1/2)$ are not necessary and the matrix may be represented as:

$$
\begin{bmatrix} X_{11} & X_{12} & X_{13} \\ X_{21} & X_{22} & X_{23} \\ X_{31} & X_{32} & X_{33} \end{bmatrix} \rightarrow \begin{bmatrix} X_1 & X_6 & X_5 \\ X_6 & X_2 & X_4 \\ X_5 & X_4 & X_3 \end{bmatrix}
\tag{2.81}
$$

so that:

$$
P_j = d_{ij} X_j \quad (i = 1,2,3; \; j = 1,2,\dots,6)
\tag{2.82a}
$$

or

$$
\begin{bmatrix} P_1 \\ P_2 \\ P_3 \end{bmatrix} = \begin{bmatrix} d_{11} & d_{12} & d_{13} & d_{14} & d_{15} & d_{16} \\ d_{21} & d_{22} & d_{23} & d_{24} & d_{25} & d_{26} \\ d_{31} & d_{32} & d_{33} & d_{34} & d_{35} & d_{36} \end{bmatrix} \begin{bmatrix} X_1 \\ X_2 \\ X_3 \\ X_4 \\ X_5 \\ X_6 \end{bmatrix}
\tag{2.82b}
$$

Putting $(1/2)$ for strain tensor reduction and no $(1/2)$ for stress tensor reduction is also related with "extensive (volume relevant)" and "intensive (volume irrelevant)" parameters, respectively. Although the matrix notation has the advantage of being more compact and tractable than the tensor notation, one must remember that the matrix coefficients, d_{ij}, do not transform like the tensor components, d_{ijk}. Applying the matrix notation in a similar manner to the *electrostrictive coefficients* M_{ijkl}, we obtain the following equation corresponding to Eq. (2.69):

$$
\begin{bmatrix} x_1 \\ x_2 \\ x_3 \\ x_4 \\ x_5 \\ x_6 \end{bmatrix} = \begin{pmatrix} d_{11} & d_{21} & d_{31} \\ d_{12} & d_{22} & d_{32} \\ d_{13} & d_{23} & d_{33} \\ d_{14} & d_{24} & d_{34} \\ d_{15} & d_{25} & d_{35} \\ d_{16} & d_{26} & d_{36} \end{pmatrix} \begin{bmatrix} E_1 \\ E_2 \\ E_3 \end{bmatrix}
$$

$$
+ \begin{pmatrix} M_{11} & M_{21} & M_{31} & M_{41} & M_{51} & M_{61} \\ M_{12} & M_{22} & M_{32} & M_{42} & M_{52} & M_{62} \\ M_{13} & M_{23} & M_{33} & M_{43} & M_{53} & M_{63} \\ M_{14} & M_{24} & M_{34} & M_{44} & M_{54} & M_{64} \\ M_{15} & M_{25} & M_{35} & M_{45} & M_{55} & M_{65} \\ M_{16} & M_{26} & M_{36} & M_{46} & M_{56} & M_{66} \end{pmatrix} \begin{pmatrix} E_1^2 \\ E_2^2 \\ E_3^2 \\ E_2 E_3 \\ E_3 E_1 \\ E_1 E_2 \end{pmatrix}
\tag{2.83}
$$

Tables 2.3 and 2.4 summarize the matrices d_{ij} (or g_{ij}) and Q_{ij} (or M_{ij}) for all crystallographic point groups.[14]

TABLE 2.3
Piezoelectric Strain Coefficient (d) Matrices[14]

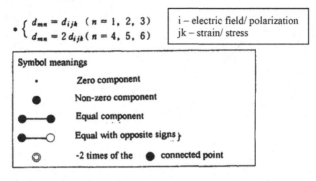

$$*\begin{cases} d_{mn} = d_{ijk} \ (n = 1, 2, 3) \\ d_{mn} = 2d_{ijk} \ (n = 4, 5, 6) \end{cases}$$

i – electric field/ polarization
jk – strain/ stress

Symbol meanings

· Zero component

● Non-zero component

●—● Equal component

●—○ Equal with opposite signs }

○ -2 times of the ● connected point

I Centro symmetric point group

Point group $\overline{1}$, $2/m$, mmm, $4/m$, $4/mmm$, $m3$, $m3m$, $\overline{3}$, $\overline{3}m$, $6/m$, $6/mmm$ All components are zero

II Non- centro symmetric point group

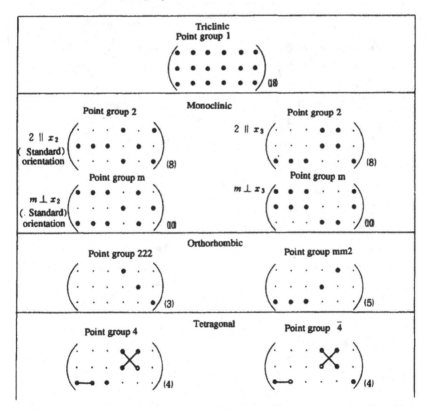

(Continued)

TABLE 2.3 (*Continued*)
Piezoelectric Strain Coefficient (d) Matrices[14]

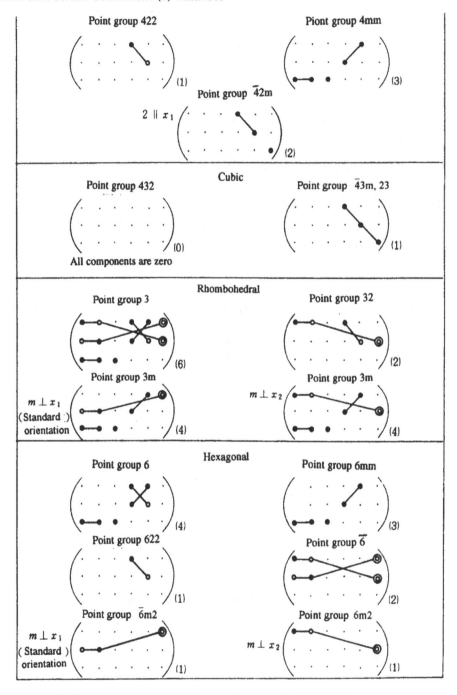

TABLE 2.4
Electrostriction Q Coefficient Matrices[14]

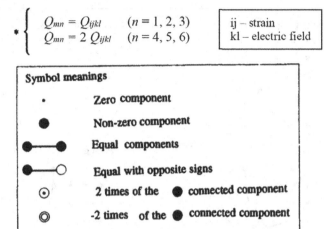

$$* \begin{cases} Q_{mn} = Q_{ijkl} & (n = 1, 2, 3) \\ Q_{mn} = 2\,Q_{ijkl} & (n = 4, 5, 6) \end{cases}$$

ij – strain
kl – electric field

Symbol meanings

·	Zero component
●	Non-zero component
●—●	Equal components
●—○	Equal with opposite signs
⊙	2 times of the ● connected component
◎	-2 times of the ● connected component
×	$(Q_{11} - Q_{12})$

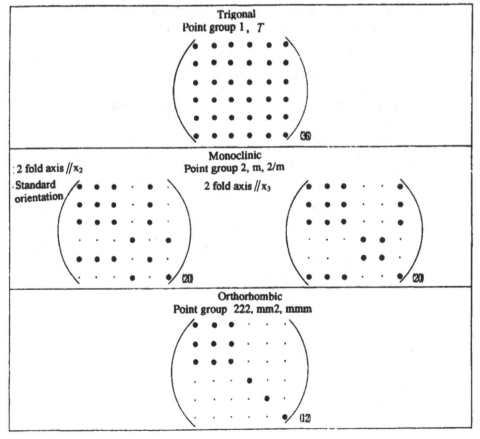

Trigonal
Point group 1, T
(36)

Monoclinic
Point group 2, m, 2/m
2 fold axis // x_2
Standard orientation
(20)

2 fold axis // x_3
(20)

Orthorhombic
Point group 222, mm2, mmm
(12)

(Continued)

TABLE 2.4 (*Continued*)
Electrostriction Q Coefficient Matrices[14]

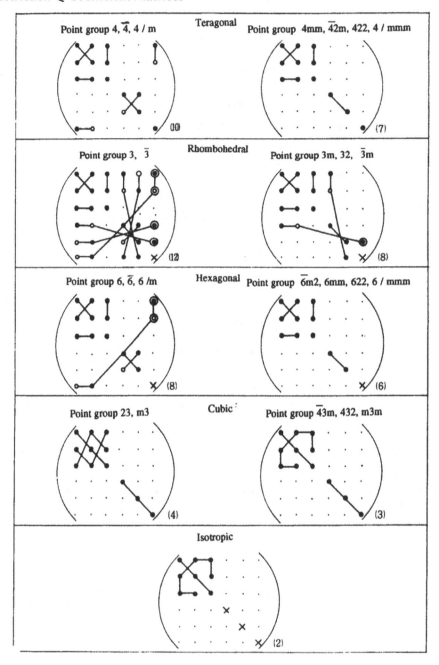

Example Problem 2.10

For a third-rank tensor such as the piezoelectric tensor, the transformation due to a change in the coordinate system is represented by

$$d'_{ijk} = \sum a_{il} a_{jm} a_{kn} d_{lmn} \qquad \text{(P2.10.1)}$$

The transformation matrix for rotation (angle θ) about a principal z axis is provided by:

$$\begin{pmatrix} a_{11} & a_{12} & a_{13} \\ a_{21} & a_{22} & a_{23} \\ a_{31} & a_{32} & a_{33} \end{pmatrix} = \begin{pmatrix} \cos\theta & \sin\theta & 0 \\ -\sin\theta & \cos\theta & 0 \\ 0 & 0 & 1 \end{pmatrix} \qquad \text{(P2.10.2)}$$

By solving the following problems step by step, the reader is requested to learn how to reduce the tensor components, according to the crystal symmetry.

a. When the crystal has a fourfold axis along z axis (i.e., $\theta = 90°$), provide the transformation matrix explicitly.
b. Calculate d'_{122} and d'_{211}, and express them with non-prime d tensor components. Step-by-step derivation process should be described.
c. Considering that $d'_{122} = d_{122}$ and $d'_{211} = d_{211}$ for a crystal with a fourfold symmetry, derive the relationship for the piezoelectric tensor components, d_{122} and d_{211} with a crystallographic reason.

Hint

- Piezoelectric third-rank tensor: the transformation due to a change in coordinate system is represented by

$$d'_{ijk} = \sum_{l,m,n} a_{il} a_{jm} a_{kn} d_{lmn}$$

- when the crystal has a fourfold axis symmetry along the z axis, the 90°-transformed d'_{ijk} should be identical to the original d_{ijk}.
- General tensor symmetry with m and n: such that $d_{123} = d_{132}$ and $d_{213} = d_{231}$

Solution

a. Taking $\cos 90° = 0$, and $\sin 90° = 1$, the rotation matrix becomes

$$\begin{pmatrix} 0 & 1 & 0 \\ -1 & 0 & 0 \\ 0 & 0 & 1 \end{pmatrix}.$$

Note that there are only three components in the rotation matrix, a_{21}, a_{12}, and a_{33}.

b. $d'_{ijk} = \sum_{l,m,n} a_{il} a_{jm} a_{kn} d_{lmn}$

$d'_{122} = \sum_{l,m,n} a_{1l} a_{2m} a_{2n} d_{lmn}$

$= a_{12} a_{21} a_{21} d_{211} = (+1)(-1)(-1)d_{211} = d_{211} \qquad \text{(P2.10.3)}$

When we see the first suffix "1" in a_{1i}, the second suffix is automatically determined as "2", because $a_{11} = a_{13} = 0$, as shown above. Similarly,

$$d'_{211} = a_{21}a_{12}a_{12}d_{211} = (-1)(+1)(+1)d_{211} = -d_{211} \tag{P2.10.4}$$

c. *Reason*: Because of the crystal symmetry of "fourfold", the 90° tensor rotation should not identify the difference between the prime and non-prime tensor components, leading to the relations: $d'_{122} = d_{122}$ and $d'_{211} = d_{211}$.
   Accordingly,

$$d'_{122} = d_{122} = d'_{211} = d_{211} = -d_{211} \tag{P2.10.5}$$

Thus, $d_{122} = d_{211} = 0 - [x = -x \rightarrow x = 0]$ is an important logic.

If you continue similar reduction processes for all 18 tensor components, you will get the followings:

$$d'_{111} = a_{12}a_{12}a_{12}d_{222} = (+1)(+1)(+1)d_{222} = d_{222}, d'_{111} \equiv d_{111} \rightarrow d_{111} = d_{222} \tag{P2.10.6}$$

$$d'_{122} = a_{12}a_{21}a_{21}d_{211} = (+1)(-1)(-1)d_{211} = d_{211}, d'_{122} \equiv d_{122} \rightarrow d_{122} = d_{211} \tag{P2.10.7}$$

$$d'_{133} = a_{12}a_{33}a_{33}d_{233} = (+1)(+1)(+1)d_{233} = d_{233}, d'_{133} \equiv d_{133} \rightarrow d_{133} = d_{233} \tag{P2.10.8}$$

$$d'_{123} = a_{12}a_{21}a_{33}d_{213} = (+1)(-1)(+1)d_{213} = -d_{213}, d'_{123} \equiv d_{123} \rightarrow d_{123} = -d_{213} = d_{132} = -d_{231} \tag{P2.10.9}$$

$$d'_{131} = a_{12}a_{33}a_{12}d_{232} = (+1)(+1)(+1)d_{232} = d_{232}, d'_{131} \equiv d_{131} \rightarrow d_{131} = d_{232} = d_{113} = d_{223} \tag{P2.10.10}$$

$$d'_{112} = a_{12}a_{12}a_{21}d_{221} = (+1)(+1)(-1)d_{221} = -d_{221}, d'_{112} \equiv d_{112} \rightarrow d_{112} = -d_{221} = d_{121} = -d_{212} \tag{P2.10.11}$$

$$d'_{211} = a_{21}a_{12}a_{12}d_{122} = (-1)(+1)(+1)d_{122} = -d_{122}, d'_{211} \equiv d_{211} \rightarrow d_{211} = -d_{122} \tag{P2.10.12}$$

$$d'_{222} = a_{21}a_{21}a_{21}d_{111} = (-1)(-1)(-1)d_{111} = -d_{111}, d'_{222} \equiv d_{222} \rightarrow d_{222} = -d_{111} \tag{P2.10.13}$$

$$d'_{233} = a_{21}a_{33}a_{33}d_{133} = (-1)(+1)(+1)d_{133} = -d_{133}, d'_{233} \equiv d_{233} \rightarrow d_{233} = -d_{133} \tag{P2.10.14}$$

$$d'_{223} = a_{21}a_{21}a_{33}d_{113} = (-1)(-1)(+1)d_{113} = d_{113}, d'_{223} \equiv d_{223} \rightarrow d_{223} = d_{113} \tag{P2.10.15}$$

$$d'_{231} = a_{21}a_{33}a_{12}d_{132} = (-1)(+1)(+1)d_{132} = -d_{132}, d'_{231} \equiv d_{231} \rightarrow d_{231} = -d_{132} \tag{P2.10.16}$$

$$d'_{212} = a_{21}a_{12}a_{21}d_{121} = (-1)(+1)(-1)d_{121} = d_{121}, d'_{212} \equiv d_{212} \rightarrow d_{212} = d_{121} \tag{P2.10.17}$$

$$d'_{311} = a_{33}a_{12}a_{12}d_{322} = (+1)(+1)(+1)d_{322} = d_{322}, d'_{311} \equiv d_{311} \rightarrow d_{311} = d_{322} \tag{P2.10.18}$$

$$d'_{322} = a_{33}a_{21}a_{21}d_{311} = (+1)(-1)(-1)d_{311} = d_{311}, d'_{322} \equiv d_{322} \rightarrow d_{322} = d_{311} \tag{P2.10.19}$$

$$d'_{333} = a_{33}a_{33}a_{33}d_{333} = (+1)(+1)(+1)d_{333} = d_{333}, d'_{333} \equiv d_{333} \rightarrow d_{333} = d_{333} \tag{P2.10.20}$$

$$d'_{323} = a_{33}a_{21}a_{33}d_{313} = (+1)(-1)(+1)d_{313} = -d_{313}, d'_{323} \equiv d_{323} \rightarrow d_{323} = -d_{313} \tag{P2.10.21}$$

$$d'_{331} = a_{33}a_{33}a_{12}d_{332} = (+1)(+1)(+1)d_{332} = d_{332}, d'_{331} \equiv d_{331} \rightarrow d_{331} = d_{332} \tag{P2.10.22}$$

$$d'_{312} = a_{33}a_{12}a_{21}d_{321} = (+1)(+1)(-1)d_{321} = -d_{321}, d'_{312} \equiv d_{312} \rightarrow d_{312} = -d_{321} \tag{P2.10.23}$$

Now we obtain the results:

$$d_{111} = d_{222} = d_{112} = d_{121} = d_{211} = d_{221} = d_{212} = d_{122}$$
$$= d_{331} = d_{313} = d_{133} = d_{332} = d_{323} = d_{233}$$
$$= d_{312} = d_{321} = 0$$

$$d_{333} \neq 0$$
$$d_{311} = d_{322}$$
$$d_{113} = d_{131} = d_{223} = d_{232}$$
$$d_{123} = d_{132} = -d_{213} = -d_{231}$$

Therefore, in a matrix notation:

$$\begin{pmatrix} 0 & 0 & 0 & d_{14} & d_{15} & 0 \\ 0 & 0 & 0 & d_{15} & -d_{14} & 0 \\ d_{31} & d_{31} & d_{31} & 0 & 0 & 0 \end{pmatrix} \qquad (P2.10.24)$$

for the point group **4**.

Example Problem 2.11

BaTiO$_3$ has a tetragonal crystal symmetry (point group **4mm**) at room temperature. Because of additional mirror symmetry to the four-fold symmetry, $d_{14} = 0$, and the piezoelectric strain coefficient matrix is, therefore, of the form:

$$d_{ij} = \begin{pmatrix} 0 & 0 & 0 & 0 & \mathbf{d_{15}} & 0 \\ 0 & 0 & 0 & \mathbf{d_{15}} & 0 & 0 \\ \mathbf{d_{31}} & \mathbf{d_{31}} & \mathbf{d_{33}} & 0 & 0 & 0 \end{pmatrix}$$

a. Determine the strain induced in a piezoelectric material when an electric field is applied along the crystallographic **c** axis.
b. Determine the strain induced in a piezoelectric material when an electric field is applied along the crystallographic **a** axis.

Hint
The matrix equation that applies in this case is:

$$\begin{bmatrix} x_1 \\ x_2 \\ x_3 \\ x_4 \\ x_5 \\ x_6 \end{bmatrix} = \begin{pmatrix} 0 & 0 & d_{31} \\ 0 & 0 & d_{31} \\ 0 & 0 & d_{33} \\ 0 & d_{15} & 0 \\ d_{15} & 0 & 0 \\ 0 & 0 & 0 \end{pmatrix} \begin{bmatrix} E_1 \\ E_2 \\ E_3 \end{bmatrix} \qquad (P2.11.1)$$

Solution
We can derive expressions for the induced strains:

$$x_1 = x_2 = d_{31}E_3, x_3 = d_{33}E_3, x_4 = d_{15}E_2, x_5 = d_{15}E_1, x_6 = 0$$

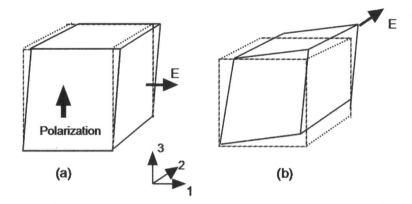

FIGURE 2.19   Induced strains considered for Problems 2.11 and 2.12: (a) shear strain induced in a BT single crystal with *tetragonal* symmetry **4mm** under E_1, and (b) electrostrictive strain induced along [111] in a PMN single crystal with *cubic* symmetry **m3m**.

so that the following determinations can be made:

    a. When E_3 is applied, elongation in the c direction ($x_3 = d_{33}E_3$, $d_{33} > 0$) and contraction in the a and b directions ($x_1 = x_2 = d_{31}E_3$, $d_{31} < 0$) occur. Note that PZTs and other oxide piezoelectrics exhibit positive d_{33}, that is, extension along the electric field direction, while polymer polyvinylidene difluoride (PVDF) shows negative d_{33}.

    b. When E_1 is applied, a shear strain x_5 (= $2x_{31}$) = $d_{15}E_1$ is induced. The case where $d_{15} > 0$ and $x_5 > 0$ is illustrated in Figure 2.19a.

Example Problem 2.12

Lead magnesium niobate (Pb(Mg$_{1/3}$Nb$_{2/3}$)O$_3$) has a cubic crystal symmetry (point group **m3m**) at room temperature and does not exhibit piezoelectricity; however, a large electrostrictive response is induced with the application of an electric field. The relationship between the induced strain and the applied electric field is given in matrix form by:

$$\begin{bmatrix} X_1 \\ X_2 \\ X_3 \\ X_4 \\ X_5 \\ X_6 \end{bmatrix} = \begin{pmatrix} M_{11} & M_{12} & M_{12} & 0 & 0 & 0 \\ M_{12} & M_{11} & M_{12} & 0 & 0 & 0 \\ M_{12} & M_{12} & M_{11} & 0 & 0 & 0 \\ 0 & 0 & 0 & M_{44} & 0 & 0 \\ 0 & 0 & 0 & 0 & M_{44} & 0 \\ 0 & 0 & 0 & 0 & 0 & M_{44} \end{pmatrix} \begin{pmatrix} E_1^2 \\ E_2^2 \\ E_3^2 \\ E_2 E_3 \\ E_3 E_1 \\ E_1 E_2 \end{pmatrix} \quad \text{(P2.12.1)}$$

Calculate the strain induced in the material when an electric field is applied along the [111] direction.

Hint
$E_1 E_2$ and $E_3 E_1$ are not independent of E_1^2, but this matrix formation handles as though independent.

Solution

Substituting the electric field applied in this case:

$$E_1 = E_2 = E_3 = E_{[111]}/\sqrt{3}$$

into Eq. (P2.12.1), we obtain:

$$x_1 = x_2 = x_3 = (M_{11} + 2M_{12})\frac{E_{[111]}^2}{3}(= x_{11} = x_{22} = x_{33})$$ (P2.12.2a)

$$x_4 = x_5 = x_6 = M_{44}\frac{E_{[111]}^2}{3}(= 2x_{23} = 2x_{31} = 2x_{12})$$ (P2.12.2b)

The resulting distortion is illustrated in Figure 2.19b. The strain, x, induced along an arbitrary direction is given by

$$x = \sum x_{ij}l_i l_j$$ (P2.12.3)

where l_i is a direction cosine defined with respect to the i axis. The strain induced along the [111] direction, $x_{[111]//}$, is therefore given by:

$$x_{[111]//} = \sum x_{ij}(1/\sqrt{3})(1/\sqrt{3})$$

$$= x_1 + x_2 + x_3 + \frac{2[(x_4/2) + (x_5/2) + (x_6/2)]}{3}$$

$$= (M_{11} + 2M_{12} + M_{44})\frac{E_{[111]}^2}{3}$$ (P2.12.4)

The strain induced perpendicular to the [111] direction, $x_{[111]\perp}$, is calculated in a similar fashion.

$$x_{[111]\perp} = \left[M_{11} + 2M_{12} - (M_{44}/2)\right]\frac{E_{[111]}^2}{3}$$ (P2.12.5)

It is important to note here that the volumetric strain ($\Delta V/V$) will be given by:

$$\frac{\Delta V}{V} = x_{[111]//} + 2x_{[111]\perp} = (M_{11} + 2M_{12})E_{[111]}^2$$ (P2.12.6)

and is independent of the applied field direction.

2.7 FERROELECTRIC MATERIALS

Quartz (SiO_2) and zinc oxide (ZnO) are popular piezoelectric but non-ferroelectric (nonpolar) materials. The "direct" piezoelectric effect was discovered in quartz by Pierre and Jacques Curie in 1880. The "converse" piezoelectric effect was discovered successively in 1881 by Gabriel Lippmann. Shipwreck of Titanic and World War I were the motivation of the undersea transducer, sonar. Paul Langevin developed the so-called Langevin-type transducer, which was originally composed of natural tiny quartz single crystals sandwiched by two metal blocks, in order to tune the transducer resonance frequency around 26 kHz which is a desired range for underwater applications.

On the other hand, the first ferroelectric discovered is Rochelle salt ($NaKC_4H_4O_6 \cdot 4H_2O$) in 1921. Though this material was studied from an academic viewpoint, it has not been widely utilized in practice because it is water-soluble and its Curie temperature is just above room temperature. KH_2PO_4 (KDP) is the second discovery in 1935, which is also water-soluble and the Curie temperature is −150°C. We needed to wait until World War II for the third and most famous ceramic ferroelectric, that is, barium titanate (BaTiO$_3$), which was actually commercialized first as a transducer material. In order to develop compact capacitors for portable "radar" systems to be used in the battle fields, TiO$_2$-based conventional "condenser materials" were widely researched by doping various ions such as CaO, SrO, BaO, MgO, and Fe$_2$O$_3$. Four groups in the United States, Russia, Germany, and Japan discovered BaTiO$_3$ almost the same time in the World War II.[15]

2.7.1 Physical Properties of Barium Titanate

An important ferroelectric material is barium titanate, BaTiO$_3$ (BT). It is often presented as a classic example of a ferroelectric that exhibits the so-called first-order behavior. BT has the perovskite crystal structure depicted in Figure 2.20. In its high temperature, paraelectric (nonpolar) phase, there is no spontaneous polarization and a cubic symmetry (O$_h$-m3m) exists. Below the transition temperature, which is designated by T_C and called the *Curie temperature* (~130°C for BaTiO$_3$), spontaneous polarization develops. Refer to Figure 2.8 to see detailed ionic shifts in BaTiO$_3$ at room temperature. The crystal structure becomes slightly elongated, assuming a tetragonal symmetry (C$_{4v}$ –4mm). There are also two lower temperature phase transitions for BaTiO$_3$, one from a tetragonal to an orthorhombic phase (~ 0°C) and the other from an orthorhombic to a rhombohedral phase (~ −90°C). Figure 2.21 illustrates these successive phase transitions from cubic, tetragonal, orthorhombic, and rhombohedral phases with a decrease in temperature. We will focus for the moment on the higher temperature paraelectric to ferroelectric transition.

Figure 2.22 shows the temperature dependence of the spontaneous polarization P_S and permittivity (dielectric constant) ε in a BT-like first-order transition ferroelectric schematically. P_S decreases with increasing temperature and vanishes at the Curie temperature, while ε tends to diverge near T_C. Also, the reciprocal (relative) permittivity $1/\varepsilon$ is known to be linear with respect to the temperature over a wide range in the paraelectric phase (so-called *Curie–Weiss law*),

$$\varepsilon = C/(T - T_0) \tag{2.84}$$

where C is the *Curie–Weiss constant* and T_0 is the *Curie–Weiss temperature*. T_0 is slightly lower than the exact transition temperature T_C.

It is also known that the spontaneous polarization P_S and the spontaneous strain x_S follow the relationship:

$$x_S = QP_S^2, \tag{2.85}$$

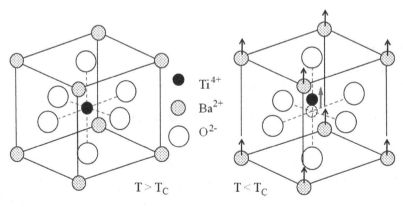

T$_C$: Curie temperature

FIGURE 2.20 Crystal structures of BaTiO$_3$: higher (left) and lower (right) than T_C.

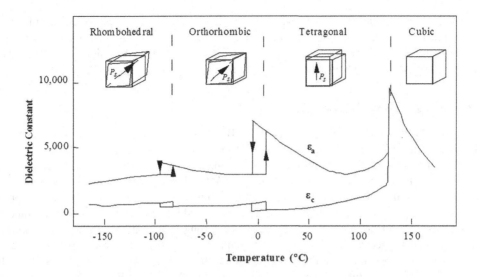

FIGURE 2.21 Various phase transitions in barium titanate (BT).

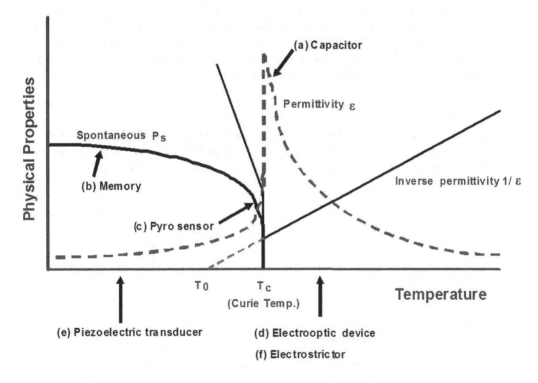

FIGURE 2.22 Temperature dependence of the spontaneous polarization and permittivity in a ferroelectric material. (a)–(f) indicate the temperature ranges for each application.

and x_s decreases almost linearly with increasing temperature. In the case of $BaTiO_3$, it exhibits the piezoelectric effect in the ferroelectric phase, while in the paraelectric phase, it is non-piezoelectric and exhibits only the electrostrictive effect. The general descriptions above are almost consistent with Section 2.5 Theory of Ferroelectric Phenomenology.

If the above consideration is rephrased from the application viewpoints, the Curie temperature for capacitor materials is designed to set around room temperature (RT); T_C for memory materials around 100°C higher than RT; T_C for pyro sensors just above RT; T_C for piezoelectric transducers typically much higher than RT, higher than 200°C; T_C for electrooptic and electrostrictive devices are lower than RT to use their paraelectric state. In other words, we will design practical materials with their Curie points suitable for each application.

2.7.2 LEAD ZIRCONATE TITANATE (PZT)

After the World War II, the ferroelectric search on BT-isomorphic perovskite materials was accelerated in the world. PZT solid solution systems [$(1 - x)$PbZrO$_3$ – xPbTiO$_3$] were discovered in 1954 by Japanese researchers, Shirane, Sawaguchi, and Takagi.[16] Figure 2.23 shows the phase diagram of PZT, where the *Morphotropic Phase Boundary (MPB)* between the tetragonal and rhombohedral phases exists around 52 PZ-48 PT compositions. However, the enormous piezoelectric properties were discovered by B. Jaffe,[17] Clevite Corporation, and Clevite took the most important PZT patents for transducer applications. Because of this strong basic patent, Japanese ceramic companies were encouraged actually to develop ternary systems to overcome the performance and, more importantly, to escape from the Clevite's patent; that is, PZT + a complex perovskite such as Pb(Mg$_{1/3}$Nb$_{2/3}$)O$_3$ (Matsushita Panasonic), Pb(Ni$_{1/3}$Nb$_{2/3}$)O$_3$ (NEC), and Pb(Zn$_{1/3}$Nb$_{2/3}$)O$_3$ (Toshiba), which are the basic compositions in recent years. Figure 2.24 plots the dependence of several piezoelectric d constants on composition near the MPB in the PZT system. Note that the maximum piezoelectric performance is obtained around the MPB composition in the pure PZT solid solution system. Typical electromechanical coupling factor k_{33} value is 70% in commercially available PZTs. Because the PZTs are the most dominant materials in piezoelectric applications, this book introduces the high-power characterization results, based on the PZTs.

2.7.3 RELAXOR FERROELECTRICS

Relaxor ferroelectrics can be prepared either in polycrystalline form or as single crystals. Different from the previously mentioned normal ferroelectrics such as BT and PZT, *relaxor ferroelectrics* exhibit a broad phase transition from the paraelectric to ferroelectric state, a strong frequency

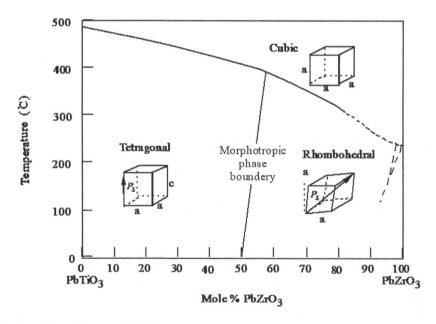

FIGURE 2.23 Phase diagram of PZT.[16]

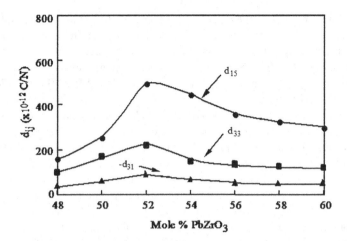

FIGURE 2.24 Dependence of several d constants on composition near the MPB in the PZT system.

dependence of the dielectric constant (i.e., *dielectric relaxation*) and a weak remnant polarization. Lead-based relaxor materials have complex disordered perovskite structures.

Simple perovskites in Figure 2.25a, such as barium titanate ($BaTiO_3$), lead titanate ($PbTiO_3$), and lead zirconate ($PbZrO_3$), and their solid solutions such as $A(B,B')O_3$, and complex perovskites such as $A^{2+}(B^{3+}_{1/2}B'^{5+}_{1/2})O_3$ and $A^{2+}(B^{2+}_{1/3} \cdot B'^{5+}_{2/3})O_3$ are all readily formed. Supercells of the complex perovskite structures listed above, in which ordering of the B-site cations exists, are pictured in Figure 2.25b and c. When the B and B' cations are randomly distributed as in $Pb(Mg_{1/3}Nb_{2/3})O_3$, the unit cell corresponds to the simple perovskite structure depicted in Figure 2.25a.

Relaxor-type electrostrictive materials, such as those from the lead magnesium niobate–lead titanate, $Pb(Mg_{1/3}Nb_{2/3})O_3$–$PbTiO_3$ (or PMN-PT), solid solution are highly suitable for actuator applications. Ceramics of PMN are easily poled when the poling field is applied near the transition temperature, but they are depoled completely when the field is removed as the macrodomain structure reverts to microdomains (with sizes on the order of several 100 Å, which is called "super-paraelectric"). This microdomain structure is believed to be the source of the exceptionally

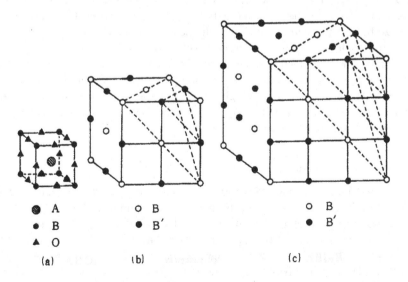

FIGURE 2.25 Complex perovskite structures with various B ion arrangements: (a) ABO_3, simple structure, (b) $A(B_{1/2}B'_{1/2})O_3$, 1:1 ordering, double cell, and (c) $A(B_{1/3}B'_{2/3})O_3$, 1:2 ordering, multiple cell.

large electrostriction exhibited in these materials. The usefulness of the material is thus further enhanced when the transition temperature is adjusted to near-room temperature. The longitudinal induced strain at room temperature as a function of applied electric field for $0.9PMN-0.1PbTiO_3$ ceramic is shown in Figure 2.10b.[18] Notice that the order of magnitude of the electrostrictive strain (10^{-3}) is similar to that induced under unipolar drive in La-doped PZT (PLZT) (7/62/38) through the piezoelectric effect (Figure 2.10a). An attractive feature of this material is the near absence of hysteresis, while the nonlinear strain behavior ($x = ME^2$) requires a sophisticated drive circuit. This discovery of giant electrostriction boosted "piezoelectric actuator" developments. The relaxor electrostrictor also exhibits an induced piezoelectric effect. That is, the electromechanical coupling factor k_t varies with the applied DC bias field. As the DC bias field increases, the coupling increases and saturates. Since this behavior is reproducible, these materials can be applied as ultrasonic transducers which are tunable by the bias field, especially useful in medical diagnostic applications.[19]

Another epoch-making discovery by the author's group is in single-crystal relaxor ferroelectrics with the MPB composition in the late 1970s, which show tremendous promise as ultrasonic transducers and electromechanical actuators. Single crystals of $Pb(Zn_{1/3}Nb_{2/3})O_3$ (PZN), $Pb(Mg_{1/3}Nb_{2/3})O_3$ (PMN), and binary systems of these materials combined with $PbTiO_3$ (PZN-PT and PMN-PT) exhibit extremely large electromechanical coupling factors.[20,21] Figure 2.26 shows changes of piezoelectric parameters with mole fraction x of PT in the $Pb(Zn_{1/3}Nb_{2/3})O_3-PbTiO_3$ solid solution system.[21] Large coupling coefficients and large piezoelectric constants have been found for crystals from the morphotropic phase boundaries of these solid solutions. PZN-8% PT single crystals were found to possess a high k_{33} value of 0.94 for the (001) crystal cuts (crystal orientation is the key!); this is very high compared to the k_{33} of conventional PZT ceramics of around 0.70~0.80. 15 years after this discovery, Yamashita's Toshiba group[22] and Shrout's Penn State group[23] reconfirmed the superiority of these relaxor-PT solid solution single crystals and demonstrated the strains as large as 1.7% induced practically for the PZN-PT crystals, then the present enthusiastic worldwide developments have arisen.

2.7.4   PVDF

Thanks to Kawai's efforts, PVDF or PVF_2 was discovered in 1969.[24] Though the piezoelectric d constant is not as high as piezo-ceramics, high piezoelectric g constant due to small permittivity ε is attractive from the sensor application viewpoint.

The PVDF is a polymer with monomers of CH_2CF_2. Since H and F have positive and negative ionization tendency, the monomer itself has a dipole moment. Crystallization from the melt forms the nonpolar α-phase, which can be converted into the polar β-phase by a uniaxial or biaxial drawing operation; the resulting dipoles are then reoriented through electric poling (see Figure 2.27).

Large sheets can be manufactured and thermally formed into complex shapes. Piezoelectric polymers have the following characteristics: (a) small piezoelectric d constants (for actuators) and large g constants (for sensors), (b) light weight and soft elasticity, leading to good acoustic impedance matching with water or the human body, and (c) a low mechanical quality factor Q_m, allowing for a broad resonance bandwidth.

2.7.5   Pb-Free Piezo-Ceramics

In 2006, European Union started RoHS (Restrictions on the use of certain Hazardous Substances), which explicitly limits the usage of lead (Pb) in electronic equipment. Basically, we may need to regulate the usage of PZT, current most famous piezoelectric ceramics, in the future. Japanese and European industrial communities may experience governmental regulation on the PZT usage in these 10 years. Pb (lead)-free piezo-ceramics have started to be developed after 1999. The Pb-free materials include (1) $(K,Na)(Ta,Nb)O_3$ based, (2) $(Bi,Na)TiO_3$, and (3) $BaTiO_3$, which reminds us "the history will repeat" (i.e., "piezoelectric Renaissance"). High-power performance of Pb-free materials is discussed in Chapter 9.

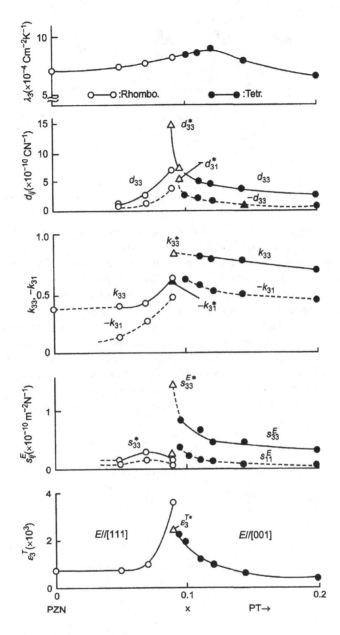

FIGURE 2.26 Changes of piezoelectric parameters with mole fraction x of PT in the $Pb(Zn_{1/3}Nb_{2/3})$ O_3-$PbTiO_3$ solid solution system.[21]

The share of the patents for bismuth compounds (bismuth layered type and $(Bi,Na)TiO_3$ type) exceeds 61%. This is because bismuth compounds are easily fabricated in comparison with other compounds. Honda Electronics, Japan developed Langevin transducers with using the BNT-based ceramics for ultrasonic cleaner applications.[25] Their composition $0.82(Bi_{1/2}Na_{1/2})TiO_3$-$0.15BaTiO_3$-$0.03(Bi_{1/2}Na_{1/2})(Mn_{1/3}Nb_{2/3})O_3$ exhibits $d_{33} = 110 \times 10^{-12}C/N$, which is only 1/3 of that of a hard PZT, but the electromechanical coupling factor $k_t = 0.41$ is larger because of much smaller permittivity ($\varepsilon = 500$) than that of the PZT. Furthermore, the maximum vibration velocity of a rectangular plate (k_{31} mode) is close to $1\,m/s$ (rms value), which is higher than that of hard PZTs.

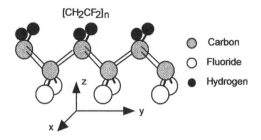

FIGURE 2.27 Molecular structure of PVDF.

(Na,K)NbO$_3$ systems exhibit the highest performance among the present Pb-free materials, because of the MPB usage. Figure 2.28 shows the current best data reported by Toyota Central Research Lab, where strain curves for oriented and unoriented (K,Na,Li)(Nb,Ta,Sb)O$_3$ ceramics are shown.[26] Note that the maximum strain reaches up to $1,500 \times 10^{-6}$, which is equivalent to the PZT strain. Drawbacks include their sintering difficulty and the necessity of the sophisticated preparation technique (topochemical method for preparing flaky raw powder).

Tungsten–bronze (TB) types are another alternative choice for resonance applications, because of their high Curie temperature and low loss. Taking into account general consumer attitude on disposability of portable equipment, Taiyo Yuden, Japan developed micro-ultrasonic motors using non-Pb multilayer (ML) piezo-actuators.[27] Their composition is based on TB ((Sr,Ca)$_2$NaNb$_5$O$_{15}$) without heavy metal. The basic piezoelectric parameters in TB$\left(d_{33} = 55 \sim 80 \text{ pC/N}, T_C = 300°C\right)$ are not very attractive. However, once the c axis oriented ceramics are prepared, the d_{33} is dramatically enhanced up to 240 pC/N. Further, since the Young's modulus Y_{33}^E=140 GPa is more than twice of that of PZT, the higher generative stress (blocking stress) is expected, which is suitable to

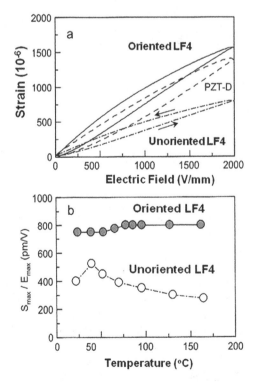

FIGURE 2.28 Strain curves for oriented and unoriented (K,Na,Li)(Nb,Ta,Sb)O$_3$ ceramics.[26]

ultrasonic motor applications. Taiyo Yuden developed a sophisticated preparation technology for oriented ceramics with a ML configuration: that is, preparation under strong magnetic field, much simpler than the flaky powder preparation.

2.8 APPLICATIONS OF FERROELECTRICS

Ferroelectric materials, especially polycrystalline ceramics, are very promising for a variety of applications such as *high-permittivity capacitors*, *ferroelectric memories*, *pyroelectric sensors*, *piezoelectric* and *electrostrictive transducers*, *electrooptic devices*, and *PTC thermistors*. Refer to Ref. [7] for the details.

For capacitor dielectrics, the peak dielectric constant around the transition (Curie) temperature is utilized, while for memory applications, the material must be ferroelectric at room temperature (refer to Figure 2.22). A large temperature dependence of the spontaneous polarization below T_C is sought for pyroelectric sensors. The converse pyroelectric effect is the electrocaloric effect (electric field generates the temperature decrease), which is becoming a hot research topic in this energy-saving age.

Piezoelectric materials are used as sensors and actuators, where the T_C should be much above room temperature. Pressure and acceleration sensors are now commercially available in addition to conventional piezo-vibrators. Precision positioners and pulse drive linear motors have already been installed in precision lathe machines, semiconductor manufacturing apparatuses, and office equipment. Recent enthusiastic development is found in ultrasonic motors, aiming at "electromagnetic and sound noise"-free and very compact motor applications. Recently, in parallel to the new energy source programs, piezoelectric energy harvesting systems became popular. Waste mechanical energy such as noise vibration, wind, and human walk can be converted into electrical energy to directly use for signal transmission or to charge up batteries for portable electronics.

Electrooptic materials will become key components in displays and optical communication systems in the near future. Optical beam scanners, light valves, and switches are urgent necessities. For thermistor applications, semiconductive ferroelectric ceramics with a positive temperature coefficient of resistivity (PTCR) based on a junction effect have been developed from barium titanate-based materials. Some recent application examples are introduced below.

2.8.1 PIEZOELECTRIC MULTILAYER ACTUATORS FOR AUTOMOBILE

Diesel engines are recommended rather than regular gasoline cars from the energy conservation and global warming viewpoint. When we consider the total energy of gasoline production, both oil well-to-tank and tank-to-wheel, the energy efficiency, measured by the total energy required to realize unit drive distance for a vehicle (MJ/km), is of course better for high-octane gasoline than diesel oil. However, since the electric energy required for purification is significant, the gasoline is inferior to diesel fuel.[28] As well known, the conventional diesel engine, however, generates toxic exhaust gases such as SO_x and NO_x. In order to solve this problem, new diesel injection valves were developed by Siemens, Bosch, and Toyota with piezoelectric multilayered actuators. Figure 2.29a shows such a common rail-type diesel injection valve with an ML piezo-actuator which produces high pressure fuel and quick injection control.[29] As shown in the diesel fuel injection timing chart (Figure 2.29b), during a cycle of about 60–100 Hz, five quick fuel injections (now seven) are required with a very sharp rise and fall pulse shape. For this purpose, piezoelectric actuators, specifically ML types, were adopted. The highest reliability of these devices at an elevated temperature (150°C) for a long period (10 years) has been achieved.[29] Despite pseudo-DC drive around 100 Hz, high voltage under continuous cyclic usage, the ML generates significant heat via the piezo-material's dielectric loss, which will be discussed in Chapter 6. The piezoelectric actuator is namely the key to increase burning efficiency and minimize the toxic exhaust gases.

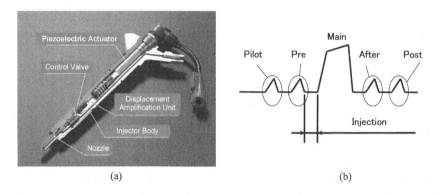

(a)   (b)

FIGURE 2.29   (a) Common rail-type diesel injection valve with a piezoelectric ML actuator. (Courtesy by Denso Corporation.)[29] (b) Diesel fuel injection timing chart in one cycle (about 60 Hz cycle).

2.8.2 ULTRASONIC MOTORS FOR CAMERA MODULES

A micromotor called "metal tube type" consisting of a metal hollow cylinder and two PZT rectangular plates was developed by the Penn State University in the late 1990s [see Figure 2.30a]. When one of the PZT plates is driven (single phase drive), Plate X, a bending vibration is excited basically along x axis. However, because of an asymmetrical mass (Plate Y), another hybridized bending mode is excited with some phase lag along y axis, leading to an elliptical locus in a clockwise direction, like a "Hula-Hoop" motion. The rotor of this motor is a cylindrical rod with a pair of stainless ferrules pressing down with a spring. The metal cylinder motor 2.4 mm in diameter and 12 mm in length was driven at 62.1 kHz in both rotation directions. No-load speed of 1,800 rpm and output torque of up to 1.8 mN·m were obtained for rotation in both directions under an applied *rms* voltage of 80 V. A rather high maximum efficiency of about 28% for this small motor is a noteworthy feature.[30,31] Various modifications were made for the stator, including a type with four PZT plates, arranged symmetrically and driven by two-phase (sine and cosine) voltages.

In collaboration with the Penn State University, Samsung Electromechanics, Korea, developed a zoom and focus mechanism with two micro-rotary motors in 2003. Two micro metal tube motors with 2.4 mm diameter and 14 mm length were installed to control zooming and focusing lenses independently in conjunction with screw mechanisms, as illustrated in Figure 2.30b.[32] A screw is rotated through a pulley, which is then transferred to the lens up-down motion. The square chip (3×3 mm^2) on the camera module in Figure 2.30c is a high-frequency drive voltage supply. Newscale Technologies (Victor, NY) integrated a screw in the metal tube motor, and commercialized "squiggle motors" worldwide for camera module applications, with a partnership with ALPS,

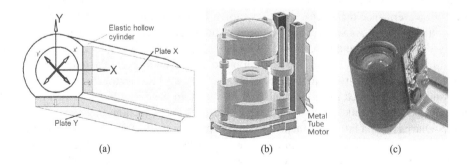

(a)   (b)   (c)

FIGURE 2.30   (a) Structure of a "Hula Hoop Metal tube" motor using a metal tube and two rectangular PZT plates. (b) Structure of the camera auto zooming/focusing mechanism with two metal tube USMs. (c) Photo of the camera module in a Samsung cellular phone.

Tamron, and TDK-EPCOS.[33] Samsung Electromechanics is now utilizing much smaller micro-ML chip linear USMs for the Galaxy series camera modules due to the thinner design necessity.[34]

In parallel to the USM usage, Konica Minolta, Japan, developed a Smooth Impact Drive Mechanism (SIDM) using an ML piezo-element.[35] The idea comes from the "stick & slick" condition of the ring object attached on a drive rod. By applying a sawtooth-shaped voltage to an ML actuator, alternating slow expansion and quick shrinkage are excited on a drive friction rod. A ring slider placed on the drive rod will "stick" on the rod due to friction during a slow expansion period, while it will "slide" during a quick shrinkage period, so that the slider moves from one end to the other end of the rod. The lens is attached to this slider. When the voltage saw shape is reversed, opposite motion can be obtained. Piezo Tech, Korea, developed a similar SIDM motor but with using a bimorph, instead of an ML actuator, that suppressed the cost significantly. The TULA (tiny ultrasonic linear actuator)[36], though a flexural bimorph is used, the driving frequency is much higher than 40 kHz (ultrasonic range) due to a small size.

2.8.3 PIEZOELECTRIC ENERGY HARVESTING SYSTEMS

One of the recent research interests is piezoelectric energy harvesting. Cyclic electric field excited in the piezoelectric plate by the environmental noise vibration is accumulated into a rechargeable battery. Originally, we consumed the converted electrical energy via Joule heat in order to damp the vibration rapidly in the 1980s.[37] After getting into the 1990s, we started to accumulate it in a rechargeable battery, which seems to be the pioneering study.[38–40] There are three major phases/steps associated with piezoelectric energy harvesting: (1) mechanical–mechanical energy transfer, including mechanical stability of the piezoelectric transducer under large stresses, and mechanical impedance matching, (2) mechanical–electrical energy transduction, relating with the electromechanical coupling factor in the composite transducer structure, and (3) electrical–electrical energy transfer, including electrical impedance matching. A suitable DC/DC converter is required to accumulate the electrical energy from a high-impedance piezo-device into a rechargeable battery (low impedance).[40]

Our application target of the "Cymbal" was set to hybrid vehicles with both an engine and an electromagnetic motor, in collaboration with Toyota Central Research Labs; reducing the engine vibration and harvesting electric energy (~ 1 W) to car batteries to increase the mileage. A cymbal with 29 mm diameter and 1~2 mm thick (0.3 mm thick stainless steel endcaps), inserted below a 4 kg engine weight (40 N bias force), was shaken under an electromagnetic shaker at 100 Hz, which generated 80 mW electric power.[38,39] By cascading nine cymbals embedded in engine damper rubber sheets, we succeeded to obtain the total electric power level close to 1 W to a rechargeable battery.

We should also point out here that there is another research school of piezo-energy harvesting; that is, small energy harvesting (mW) for signal transfer applications, where the efficiency is not a primary objective. This school usually treats an impulse/snap action load to generate instantaneous electric energy for transmitting signals for a short period (100 ms – 10 s), without accumulating the electricity in a rechargeable battery. NEC-Tokin developed an LED traffic light array system driven by a piezoelectric windmill, which is operated by wind generated effectively by passing automobiles. Successful products (million sellers) in the commercial market belong mostly to this category at present, including "Lightning Switch"[41] and the 25 mm caliber "Programmable Ammunition"[42]. The former by PulseSwitch Systems, VA, is a remote switch for room lights, with using a unimorph piezoelectric component [Figure 2.31a]. In addition to the living convenience, Lightning Switch can reduce the housing construction cost drastically, due to a significant reduction of the copper electric wire and the aligning labor. To the contrary, the 25-mm caliber "Programmable Ammunition"[42] is based on electricity generation with an ML piezo-actuator under gun-shot impact to maneuver the bullet via an operational amplifier [Figure 2.31b], developed by Micromechatronics Inc., PA, and ATK Integrated Weapon Systems, AZ. A piezo energy harvesting device was installed because the original button battery was decayed in 2 months under high-temperature atomosphere in the Afghanistan battle field.

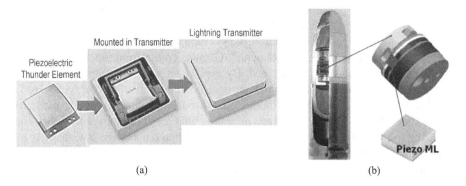

FIGURE 2.31 (a) Lightning Switch with piezoelectric Thunder actuator. (Courtesy by Face Electronics.) (b) Programmable air-bust munition (PABM, 25 mm caliber) developed by Micromechatonics.

CHAPTER ESSENTIALS

1. *Functional classification*: Dielectrics > Piezoelectrics > Pyroelectrics > Ferroelectrics
2. *Origin of the spontaneous polarization*: the dipole coupling with the local field (i.e., Lorentz factor), in conjunction with nonlinear atomic elasticity.
3. *Piezoelectricity*: can be modeled with "two" ideal springs.
   Piezostriction: can be described in terms of the difference between the harmonic terms of the two equivalent springs.
   Electrostriction: can be modeled with a "single" anharmonic equivalent spring.
4. Origins of the field-induced strains:
   a. *Inverse piezoelectric effect*: $x = d \cdot E$
   b. *Electrostriction*: $x = M \cdot E^2$
   c. *Domain reorientation*: strain hysteresis – this is mostly discussed in this textbook.
   d. *Phase transition* (antiferroelectric↔ferroelectric): strain "jump". Refer to Ref. [7]
5. *Shear strain*: $x_5 = 2x_{31} = 2\phi$, taken as positive for smaller angle.
6. The electrostriction equation:

$$x = \underset{\text{(spontaneous strain)}}{Q P_S^2} + \underset{\text{(piezostriction)}}{2Q\varepsilon_0\varepsilon P_S E} + \underset{\text{(electrostriction)}}{Q\varepsilon_0^2\varepsilon^2 E^2}$$

7. The piezoelectric constitutive equations:

$$x = s^E X + d E$$

$$P = d X + \varepsilon_0 \varepsilon^X E$$

8. Electromechanical coupling factor (k):

$$k^2 = \frac{d^2}{s^E\left(\varepsilon^T \varepsilon_0\right)} = \frac{h^2}{c^D\left(\beta^s / \varepsilon_0\right)}$$

9. $\varepsilon^x = \varepsilon^X(1 - k^2)$; $s^D = s^E(1 - k^2)$ – constraint permittivity and elastic compliance are smaller than those under free conditions.
10. PZTs are the major ceramics for piezoelectric device applications.
11. Pb-free piezo-ceramics are being developed for overcoming the future RoHS regulation to PZTs.

CHECK POINT

1. The local field is the driving force for the spontaneous polarization. How is γ called, which enhances the applied electric field E.
2. (T/F) If the atomic spring constant is purely linear, no ferroelectricity should occur. True or False?
3. (T/F) A polycrystalline piezoelectric PZT has three independent piezoelectric d matrix components, d_{33}, d_{13}, and d_{15}. True or False?
4. (T/F) The following two force configurations are equivalent mathematically. True or False?

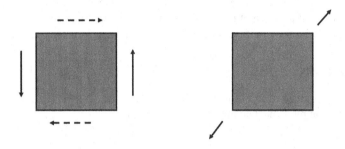

5. Elastic Gibbs energy (in 1D expression) is given by:

$$G_1 = (1/2)\alpha P^2 + (1/4)\beta P^4 + (1/6)\gamma P^6 + \cdots - (1/2)sX^2 - QP^2X$$

Why don't we include the "odd-number" power terms of polarization P? Answer simply.
6. Elastic Gibbs energy is given by:

$$G_1 = (1/2)\alpha P^2 + (1/4)\beta P^4 + (1/6)\gamma P^6 + \cdots - (1/2)sX^2 - QP^2X$$

Where can you find the primary expansion term in respect of "temperature"? Answer simply.
7. (T/F) The phenomenology suggests that the spontaneous polarization of a ferroelectric changes with temperature linearly. True or False?
8. (T/F) The phenomenology suggests that the polarization curve of a ferroelectric under a large electric field cycle shows a hysteresis phenomenon. True or False?
9. (T/F) The phenomenology suggests that the spontaneous strain in a ferroelectric changes with temperature linearly. True or False?
10. (T/F) The phenomenology suggests that the permittivity of a ferroelectric material exhibits the maximum at its Curie temperature. True or False?
11. (T/F) The phenomenology suggests that the piezoelectric constant of a ferroelectric material exhibits the maximum just below its Curie temperature. True or False?
12. (T/F) The phenomenology suggests that the electrostrictive coefficient M of a paraelectric material is almost temperature-insensitive constant. True or False?
13. (T/F) The phenomenology suggests that the Curie–Weiss temperature of a ferroelectric material is always higher than (or equal to) the Curie temperature. True or False?
14. How is the piezoelectric coefficient d related with the electrostrictive coefficient Q, spontaneous polarization P_S, and relative permittivity ε in the Devonshire phenomenology? Using the air permittivity ε_0, provide a simplest formula.
15. (T/F) There is a highly resistive (no electric carrier/impurity in a crystal) piezoelectric single crystal (spontaneous polarization P_S) with a mono-domain state without surface

electrode. The "depolarization electric field" in the crystal is given by $E = -\left(\dfrac{P_S}{\varepsilon_0}\right)$.
True or False?

16. What is a typical number for k_{33} in soft PZT ceramics? 1%, 10%, 50%, or 70%?
17. There is a PZT ML actuator with a cross-section area $5 \times 5\,mm^2$. Provide a generative (blocking force) roughly. 1 N, 10 N, 100 N, or 1 kN?
18. What is the fundamental longitudinal resonance frequency of a PZT ML actuator with a length 10 mm? 1 kHz, 10 kHz, 100 kHz, or 1 MHz?
19. What is the MPB composition of the PZT system at room temperature? $PbZrO_3:PbTiO_3 = (48{:}52)$, (50:50), (52:48), or (none of these)?
20. (T/F) The MPB composition of the PZT system exhibits the maximum electromechanical coupling k, piezoelectric coefficient d, and the minimum permittivity ε. True or False?

CHAPTER PROBLEMS

2.1 The room temperature form of quartz belongs to class **32**.
   a. Show that the piezoelectric matrix (d_{ij}) is given by:

$$
\begin{pmatrix}
d_{11} & -d_{11} & 0 & d_{14} & 0 & 0 \\
0 & 0 & 0 & 0 & -d_{14} & -2d_{11} \\
0 & 0 & 0 & 0 & 0 & 0
\end{pmatrix}
$$

   Notice that the piezoelectric tensor must be invariant for a 120° rotation around the 3 axis and for a 180° rotation around the 1 axis. The transformation matrices are, respectively:

$$
\begin{pmatrix}
-\tfrac{1}{2} & \tfrac{\sqrt{3}}{2} & 0 \\
-\tfrac{\sqrt{3}}{2} & -\tfrac{1}{2} & 0 \\
0 & 0 & 1
\end{pmatrix}
\text{ and }
\begin{pmatrix}
1 & 0 & 0 \\
0 & -1 & 0 \\
0 & 0 & -1
\end{pmatrix}
$$

   b. The measured values of the d_{ij} for right-handed quartz are:

$$
\begin{pmatrix}
-2.3 & 2.3 & 0 & -0.67 & 0 & 0 \\
0 & 0 & 0 & 0 & 0.67 & 4.6 \\
0 & 0 & 0 & 0 & 0 & 0
\end{pmatrix} \times 10^{-12}\ (\text{C/N})
$$

      i. If a compressive stress of 1 kgf/cm^2 is applied along the 1 axis of the crystal, find the induced polarization. (kgf = kilogram force = 9.8 N)
      ii. If an electric field of 100 V/cm is applied along the 1 axis, find the induced strains.

2.2 In the case of a first-order phase transition, the Landau free energy is expanded as:

$$
F(P,T) = (1/2)\alpha P^2 + (1/4)\beta P^4 + (1/6)\gamma P^6 \quad \left(\alpha = (T - T_0)/\varepsilon_0 C\right)
$$

   Calculate the inverse permittivity near the Curie temperature, and verify that the slope $[\partial(1/\varepsilon)/\partial T]$ just below T_C is eight times larger than the slope just above T_C.

Hint

In the first-order phase transition, P_S satisfies the following equation in the temperature range of $T < T_C$:

$$E = \alpha + \beta P_S^2 + \gamma P_S^4 = 0.$$

The permittivity is given by $1/\varepsilon_0\varepsilon = \alpha + 3\beta P_S^2 + 5\gamma P_S^4$. Thus,

$$1/\varepsilon_0\varepsilon = \alpha + 3\beta P_S^2 + 5\left(-\alpha - \beta P_S^2\right) = -4\alpha - 2\beta P_S^2$$

Since $a = (T - T_0)/\varepsilon_0 C$, $P_S^2 = \left(\sqrt{\beta^2 - 4\alpha\gamma} - b\right)\!\big/2g$ and

$$(T - T_0)/\varepsilon_0 C = (3/16)\left(\beta^2/\gamma\right) - (T_C - T)/\varepsilon_0 C,$$

we can obtain

$$1/\varepsilon_0\varepsilon = -4\alpha - 2\beta P_S^2$$

$$= -4\left[(3/16)\left(\alpha^2/\gamma\right) - (T_C - T)/\varepsilon_0 C\right] + \left(\beta^2/\gamma\right) - (\beta/\gamma)\sqrt{\beta^2 - 4\gamma\left[\left(\frac{3}{16}\right)\left(\frac{\beta^2}{\gamma}\right) - \frac{T_C - T}{\varepsilon_0 C}\right]}$$

Considering $(T_C - T) \ll 1$, obtain the approximation of this equation.

REFERENCES

1. J. Valasek, Piezoelectric and allied phenomena in Rochelle salt, *Physical Review*. **15**(6) 537 (1920).
2. K. Aizu, *Journal of the Physical Society of Japan*. **38**, 1592 (1975).
3. C. Kittel, *Introduction to Solid State Physics,* 6th Edition, Chap.13, (John Wiley & Sons, New York, 1986).
4. W. Kinase, Y. Uemura and M. Kikuchi, *Journal of Physics and Chemistry of Solids*. **30**, 441 (1969).
5. K. Uchino, S. Nomura, *Japanese Journal of Applied Physics Bulletin*. **52**, 575 (1983); K. Uchino, S. Nomura, L.E. Cross, R.E. Newnham, S.J. Jang, Electrostrictive effect in perovskites and its transducer applications, *Journal of Material Science*. **16**, 569 (1981).
6. K. Uchino, Electrostrictive actuators: Materials and applications, *American Ceramic Society Bulletin*. **65**(4), 647 (1986).
7. K. Uchino, *Ferroelectric Devices*, 2nd Edition, (CRC Press, New York, 2009).
8. A. F. Devonshire, *Philosophical Magazine*. **40**, 1040 (1949).
9. A. F. Devonshire, *Advances in Physics*. **3**, 85 (1954).
10. H. F. Kay, *Reports on Progress in Physics***43**, 230 (1955).
11. K. Uchino, Barium Titanate Study Committee Paper XXXI-171-1067 (1983).
12. K. Uchino, S. Nomura, L.E. Cross, S.J. Jang and R.E. Newnham, Electrostrictive effect in lead magnesium niobate single crystals, *Journal of Applied Physics*. **51**(2), 1142–1145 (1980).
13. D. R. Tobergte, S. Curtis, IUPAC. Compendium of Chemical Terminology, (the "Gold Book") 53 (2013).
14. J. F. Nye, *Physical Properties of Crystals*, (Oxford University Press, London, 1972), p. 123, p. 140.
15. K. Uchino, *Advanced Piezoelectric Materials*, 2nd Edition, edited by Kenji Uchino, Chapter 1 The development of piezoelectric materials and the new perspective, (Woodhead Publishing/Elsevier, Cambridge, UK, 2017).
16. E. Sawaguchi, G. Shirane, Y. Takagi, Phase transition in lead zirconate, *Journal of the Physical Society of Japan*, **6**, 333–339 (1951).
17. B. Jaffe, Piezoelectric transducers using lead titanate and lead zirconate, US Patent 2,708,244, May (1955).
18. L.E. Cross, S.J. Jang, R.E. Newnham, S. Nomura, K. Uchino, *Ferroelectrics*. **23**, 187 (1980).

19. H. Takeuchi, H. Masuzawa, C. Nakaya, Y. Ito, *Proceedings of IEEE 1990 Ultrasonics Symposium*, 697 (1990).
20. J. Kuwata, K. Uchino, S. Nomura, *Ferroelectrics.* **37**, 579 (1981).
21. J. Kuwata, K. Uchino, S. Nomura, *Japanese Journal of Applied Physics.* **21**, 1298 (1982).
22. K. Yanagiwawa, H. Kanai, Y. Yamashita, *Japanese Journal of Applied Physics.* **34**, 536 (1995).
23. S.E. Park, T.R. Shrout, *Material Research Innovations.* **1**, 20 (1997).
24. H. Kawai, The piezoelectricity of poly (vinylidene fluoride), *Japanese Journal of Applied Physics.* **8**, 975 (1969).
25. T. Tou, Y. Hamaguchi, Y. Maida, H. Yamamori, K. Takahashi, Y. Terashima, *Japanese Journal of Applied Physics.* **48**, 07GM03, (2009).
26. Y. Saito, H. Takao, T. Tani, T. Nonoyama, K. Takatori, T. Homma, T. Nagaya, M. Nakamura, *Nature.* **432**, 84–87 (2004).
27. Y. Doshida, *Proceedings of 81st Smart Actuators/Sensors Study Committee*, Japanese Technology Transfer Association, December 11, Tokyo (2009).
28. www.marklines.com/ja/amreport/rep094_200208.jsp.
29. A. Fujii, *Proceedings of JTTAS Meeting*, December 2, 2005, Tokyo.
30. B. Koc, S. Cagatay, K. Uchino, *IEEE Transaction on Ultrasonic, Ferroelectric, Frequency Control.* **49**(4), 495–500 (2002).
31. S. Cagatay, B. Koc, K. Uchino, *IEEE Transaction on UFFC.* **50**(7), 782–786 (2003).
32. K. Uchino, *Proceedings of New Actuator 2004*, Bremen, June14–16, p. 127 (2004).
33. https://www.newscaletech.com/about-us/.
34. B. Koc, J. Ryu, D. Lee, B. Kang, B.H. Kang, *Proceedings on New Actuator 2006*, Bremen, June 14–16, p. 58 (2006).
35. Y. Okamoto, R. Yoshida, H. Sueyoshi, *Konica Minolta Technology Report.* **1**, 23 (2004).
36. http://www.piezo-tech.com/eng/product/ (2008).
37. K. Uchino, T. Ishii, *Journal of the Ceramic Society of Japan.* **96**(8), 863–867 (1988).
38. H.W. Kim, A. Batra, S. Priya, K. Uchino, D. Markley, R.E. Newnham, H.F. Hofmann, Energy harvesting using a piezoelectric "Cymbal" transducer in dynamic environment, *Japanese Journal of Applied Physics.* **43**(9A), 6178–6183 (2004).
39. H.W. Kim, S. Priya, K. Uchino, R.E. Newnham, *Journal of Electroceramics.* **15**, 27–34 (2005).
40. K. Uchino, *Energy Harvesting with Piezoelectric and Pyroelectric Materials*, edited by Nantakan Muensit, Partial Charge Chapter 4 Energy flow analysis in piezoelectric harvesting systems, (Trans Tech Publications, Zuerich, Switzerland, 2011).
41. K. Uchino, *Proceedings of 12th International Conference on New Actuators*, A3.0, Bremen, Germany, June 14–16 (2010). http://www.lightningswitch.com/.
42. http://www.atk.com/MediaCenter/mediacenter_video gallery.asp.
43. K. Uchino, Piezoelectric actuator renaissance, *Journal of Energy Harvesting and Systems.* **1**(1–2), 45–56 (2014).

3 Fundamentals of Losses

ABSTRACT

Though losses in piezoelectrics are considered to consist of three mechanisms, namely, dielectric, elastic, and piezoelectric losses, we discuss primarily on general electric and mechanical losses in this chapter without considering the coupling loss. In the electric loss (high frequency, Peta Hertz), electromagnetic wave is absorbed by the material's conducting loss. Absorption coefficient is frequency dependent; while, the electric loss at low frequency, we consider (1) inductor components inevitably accompanied by resistive loss and (2) dielectric loss. Though electronic polarization responds up to Peta Hertz, ionic and dipolar polarization does have response limit, which provides the intrinsic dielectric loss. Small conductance in the material contributes as additional effective dielectric loss. We handle three mechanical loss models: (a) Solid damping – damping force proportional to the displacement $F = \zeta \cdot cu_{max}$, (b) Coulomb damping – damping force constant $\pm F$ – linear decay of the displacement amplitude, and (c) viscous damping – damping force proportional to velocity $F = \xi \cdot \dot{u}$ – exponential decay of the displacement amplitude, in this chapter.

3.1 ELECTRIC/DIELECTRIC LOSSES

The most popular loss for the reader may be Joule heat in the electrical circuit, since your elementary school age. We consider first electric and dielectric losses in conducting and insulative materials, respectively.

3.1.1 ELECTRIC LOSSES

Let us start from the Maxwell electromagnetic (EM) equations:

$$\nabla \cdot \boldsymbol{D} = \rho \ (\rho : \text{charge}) \quad : \text{Gauss's Law} \tag{3.1}$$

$$\nabla \cdot \boldsymbol{B} = 0 \quad : \text{Gauss's Law for magnetism} \tag{3.2}$$

$$\nabla \times \boldsymbol{E} = -\frac{\partial \boldsymbol{B}}{\partial t} \quad : \text{Faraday's Law} \tag{3.3}$$

$$\nabla \times \boldsymbol{H} = \frac{\partial \boldsymbol{D}}{\partial t} + \boldsymbol{J} \quad : \text{Ampère-Maxwell Law} \tag{3.4}$$

When we consider first a conducting material, which obeys Ohm's law, we can write

$$\boldsymbol{J} = \sigma \cdot \boldsymbol{E}, \tag{3.5}$$

where σ is the conductivity. Then, the above third and fourth dynamic equations become

$$\nabla \times \boldsymbol{E} = -\mu_0 \mu \, \frac{\partial \boldsymbol{H}}{\partial t} \tag{3.6}$$

$$\nabla \times \boldsymbol{H} = \varepsilon_0 \varepsilon \frac{\partial \boldsymbol{E}}{\partial t} + \sigma \cdot \boldsymbol{E}, \tag{3.7}$$

where we assume isotropic dielectric and magnetic properties as $\boldsymbol{D} = \varepsilon_0 \varepsilon \boldsymbol{E}$ and $\boldsymbol{B} = \mu_0 \mu \boldsymbol{H}$. Here $\varepsilon_0 = 8.854 \times 10^{-12}$F/m and $\mu_0 = 4\pi \times 10^{-7}$H/m, and $\varepsilon_0 \cdot \mu_0 = 1/c^2$ (c: light velocity in vacuum) in the MKS unit system. The usual method of taking "curl" of the first and using the second equations, and $\nabla \times \nabla \times \boldsymbol{E} = \nabla(\nabla \cdot \boldsymbol{E}) - \nabla^2 \boldsymbol{E}$ now yields

$$\nabla^2 \boldsymbol{E} = \mu_0 \mu \left(\nabla \times \frac{\partial \boldsymbol{H}}{\partial t} \right) = \mu_0 \mu \left(\varepsilon_0 \varepsilon \frac{\partial^2 \boldsymbol{E}}{\partial t^2} + \sigma \frac{\partial \boldsymbol{E}}{\partial t} \right) = \frac{\mu \cdot \varepsilon}{c^2} \ddot{\boldsymbol{E}} + \mu_0 \mu \sigma \dot{\boldsymbol{E}}. \tag{3.8}$$

Here, we can assume $\nabla \cdot \boldsymbol{E} = 0$ from Gauss's law only when free charge $\rho = 0$ (such as in reasonably resistive medium with low σ). A similar equation for $\boldsymbol{H}$,

$$\nabla^2 \boldsymbol{H} = -\varepsilon_0 \varepsilon \left(\nabla \times \frac{\partial \boldsymbol{E}}{\partial t} \right) - \sigma \left(\nabla \times \boldsymbol{E} \right) = \mu_0 \mu \left(\varepsilon_0 \varepsilon \frac{\partial^2 \boldsymbol{H}}{\partial t^2} + \sigma \frac{\partial \boldsymbol{H}}{\partial t} \right) = \frac{\mu \cdot \varepsilon}{c^2} \ddot{\boldsymbol{H}} + \mu_0 \mu \sigma \dot{\boldsymbol{H}}. \tag{3.9}$$

3.1.1.1 Electromagnetic Wave (High Frequency)

Let us initially consider an EM wave at a high frequency such as Peta Hertz (i.e., visible light). If we focus on plane waves by assuming that the components of $\boldsymbol{E}$ and $\boldsymbol{H}$ are functions of the propagation direction z alone, E_z and H_z turn out to be constants which may appropriately be put equal to zero.[1] Then we have to deal with equations of the form of E_x and H_y (electric and magnetic waves are orthogonal each other) as illustrated in Figure 3.1:

$$\frac{\partial^2 E_x}{\partial z^2} = \frac{\ddot{E}_x}{v_c^2} + \frac{\sigma}{\varepsilon v_c^2} \dot{E}_x, \tag{3.10}$$

$$\frac{\partial^2 H_y}{\partial z^2} = \frac{\ddot{H}_y}{v_c^2} + \frac{\sigma}{\varepsilon v_c^2} \dot{H}_y, \tag{3.11}$$

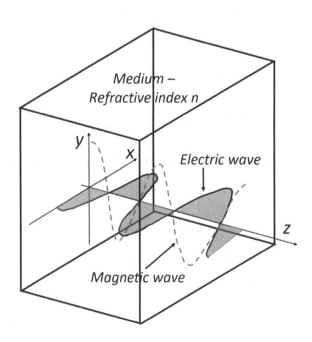

FIGURE 3.1 EM wave propagation.

where $v_c = c/n_0$ (c: light velocity in vacuum), and the refractive index $n_0 = \sqrt{\varepsilon\mu}$. Note that ε here is the relative permittivity (only electronic polarizability!) at Peta Hertz range, and the value is less than 10 (not as high as 1,000 in a ferroelectric at 1 kHz). Also note the famous relationship: $\varepsilon = n_0^2$. We consider harmonic time dependence and a solution of the form

$$E_x = f(z) \cdot e^{-j\omega t}. \tag{3.12}$$

Substitution yields, in the case of wave propagation in the positive z direction,

$$f(z) = A e^{j\sqrt{\omega^2 - \frac{i\sigma\omega}{\varepsilon}} \cdot z/v_c} \tag{3.13}$$

where A is an arbitrary constant. If we write

$$\sqrt{\varepsilon\mu\omega^2 - j\mu\sigma\omega} = \omega(n' - jn''), \tag{3.14}$$

whence

$$n' = \sqrt{\frac{\mu}{2}\left(\sqrt{\varepsilon^2 + 4\sigma^2\left(\frac{2\pi}{\omega}\right)^2} + \varepsilon\right)}, \tag{3.15}$$

$$n'' = \sqrt{\frac{\mu}{2}\left(\sqrt{\varepsilon^2 + 4\sigma^2\left(\frac{2\pi}{\omega}\right)^2} - \varepsilon\right)}. \tag{3.16}$$

This allows us to write, with appropriate attention to signs, for the part of E_x corresponding to wave propagation in the positive z direction

$$E_x = A e^{-\frac{\omega n'' x}{c}} e^{-j\omega\left(t - \frac{n'}{c}z\right)} \tag{3.17}$$

which is a wave progressing with phase velocity

$$V = c/n' \tag{3.18}$$

Here n' appears as the effective index of refraction of the conducting medium (larger than the above n_0), and V is smaller than v_c in Eqs. (3.10) and (3.11). When the conductance is small, effective refractive index approaches to $n' = n_0$ and $n'' = 2\pi\sqrt{\frac{\mu}{\varepsilon}}\sigma/\omega$. The wave is damped, and the decrease of amplitude with distance z is measured by $e^{-\frac{\omega n'' z}{c}}$, so that the effective absorption coefficient is

$$\alpha = \frac{\omega n''}{c} \tag{3.19}$$

An electrical conductor absorbs EM radiation with a damping coefficient depending on the frequency, which corresponds to an "EM loss". Moreover, the velocity itself will in general depend on the frequency [see Eq. (3.15)], and therefore, the conductor will appear as a dispersive medium. This is a major reason why a metal plate case is occasionally used (rather than light-weight plastic case) for EM shield to protect the electronic circuit from the external magnetic noise.

3.1.1.2 Electrical Circuit (Low Frequency)

Now we consider a low-frequency (< 1 GHz or EM wave length [m] or longer) case. In this case, we do not need to consider EM waves, and the electronic components such as inductance L, capacitance C, and resistance R can be handled as discrete components. Figure 3.2a and b shows example structures of inductor (helical coil) and of capacitor (multilayer type), respectively. A simple solenoid coil with dimensions, radius R_c, length L, and coil wire radius r, length l, and the number of turn N, provides inductance and resistance of

$$L = \mu\mu_0 \frac{N^2 \pi R_c^2}{l}, \tag{3.20}$$

and

$$R = \rho \frac{l}{r^2}, \tag{3.21}$$

respectively, where μ is the relative permeability of the coil core material (if we use. if not $\mu = 1$ in the air) and ρ is the resistivity of the coil wire. Solenoid electric energy under current I is given by $\frac{1}{2}LI^2$, and the Joule heat can be calculated as RI^2 (usually impedance $\omega L \gg R$). Thus, an inductor is described with the circuit component notations both L and R, as in Figure 3.2c.

On the other hand, a multilayer capacitor is composed of a high-permittivity dielectric material with high resistivity and internal electrode metal sheets laminated alternatively in order to increase the capacitance by a factor of n^2, where n is the number of layers (by neglecting unelectroded inactive side region). Though highly resistive, since the layer thickness is rather thin (typically 3–10 μm of dielectric layers), the actual conductance cannot be neglected. Thus, a capacitor is described with the circuit component notations both C and conductor G, as in Figure 3.2d. Different from the inductor controlled by current (i.e., impedance-related $j\omega L$), the capacitor is controlled by voltage (i.e., admittance-related $j\omega C$). Because of this reason, we occasionally use the parallel connection circuit of C and conductance G (usually $\omega C \gg G$), in comparison with the series connection circuit of L and resistance R (usually impedance $\omega L \gg R$). Refer to Section 3.1.2 on the microscopic origin of dielectric loss.

Regarding a resistor, its basic structure is a helical coil similar to Figure 3.2a. However, by using a film strip of high-resistivity metal or carbon, the resistance is increased significantly (usually

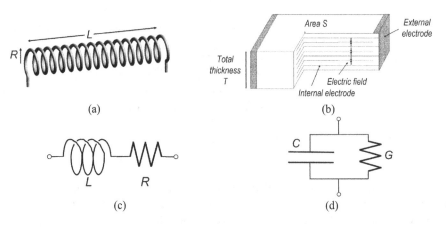

FIGURE 3.2 (a) Structure of inductor and helical coil; (b) structure of capacitor, multilayer type; (c) circuit component notation for inductor L; and (d) circuit component notation for capacitor C.

impedance $R \gg \omega L$). In practice, temperature stability of the resistance material is important as an electronic component, even the self-heating occurs due to Joule loss during usage.

3.1.2 DIELECTRIC LOSSES

In *dielectric* materials with high resistivity, the constituent atoms are considered to be ionized to a certain degree and are either positively or negatively charged. When an electric field is applied, cations are attracted to the cathode and anions to the anode due to electrostatic interaction. This phenomenon is known as *ionic polarization* of the dielectric, and the polarization is expressed quantitatively as the sum of the electric dipoles per unit volume $[C/m^2]$. Refer to Figures 2.1 and 2.2. Electronic polarization can follow alternating fields with frequencies up to THz–PHz (10^{12}–10^{15} cycle/second, light wave), which corresponds to EM wave propagation range discussed in the previous section, while ionic polarization responds up to GHz–THz (10^9–10^{12} cycle/second, microwave region). Permanent dipole reorientation can follow only up to MHz–GHz (10^6–10^9 cycle/second). Polar water molecule (H_2O) is excited in a microwave oven at around 2 GHz and absorbs the energy to increase the temperature. This is why ferroelectric materials with permanent dipoles cannot be used for microwave dielectric materials; their permittivity is typically high at low frequencies (kHz) but decreases significantly with increasing applying electric field frequency.

Compared with air-filled capacitors, dielectric capacitors can store more electric charge due to the dielectric polarization P induced by the external field, as schematically shown in Figure 3.3 (Figure 2.3 is reused). The physical quantity corresponding to the total stored electric charge per unit area is called the *electric displacement D* and is related to the externally applied electric field E by the following expression:

$$D = \varepsilon_0 E + P = \varepsilon \varepsilon_0 E. \tag{3.22}$$

Here, ε_0 is the vacuum permittivity ($= 8.854 \times 10^{-12}$F/m), ε is the material's *relative permittivity* (also simply called permittivity or *dielectric constant*, and in general is a tensor property). The subscript "*b*" of the charge in Figure 3.3 stands for "bound" charge originated from charge on the top of the induced dipole moment, and "*f*" is for "free" charge supplied from the coated electrode metal via the power supply, which usually screens/compensates the bound charge.

The magnitude of induced polarization P depends on the crystal structure. As we discussed in Section 2.3, dielectric materials with high *Lorentz polarization factor γ* should induce high polarization, leading to high permittivity. As imagined from the response speed limitation of the ionic polarizability (GHz–THz) (Figure 2.2), polarization induction requires some delay time after the

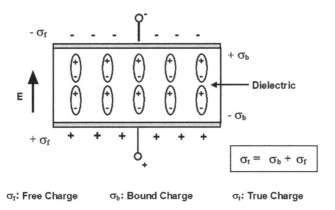

σ$_f$: Free Charge   σ$_b$: Bound Charge   σ$_t$: True Charge

FIGURE 3.3 Charge accumulation in a dielectric capacitor.

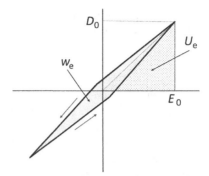

FIGURE 3.4 *D* vs. *E* (stress-free) curve with hysteresis.

external electric field *E* application. In order to express the delay, we adopt the complex permittivity ε^* such that

$$D = \varepsilon^* \varepsilon_0 E = \varepsilon_0 \left(\varepsilon' - j\varepsilon''\right) E$$

$$= \varepsilon_0 \varepsilon'(1 - j\tan\delta)E. \tag{3.23}$$

Here $j = \sqrt{-1}$, and $\tan\delta = \varepsilon''/\varepsilon'$ is called *dielectric loss factor* or *dissipation factor*.

Figure 3.4 illustrates electric displacement *D* (almost the same as polarization *P* in ferroelectrics) vs. electric field *E* curve under stress-free condition with some hysteresis. In a typical "delay" case of the ***D*** generation after the electric field ***E*** application, the hysteresis is the counterclockwise rotation (as indicated by arrows) on this narrow elliptic locus. Note also that the hysteresis shape should be an "ellipse" in this complex notation; that is, a rounded edge at the maximum field point, which suggests a discrepancy from the actual experimental result, and the limitation of this complex parameter approach. Example Problem 3.1 is for further learning the numerical derivation of the loss factor from the observed *D–E* hysteresis curve.

Example Problem 3.1

When the observed variation in electric displacement, *D*, can be represented as if it had a slight phase lag with respect to the applied electric field (i.e., harmonic approximation), we can describe the relation *D* vs. *E* as

$$E^* = E_0 e^{j\omega t} \tag{P3.1.1}$$

$$D^* = D_0 e^{j(\omega t - \delta)} \tag{P3.1.2}$$

We can rewrite the relationship between *D* and *E* by squeezing the phase lag into the complex permittivity:

$$D = \varepsilon^* \varepsilon_0 E \tag{P3.1.3}$$

where the *complex dielectric constant, ε^*,* is

$$\varepsilon^* = \varepsilon' - j\varepsilon'', \tag{P3.1.4}$$

$$\varepsilon''/\varepsilon' = \tan\delta \tag{P3.1.5}$$

The integrated area inside the hysteresis loop, labeled w_e in Figure 3.4, is equivalent to the energy loss per cycle per unit volume of the dielectric. It is defined for an isotropic dielectric as:

$$w_e = -\int D\, dE = -\int_0^{\frac{2\pi}{\omega}} D\,\frac{dE}{dt}\, dt \tag{P3.1.6}$$

1. Substituting the electric field, E, and electric displacement, D, into Eq. (P3.1.6), obtain the following equations:

$$w_e = \pi\varepsilon''\varepsilon_0 E_0^2 = \pi\varepsilon''\varepsilon_0 E_0^2 \tan\delta \tag{P3.1.7}$$

2. Verify an alternative expression (i.e., visual calculation from the hysteresis loop) for the dissipation factor:

$$\tan\delta = (1/2\pi)\left(w_e/U_e\right), \tag{P3.1.8}$$

where U_e, the integrated area so labeled in Figure 3.4, represents the energy stored during a quarter cycle.

Solution

The integrated area inside the hysteresis loop, labeled w_e in Figure 3.4, is equivalent to the energy loss per cycle per unit volume of the dielectric. It is defined as $w_e = -\int D\, dE = -\int_0^{\frac{2\pi}{\omega}} D\,\frac{dE}{dt}\, dt$.

Substituting the real parts of the electric field, $E*$ (i.e., $E_0 \cos(\omega t)$), and electric displacement, $D_0 \cos(\omega t - \delta)$, into Eq. (P3.1.6) yields:

$$w_e = \int_0^{\frac{2\pi}{\omega}} D_0 \cos(\omega t - \delta)\left[E_0\omega \cdot \sin(\omega t)\right] dt = E_0 D_0\ \omega \sin(\delta) \cdot \int_0^{\frac{2\pi}{\omega}} \sin^2(\omega t) dt$$

$$= \pi E_0 D_0 \sin(\delta) \tag{P3.1.9}$$

so that

$$w_e = \pi\varepsilon''\varepsilon_0 E_0^2 = \pi\varepsilon'\varepsilon_0 E_0^2 \tan\delta \tag{P3.1.10}$$

When there is a phase lag, an energy loss (or non-zero w_e) will occur for every cycle of the applied electric field, resulting in the heat generation in the dielectric material. The quantity $\tan\delta$ is referred to as the *dissipation factor*. The electrostatic energy stored during a half-cycle of the applied electric field is $2U_e$, where U_e, the integrated area so labeled in Figure 3.4, represents the energy stored during a quarter cycle.

$$2U_e = 2\left[(1/2)\left(E_0 D_0 \cos\delta\right)\right] = \left(E_0 D_0\right)\cos\delta \tag{P3.1.11}$$

Knowing that $\varepsilon'\varepsilon_0 = (D_0/E_0) \cos\delta$, Eq. (P3.1.11) may be rewritten in the form:

$$2U_e = \varepsilon'\varepsilon_0 E_0^2 \tag{P3.1.12}$$

Then, an alternative expression for the dissipation factor can be obtained:

$$\tan\delta = (1/2\pi)\left(w_e/U_e\right) \tag{P3.1.13}$$

Note that the factor 2π comes from the integration process for one cycle. In conclusion, the loss dissipation factor $\tan \delta$ is obtained by getting the area ratio on the experimentally obtained P–E hysteresis curve.

Ferroelectric materials are occasionally doped by donor or acceptor ions in order to enhance the piezoelectric performance, and polycrystalline (ceramic) specimens possess grain boundaries with various crystal dislocations/deficits. In these cases, the material's resistivity is occasionally degraded. Let us discuss a slightly conductive (or semiconductive) dielectric material's case, where the hysteresis becomes enhanced apparently.

We start from Eq. (3.1), pseudo-static Gauss Law $\nabla \cdot \boldsymbol{D} = \rho$. Here ρ is the true charge density. Taking a *continuity equation*,

$$\frac{\partial \rho}{\partial t} + div(\boldsymbol{i}) = 0 \tag{3.24}$$

and the relations, $\dfrac{\partial}{\partial t} \to j\omega$ (harmonic approximation) and $\boldsymbol{i} = \sigma \boldsymbol{E}$, we obtain

$$\nabla \cdot \left(\boldsymbol{D} - \left(\frac{j\sigma}{\omega} \right) \boldsymbol{E} \right) = 0. \quad \text{(Extended Gauss's law)} \tag{3.25}$$

If we adopt $\boldsymbol{D} = \varepsilon * \varepsilon_0 \boldsymbol{E}$, we rewrite the equation as

$$\varepsilon_0 \left(\varepsilon' - j \left[\varepsilon'' + \left(\frac{\sigma}{\varepsilon_0 \omega} \right) \right] \right) \nabla \cdot \boldsymbol{E} = 0 \tag{3.26}$$

Thus, conductance enhances the imaginary loss part ε'', leading to the phase lag increase and a wider hysteresis curve apparently. This is the main reason why we observe a rounded large hysteresis in the $P - E$ curve measurement in a lossy ferroelectric/dielectric material. Note that the intrinsic dielectric loss $\tan \delta$ is not directly related with the measuring frequency, but the conductivity-related "apparent" dielectric loss exhibits a significant frequency dependence (i.e., it diminishes with increasing the frequency ω). This is the experimental base to distinguish the conductance loss in a piezoelectric material.

3.1.3   LCR Circuit

The reader is probably familiar with a resonance circuit composed of an inductor L and a capacitor C. We discuss on a series LCR circuit (by adding a resistance R), as shown in Figure 3.5. Under a voltage $V(t)$ applied, denoting the common circuit current $I(t)$ and charge on a capacitor Q, we obtain the equation

$$L\left(\frac{dI}{dt} \right) + \frac{Q}{C} + RI = V(t). \tag{3.27}$$

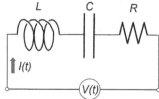

FIGURE 3.5   LCR circuit.

Since $I = \dfrac{dQ}{dt}$, Eq. (3.27) can be rewritten as

$$L\left(\frac{d^2Q}{dt^2}\right) + R\left(\frac{dQ}{dt}\right) + \frac{Q}{C} = V(t), \tag{3.28}$$

or

$$L\left(\frac{d^2I}{dt^2}\right) + R\left(\frac{dI}{dt}\right) + \frac{I}{C} = \frac{dV}{dt}. \tag{3.29}$$

When we apply a regular AC voltage $V(t) = V_0 \sin \omega t$ and assume the steady current with the same frequency ω and the phase delay ϕ, $I(t) = I_0 \sin(\omega t - \phi)$,

$$-\omega^2 LI_0 \sin(\omega t - \phi) + \omega RI_0 \cos(\omega t - \phi) + I_0 \sin(\omega t - \phi)/C = \omega V_0 \cos \omega t.$$

The solution is

$$I(t) = \frac{V_0}{Z} \sin(\omega t - \phi) \tag{3.30}$$

$$Z = \sqrt{R^2 + \left(L\omega - \frac{1}{c\omega}\right)^2} \tag{3.31}$$

$$\tan \phi = -\frac{L\omega - \dfrac{1}{c\omega}}{R} \tag{3.32}$$

When $\left(L\omega - \dfrac{1}{C\omega}\right) = 0$, or $\omega = 1/\sqrt{LC}$, the impedance $Z = R$, which is the minimum in Z, and the current shows the maximum. Frequency dependence of the impedance Z and phase $\tan \phi$ is discussed in Section 3.3.

3.2 MECHANICAL LOSS/DAMPING MODELS

Mechanical/elastic losses are associated with energy dissipation from the mechanical system and the heat generation, in parallel to the vibration damping in mechanical vibrations. When the reader was a kid, you enjoyed significant temperature rise by cyclically extending and releasing a thick rubber band.

3.2.1 MECHANICAL ENERGY LOSS CALCULATION

Vibration amplitude decay of a free vibration is originated from the energy loss. Though the vibration damping is not harmonic in general, since it is occasionally close to "harmonic" when the damping factor is not large, we assume *harmonic oscillation* from a mathematical simplicity viewpoint.

Let us consider a general free-vibration case that the harmonic displacement x under a harmonic force F with a phase lag ϕ:

$$F = F_0 \cos(\omega t) \tag{3.33}$$

$$x = x_0 \cos(\omega t - \phi) \tag{3.34}$$

We calculate the mechanical "work" (i.e., the product of force F and displacement x) done by this force during one cycle:

$$\Delta W = \int x\, dF = \int_0^{\frac{2\pi}{\omega}} x \cdot \frac{dF}{dt}\, dt = \omega x_0 F_0 \int_0^{\frac{2\pi}{\omega}} \cos(\omega t - \phi) \cdot \left[-\sin(\omega t) \right] dt$$

$$= \omega x_0 F_0 \int_0^{\frac{2\pi}{\omega}} \left[\cos(\omega t)\cos(\phi) + \sin(\omega t)\sin(\phi) \right] \cdot \left[-\sin(\omega t) \right] dt \qquad (3.35)$$

$$= -\omega x_0 F_0 \left[\cos(\phi) \int_0^{\frac{2\pi}{\omega}} \cos(\omega t)\sin(\omega t)\, dt + \sin(\phi) \int_0^{\frac{2\pi}{\omega}} \sin^2(\omega t)\, dt \right]$$

Since one cycle integration of the first part becomes zero, while the second part provides π/ω, we obtain

$$\Delta W = -\pi x_0 F_0 \sin(\phi) \qquad (3.36)$$

Negative work stands for energy loss per cycle from the system and the loss becomes maximum when the phase lag is 90° or $\pi/2$.

As you may know, in the "forced vibration" case, the energy loss above is compensated only by the force component 90° ahead of the displacement.

3.2.2 LAPLACE TRANSFORM

In order to detail the transient vibration damping, we had better review an important mathematical tool, the *Laplace transform*. The Laplace transform is generally employed for handling the transient response to a pulse-like input; while the Fourier transform is preferred for cases where a steady output is discussed under a continuous sinusoidal (harmonic) input applied. The latter steady-state discussion is held in Section 3.3.

3.2.2.1 Principle of the Laplace Transform

We consider a function $u(t)$ which is defined for $t \geq 0$ ($u(t) = 0$ for $t < 0$) and satisfies $|u(t)| \leq k\, e^{\delta t}$ for all δ not less than a certain positive real number δ_0. When these conditions are satisfied, $e^{-st}\, u(t)$ is absolutely "integrable" for $\mathrm{Re}(s) \geq \delta_0$. We define the Laplace transform:

$$U(s) = L[u(t)] = \int_0^\infty e^{-st} u(t)\, dt \qquad (3.37)$$

The inverse Laplace transform is represented as $L^{-1}[U(s)]$. Application of the useful theorems for the Laplace transform reduces the work of solving certain differential equations by reducing them to simpler algebraic forms. The procedure is applied as follows:

1. Transform the differential equation to the s-domain by means of the appropriate Laplace transform.
2. Manipulate the transformed algebraic equation and solve for the output variable.
3. Obtain the inverse Laplace transform from a list in Table 3.1.

TABLE 3.1
Some Common Forms of the Laplace Transform

	$H(t)$	$G(s)$
1	$1(t)$: Heaviside Step function $1(t) = 1, t > 0$; $1(t) = 0, t < 0$	$1/s$
2	$\delta(t)$: Dirac impulse function $\delta(t) = \infty, t = 0$; $\delta(t) = 0, t \neq 0$	1
3	$t^n/n!$ (n : positive integer)	$1/s^{n+1}$
4	e^{-at} (a : complex)	$1/(s + a)$
5	$\dfrac{t^{n-1}e^{-at}}{(n-1)!}$ $(0! = 1)$	$1/(s + a)^n$ (n: positive integer)
6	$\cos(at)$	$s/(s^2 + a^2)$
7	$\sin(at)$	$a/(s^2 + a^2)$
8	$e^{-bt}\cos(at)\, a^2 > 0$	$\dfrac{s + b}{(s+b)^2 + a^2}$
9	$e^{-bt}\sin(at)\, a^2 > 0$	$\dfrac{a}{(s+b)^2 + a^2}$
10	$\cosh(at)$	$s/(s^2 - a^2)$
11	$\sinh(at)$	$a/(s^2 - a^2)$
12	$e^{-bt}\cosh(at)\, a^2 > 0$	$\dfrac{s + b}{(s+b)^2 - a^2}$
13	$e^{-bt}\sinh(at)\, a^2 > 0$	$\dfrac{a}{(s+b)^2 - a^2}$
14		$\dfrac{1}{s}\left(e^{-as} - e^{-bs}\right)$
15		$\dfrac{1}{s}\tanh\left(\dfrac{as}{2}\right)$
16		$\dfrac{m}{s^2} - \dfrac{ma}{2s}\left[\coth\left(\dfrac{as}{2}\right) - 1\right]$

Useful theorems for the Laplace transform are summarized below.

Useful Theorems for the Laplace Transform:

a. **Linearity:**

$$L\left[a u_1(t) + b u_2(t)\right] = a U_1(s) + b U_2(s) \tag{3.38}$$

$$L^{-1}\left[a U_1(s) + b U_2(s)\right] = a u_1(t) + b u_2(t) \tag{3.39}$$

b. **Differentiation with respect to t:**

$$L\left[\frac{du(t)}{dt}\right] = sU(s) - u(0) \tag{3.40}$$

$$L\left[\frac{d^2u(t)}{dt^2}\right] = s^2U(s) - su(0) - u'(0) \qquad (3.41)$$

$$L\left[\frac{d^nu(t)}{dt^n}\right] = s^nU(s) - \sum s^{n-k}u^{k-1}(0) \qquad (3.42)$$

c. **Integration:**

$$L\left[\int u(t)\,dt\right] = U(s)/s + (1/s)\left[\int u(t)\,dt\right]_{t=0} \qquad (3.43)$$

d. **Scaling formula:**

$$L[u(t/a)] = aU(sa) \quad (a > 0) \qquad (3.44)$$

e. **Shift formula with respect to** *t*: $u(t - k) = 0$ for $t < k$ [*k*: positive real number]. The $u(t)$ curve shifts by k along the positive t axis.

$$L[u(t - k)] = e^{-ks}U(s) \qquad (3.45)$$

f. **Differentiation with respect to an independent parameter:**

$$L\left[\frac{\partial u(t,x)}{\partial x}\right] = \frac{\partial U(s,x)}{\partial x} \qquad (3.46)$$

g. **Initial and final values:**

$$\lim_{t\to 0}[u(t)] = \lim_{|s|\to\infty}[sU(s)] \qquad (3.47)$$

$$\lim_{t\to\infty}[u(t)] = \lim_{|s|\to 0}[sU(s)] \qquad (3.48)$$

3.2.2.2 Common Forms of the Laplace Transform

Table 3.1 lists some common forms of the Laplace transform to be referred in solving the damping vibration transient response in the following sections.

3.2.3 MECHANICAL LOSS CLASSIFICATION

The reader may remember the temple bell timbre. When hit suddenly, the sound pressure level decays gradually without changing its timbre, that is, all frequency components are damped in a similar fashion. On the contrary, in underwater acoustics, the acoustic absorption by water medium is accelerated with increasing the frequency. A lower frequency "sonar" (an acronym for sound navigation ranging) system is required for detecting an enemy submarine as far as possible. We introduce three loss models in the following subsections: solid damping, Coulomb damping, and viscous damping.

Figure 3.6 shows a simple mass–spring (*m*: mass, *c*: spring constant) harmonic vibration model (a) and with a damper (ζ: damping factor) (b). Without a damper, using notations, displacement *u*, and force *F*, we can describe the Newton equation as

$$m\frac{d^2u}{dt^2} + cu = F(t) \qquad (3.49)$$

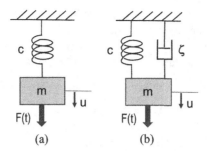

FIGURE 3.6 A simple mass–spring harmonic vibration model (a) and with a damper (b).

When $F = 0$, assuming the harmonic solution $u(t) = u_0\, e^{-i\omega t}$, Eq. (3.49) can be solved as

$$\left(-m\omega^2 + c\right)u = 0 \rightarrow \omega = \sqrt{c/m} \tag{3.50}$$

By keeping the initial displacement amplitude u_0, sinusoidal vibration with the above frequency ω will continue forever without damping. The above $\omega = \sqrt{c/m}$ is called the *resonance frequency* of this mechanical system.

Now we consider to integrate a damping factor into Eq. (3.49) as Figure 3.6b.[2] When the damping mechanism is associated directly with the displacement u, the damping factor is irrelevant to the frequency, which corresponds to the above temple bell case, and is called "solid damping". On the other hand, as you know, in the surface friction, the force direction changes sign (with keeping the force magnitude constant), according to the vibration direction. The vibration damping in this case does not behave in an exponential fashion, which is called "Coulomb damping". "Viscous damping" is observed occasionally for an object moving in liquid at relatively low speed. The damping term is represented as proportional to the object velocity, $\left(\dfrac{du}{dt}\right)$. Because of its mathematical simplicity, the viscous damping formula is most popularly used even for solid materials. More details on these three models are described in the following subsections.

3.2.4 Solid Damping

3.2.4.1 Solid Damping/Structural Damping

The solid damping is sometimes called "structural damping" and is originated from the "internal friction" in the material, which is different from the "surface friction" introduced in Section 3.2.5. The solid damping does not change with the vibration frequency but is proportional to the *maximum stress* generated during the vibration cycle, which is also different from the "viscous damping" in Section 3.2.6. When we impact a solid suddenly, various vibration modes are excited, and their sound level decreases monotonously without changing the timbre. That is, all frequency modes damp under a similar damping rate. Since the stress is almost proportional to the strain in the elastic deformation range, we can rephrase that the solid damping is proportional to the *maximum strain*.

We introduce here the damping force in proportion to the strain, irrelevant to the operating frequency:

$$F = \zeta\, cx \tag{3.51}$$

where ζ is the non-dimensional damping factor and c is the spring constant. It is important to distinguish the situation from the modification of Eq. (3.49) with an effective force constant $c(1 - \zeta)$. The vibration will diminish according to the damping force with keeping a similar resonance frequency of Eq. (3.38). Since the sign of the damping force is not clear, different from the case in

"Coulomb damping", it is difficult to analyze the damping procedure with normal kinetic equations but is better to use the energy consumption method.

Though the vibration is not perfectly harmonic, the damping vibration is supposed to be approximated by a harmonic model, if the damping factor ξ is reasonably small. The energy loss per cycle is approximated by

$$\Delta W = \pi x_0 F_0 = \pi \zeta c x_0^2 \tag{3.52}$$

From the experiments, the solid damping energy loss is irrelevant to the operating frequency and is proportional to the square of the maximum displacement x_0. The above harmonic assumption seems to be true in the case of small ζ of the system. The logarithmic damping rate δ [*logarithmic decrement* is defined as $\ln\left(\dfrac{x_1}{x_2}\right)$, where x_1 and x_2 are the successive vibration amplitudes] of a free vibration via the solid damping mechanism can be estimated from the energy loss ΔW [refer to Eq. (3.90) later] as

$$\delta = \frac{\Delta W}{2W} = \frac{\pi \zeta c x_0^2}{2 \times \dfrac{1}{2} c x_0^2} = \pi \zeta \tag{3.53}$$

Equation (3.53) is occasionally utilized for determining the damping rate of structural components of an airplane, such as wing, tail-wing, and body structure. The logarithmic damping factor δ is irrelevant to the frequency, similar to the case of viscous damping (refer to Section 3.2.6).

3.2.4.2   Kaiser Effect – Material's Memory of History

The above solid damping is proportional to the maximum strain generated during the vibration cycle, which reminds us another intriguing material's memory function on how much maximum strain has been applied historically on the material. This memory function may not be directly related with "solid damping" mechanism but is introduced for the reader's reference. Acoustic emission (AE) method is a non-destructive technique used to detect pulses of released elastic strain energy caused by deformation, crack growth, and phase change in a solid.[3] The AE method has also been used to investigate fundamental domain motions in ferromagnetic and ferroelectric materials,[4,5] as well as mechanical and fatigue properties of materials.[6] Figure 3.7a shows the measurement system in International Center for Actuators and Transducers (ICAT), Penn State University. Refer to Ref. [7] for the details. The electric field-induced AE and displacement can be observed simultaneously. The AE signal is detected by an AE sensor with the sensor PZT resonant frequency of 450 kHz. The AE signal is amplified by 40 dB through a pre-amplifier and again up to 60 dB with a main amplifier. AE event is counted (by neglecting the AE signal intensity) with applying the electric field.

The electric field-induced AE generations were measured in the piezoelectric PZT ceramics under unipolar fields (only positve field drive 0–3 kV/mm not to generate the major domain reversal). Figure 3.7b demonstrates so-called "Kaiser effect" in a PZT disk sample. From the initial "unpoled" state, unipolar field application increases the electrically poled volume in the specimen, and generates the displacement. The field-induced AE is observed only in the field "increasing" process (no signal in the field decreasing process) but is not present after poling as long as the applied field is maintained lower than that of the poling field, which is called *Kaiser effect*. The piezoelectric ceramic material seems to memorize experienced electric field or previous strain history.[7,8] In other words, input electric and converted energy loses energy into AE acoustic energy only in the field/ strain "increasing" process, which is very suggestive results to consider the solid damping mechanism. Fractal analysis of AE during the domain reversal under bipolar drive will be introduced in Chapter 8 for discussing the domain wall dynamic mechanism.

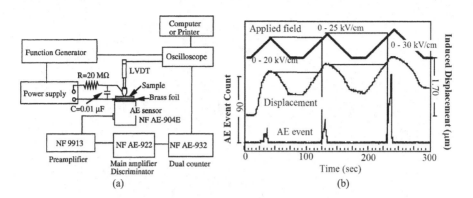

FIGURE 3.7   (a) AE system. (b) Kaiser effect observed in a piezoelectric ceramic.[7]

3.2.4.3   Piezoelectric Passive Damper

Though the mechanism is different from the solid damping, since the analytical process of energy loss per vibration cycle is analogous, the passive piezoelectric damper is introduced here.[9,10]

The principle of the piezoelectric vibration damper is explained based on a piezoelectric ceramic single plate in Figure 3.8a. When an external pulse force is applied to the piezo-plate, an electric charge is produced by the direct piezoelectric effect (Figure 3.8b).[9] Accordingly, the vibration remaining (i.e., ring-down vibration) after the removal of the external force induces an alternating voltage, which corresponds to the intensity of that vibration, across the terminals of the single plate. The electric charge produced is allowed to flow and is dissipated as Joule heat, when a resistor is put between the terminals (see Figure 3.8c). When the external resistance is too large or small, the vibration intensity is not readily reduced, and we need to tune the resistance to match exactly to the piezo-plate impedance, that is, $1/j\omega C$, where ω is the cyclic frequency (i.e., the fundamental mechanical resonance of the piezo-plate), and C is the piezo-plate capacitance.

The bimorph piezoelectric component, which consists of an elastic beam sandwiched with two sheets of piezoelectric ceramic plates (Figure 3.9a), was utilized for the mechanical damping

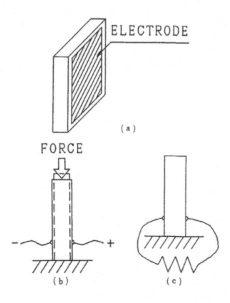

FIGURE 3.8   Mechanical damping concept with a piezoelectric. (a) A single-plate piezoelectric sample; (b) direct piezoelectric effect; and (c) electric energy dissipation through a resistance.

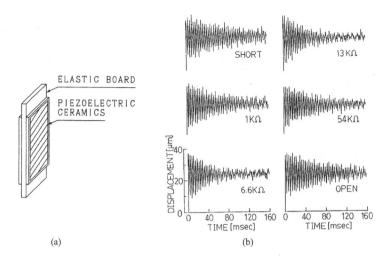

(a)

(b)

FIGURE 3.9 Vibration damping change associated with external resistance change. (a) Bimorph transducer for this measurement. (b) Damped vibration with external resistor.[9]

demonstration. The bimorph edge was hit by an impulse force, and the transient vibration displacement decay was monitored by an eddy current-type non-contact displacement sensor. Figure 3.9b shows the measured displacement data which vibrates at the bimorph resonance frequency (295 Hz), and Figure 3.10 shows the relationship between the damping time constant and an external resistance. It can be seen in the figure that the damping time constant was minimized in the vicinity of 6.6 kΩ, which is close to the impedance $1/\omega C$.

Let us evaluate the damping constant theoretically. The electric energy U_E generated can be expressed by using the electromechanical coupling factor k and the mechanical energy U_M:

$$U_E = U_M \times k^2. \tag{3.54}$$

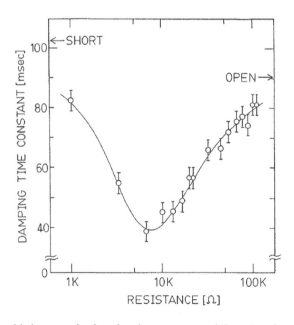

FIGURE 3.10 Relationship between the damping time constant and the external resistance.[9]

The piezoelectric damper transforms electric energy into heat energy via Joule loss when the resistor R is connected, and transforming rate of the damper can be raised to a level of up to 50%, when the electrical impedance is exactly matched. Accordingly, the vibration energy is transformed at a rate of $(1 - k^2/2)$ times with energy vibration repeated, since $k^2/2$ multiplied by the amount of mechanical vibration energy is dissipated as heat energy. As the square of the amplitude is equivalent to the amount of energy, the amplitude decreases at a rate of $(1 - k^2/2)^{1/2}$ times with every vibration repeated. If the resonance period is taken to be T_0, the number of vibrations for t sec is $2t/T_0$. Consequently, the amplitude in t sec is $\left(1 - k^2/2\right)^{t/T_0}$. If the residual vibration period is taken to be T_0, the damping in the amplitude of vibration is t sec can be expressed as follows:

$$\left(1 - k^2/2\right)^{t/T_0} = e^{-t/\tau} \tag{3.55}$$

Thus, the following relationship for the time constant of the vibration damping is obtained.

$$\tau = -\frac{T_0}{\ln\left(1 - k^2/2\right)} \tag{3.56}$$

Now let us examine the time constant of the damping using the results for the bimorph. Substitution in Eq. (3.56) of $k = 0.28$ (dynamic electromechanical coupling factor at this fundamental bending resonance mode) and $T_0 = 3.4$ m s produces $\tau = 85$ m s, which seems to be considerably larger than the value of approximately 40 m s obtained experimentally for τ (Figure 3.10). This is because the theoretical derivation Eq. (3.56) was conducted under the assumption of a loss-free (high Q_M) bimorph. In practice, however, it involved originally mechanical loss, the time constant of which can be obtained as the damping time constant under a short-circuited condition, i.e., $\tau_s = 102$ m s. The total vibration displacement can then be expressed as $e^{-t/\tau_{\text{total}}} = e^{-t/\tau_s} \times e^{-t/\tau}$. Accordingly,

$$\frac{1}{\tau_{\text{total}}} = \frac{1}{\tau_s} + \frac{1}{\tau} \tag{3.57}$$

Substitution in Eq. (3.57) of $\tau_s = 102$ m s and $\tau = 85$ m s produces $\tau_{\text{total}} = 46$ ms. This conforms to the experimental results shown in Figures 3.9b and 3.10.

3.2.5 COULOMB (FRICTION) DAMPING

Coulomb damping occurs when the mechanical object is contacted on a "dry surface". As learned in the high-school physics, the Coulomb friction force F is almost constant (irrelevant to the object speed) and expressed by the product of the normal force N and the friction constant μ:

$$F = \mu N \tag{3.58}$$

The friction constant changes from 0 to 1, but typical values are around 0.2–0.4.

3.2.5.1 Transient Response Analysis with Laplace Transform

Figure 3.11a and b shows a commercial friction damper and a schematic model with mass m, spring constant c, and friction contact, respectively. We can start from the differential equation by using mass displacement u

$$m\ddot{u} + cu = \pm F \tag{3.59}$$

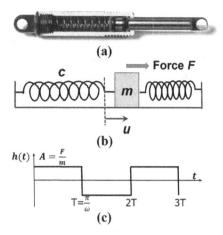

FIGURE 3.11 (a) Commercial friction damper; (b) schematic model with mass, spring, and friction contact; and (c) rectangular wave function representing friction force.

We can rewrite Eq. (3.59), using $\omega = \sqrt{c/m}$ and a rectangular wave function $h(t)$ visualized in Figure 3.11c

$$\ddot{u} + \omega^2 u = h(t) \tag{3.60}$$

Here, the amplitude A of $h(t)$ is set $F/m = \mu \cdot N/m$, and a cyclic period ($2T$) is taken as $2\pi/\omega$, corresponding to the resonance period of the original mass–spring system. We will adopt the initial conditions as:

$$u(t = 0) = a, \dot{u}(t = 0) = 0 \tag{3.61}$$

The initial mass position a (spring force $c \cdot a$) should be taken reasonably large, so that the motion will start by competing the frictional force $\mu \cdot N$:

$$a > A/\omega^2 \tag{3.62}$$

Let us solve the differential equation Eq. (3.60), using *Laplace transform*, which is a useful mathematical tool for this sort of non-sinusoidal, anharmonic input force. We denote the Laplace transforms of the displacement $u(t)$ and friction force $h(t)$ as $U(s)$ and $H(s)$: $U(s) = Lu(t)$, $H(s) = Lh(t)$. Equation (3.60) can be written as

$$L\ddot{u} + \omega^2 U(s) = H(s). \tag{3.64}$$

Taking into account the initial condition $u(t = 0) = a$ and $\dot{u}(t = 0) = 0$ [from theorem Eq. (3.41)],

$$L\ddot{u} = s^2 U - sa \tag{3.65}$$

and the Laplace transform of a rectangular wave expressed by [from #15 in Table 3.1]

$$H(s) = \frac{A}{s} \tanh\left(\frac{Ts}{2}\right) = \frac{A}{s}\left(1 - 2e^{-Ts} + 2e^{-2Ts} - 2e^{-3Ts} + \cdots\right), \tag{3.66}$$

we obtain the following equation:

$$\left(s^2 + \omega^2\right)U - sa = \frac{A}{s}\left(1 - 2e^{-Ts} + 2e^{-2Ts} - 2e^{-3Ts} + \cdots\right). \tag{3.67}$$

We can now solve it in terms of $U(s)$

$$U = \frac{s \cdot a}{\left(s^2 + \omega^2\right)} + \frac{A}{s\left(s^2 + \omega^2\right)}\left(1 - 2e^{-Ts} + 2e^{-2Ts} - 2e^{-3Ts} + \cdots\right). \tag{3.68}$$

Remember $T = \pi/\omega$ (a half of the resonance period).

Knowing $\dfrac{1}{s\left(s^2 + \omega^2\right)} = \dfrac{1}{\omega^2}\left[\dfrac{1}{s} - \dfrac{s}{\left(s^2 + \omega^2\right)}\right]$, and the famous inverse Laplace transforms as follows (Table 3.1):

$$L^{-1}\frac{1}{s} = 1 \text{ (step function, } t > 0\text{); } L^{-1}\frac{s}{\left(s^2 + \omega^2\right)} = \cos(\omega t);$$

and the "time-shift theorem" $L[u(t - a)] = e^{-as} \cdot U(s)$, we can obtain the displacement $u(t)$ solution for successive time intervals, $0 < t < T$, $T < t < 2T$, $2T < t < 3T$, …, where T is a half of the resonance period (refer to Ref. [11]):

$$u = a\cos(\omega t) + \frac{A}{\omega^2}\left(1 - \cos(\omega t)\right) \text{ for } 0 < t < T; u(T) = -a + \frac{2A}{\omega^2} \tag{3.69}$$

$$u = a\cos(\omega t) + \frac{A}{\omega^2}\left(1 - \cos(\omega t)\right) - \frac{2A}{\omega^2}\left(1 - \cos\left(\omega(t - T)\right)\right) \text{ for } T < t < 2T; u(2T) = a - \frac{4A}{\omega^2} \tag{3.70}$$

$$\cdots$$

We find that

1. The system has the resonance frequency provided by $\omega = \sqrt{c/m}$, determined by the original mass and spring.
2. Each successive swing is $\left(\dfrac{2A}{\omega^2}\right)$ shorter than the preceding one, until inside the dead region, that is, linearly decay with time, different from popular exponential decay in viscous damping in the next section. Figure 3.12 shows the linear vibration amplitude decay for the Coulomb damping, in comparison with no damping-free vibration.
3. There is the critical stop point of the vibration; that is, the minimum displacement $u(t) = A/\omega^2$, below which the spring force cannot compete the friction force.

3.2.5.2 Energy Consumption Analysis

We consider here the vibration amplitude decrease from the work by the friction force. When we consider a half-cycle from the mass velocity zero to the next zero state ($t = 0$ to $t = T$), the kinetic energy change should be zero. Supposing that the vibration amplitude is reduced by Δu from the initial $u_0 = a$, the spring potential energy change should equate to the work by the friction force times moving distance $(2u_0 - \Delta u)$:

$$\frac{1}{2}c\left[u_0^2 - \left(u_0 - \Delta u\right)^2\right] = F\left(2u_0 - \Delta u\right) \tag{3.71}$$

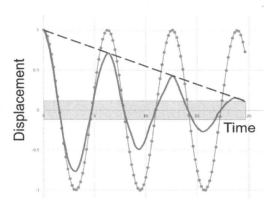

FIGURE 3.12   Linear vibration amplitude decay for Coulomb damping.

Thus,

$$\Delta u = 2F/c. \tag{3.72}$$

This implies that the vibration amplitude is reduced by $2F/c$ per a half cycle, which is exactly the same conclusion from the previous section, taking into account $2F/c = \left(\dfrac{2A}{\omega^2} \right)$: linear vibration amplitude decay for the Coulomb damping!

3.2.6   Viscous Damping

Different from the Coulomb damping for a dry surface (i.e., constant force), *viscous damping* is applied for "lubricated surface" friction (i.e., time-dependent force); so-called "dashpot" (i.e., shock absorber, buffer), illustrated in Figure 3.6b, exhibits this behavior. It can also be applied for an object moving in viscous oil, or electro-active object moving in magnetic field with its damping force in proportion to the object speed. The damping force is introduced in proportion to the velocity v of the mass m as

$$F = -\xi \cdot v = -\xi \cdot \frac{du}{dt}, \tag{3.73}$$

where ξ is called viscous damping coefficient. Thus, the dynamic equation of the mass without external force in Figure 3.6b is described as

$$m\ddot{u} = -cu - \xi \cdot \dot{u} \tag{3.74}$$

Taking the following notations,

$$\omega_0 = \sqrt{c/m} \quad \left(\text{base resonance frequency for zero damping}\right) \tag{3.75}$$

$$\zeta = \xi/2m\omega_0 \quad \left(\text{normalized damping factor}\left(\text{no dimension}\right)\right) \tag{3.76}$$

we obtain the normalized equation,

$$\ddot{u} + 2\zeta\omega_0 \cdot \dot{u} + \omega_0^2 u = 0. \tag{3.77}$$

Because of the obedient characteristic of the damping formula, we can easily solve this differential equation using the Laplace transform. Taking $L[u(t)] = U(s)$, and theorems (3.40) and (3.41) with the initial conditions, $\dot{u}(t = 0) = 0$, $u(t=0) = u_0$,

$$\left[s^2 U - su_0\right] + 2\zeta\omega_0\left[sU - u_0\right] + \omega_0^2 U = 0 \tag{3.78}$$

Then,

$$\left[s^2 + (2\zeta\omega_0)s + \omega_0^2\right]U(s) = (s + 2\zeta\omega_0)u_0 \tag{3.79}$$

or

$$U(s) = \frac{(s + 2\zeta\omega_0)u_0}{\left[s^2 + (2\zeta\omega_0)s + \omega_0^2\right]} \tag{3.80}$$

Equation (3.80) is a general solution for $U(s)$. In order to apply the inverse Laplace transform, we need to consider three cases: $0 \leq \zeta < 1$, $\zeta = 1$, and $1 < \zeta$.

3.2.6.1   Under Damping ($0 \leq \zeta < 1$)

Rewriting Eq. (3.80) as

$$U(s) = u_0\left[\frac{(s + \zeta\omega_0)}{(s + \zeta\omega_0)^2 + (1 - \zeta^2)\omega_0^2} + \frac{\zeta}{\sqrt{1 - \zeta^2}}\frac{\sqrt{1 - \zeta^2}\,\omega_0}{(s + \zeta\omega_0)^2 + (1 - \zeta^2)\omega_0^2}\right], \tag{3.81}$$

then using the inverse Laplace transforms (#8 and 9 in Table 3.1) for the first and second terms, we can obtain the solution:

$$u(t) = u_0\left[\exp(-\zeta\omega_0 t)\cos\left(\sqrt{1 - \zeta^2}\,\omega_0 t\right) + \frac{\zeta}{\sqrt{1 - \zeta^2}}\exp(-\zeta\omega_0 t)\sin\left(\sqrt{1 - \zeta^2}\,\omega_0 t\right)\right] \tag{3.82}$$

We can understand that only when $1 - \zeta^2 > 0$, sinusoidal vibration can be observed, which is called "light damping" or "under damping". It is important to note that the resonance frequency of this system is not ω_0, but $\sqrt{1 - \zeta^2}\,\omega_0$, as you can find it in Eq. (3.82). Refer to Figure 3.13 for $u(0) = u_0$. $\zeta = 0$ shows the calculated result for a simple cosine curve, $u(t) = \cos(\omega_0 t)$, and $\zeta = 0.1$ is for

$$u(t) = \exp(-0.1\omega_0 t)\cos\left(\sqrt{0.99}\omega_0 t\right) + \frac{0.1}{\sqrt{0.99}}\exp(-0.1\omega_0 t)\sin\left(\sqrt{0.99}\omega_0 t\right)$$

3.2.6.2   Critical Damping ($\zeta = 1$)

Sinusoidal vibration can be observed only when $1 - \zeta^2 > 0$ is satisfied. When $\zeta = 1$, the vibration status is called "critical damping". The response speed is not so high as the "under damping" condition, but the quickest response without any ringing vibration associated (i.e., "aperiodic" motion). Equation (3.80) is transformed as

$$U(s) = \frac{(s + 2\omega_0)u_0}{\left[s^2 + (2\omega_0)s + \omega_0^2\right]} = u_0\left[\frac{1}{s + \omega_0} + \frac{\omega_0}{(s + \omega_0)^2}\right]. \tag{3.83}$$

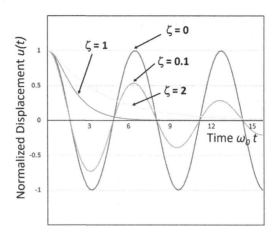

FIGURE 3.13 Free vibration amplitude decay for viscous damping. $\zeta = 1$ corresponds to the critical damping.

Then, using the inverse Laplace transforms (#4 and 5 in Table 3.1) for the first and second terms, we can obtain the solution:

$$u(t) = u_0 \left[\exp(-\omega_0 t) + \omega_0 t \cdot \exp(-\omega_0 t) \right] = u_0 \left(1 + \omega_0 t \right) \cdot \exp(-\omega_0 t) \tag{3.84}$$

The curve denoted $\zeta = 1$ in Figure 3.13 corresponds to the calculated result for $u(t) = (1 + \omega_0 t) \exp(\omega_0 t)$.

3.2.6.3 Over Damping ($\zeta > 1$)

When the damping factor $\zeta > 1$, "large damping" or "over damping" is observed. Rewriting Eq. (3.80) as

$$U(s) = u_0 \left[\frac{(s + \zeta \omega_0)}{(s + \zeta \omega_0)^2 - (\zeta^2 - 1)\omega_0^2} + \zeta \omega_0 \frac{1}{(s + \zeta \omega_0)^2 - (\zeta^2 - 1)\omega_0^2} \right], \tag{3.85}$$

then using the inverse Laplace transforms (#12 and 13 in Table 3.1) for the first and second terms, we can obtain the solution:

$$u(t) = u_0 \left[\exp(-\zeta \omega_0 t) \cosh\left(\sqrt{\zeta^2 - 1}\,\omega_0 t \right) + \frac{\zeta}{\sqrt{\zeta^2 - 1}} \exp(-\zeta \omega_0 t) \sinh\left(\sqrt{\zeta^2 - 1}\,\omega_0 t \right) \right] \tag{3.86}$$

This motion is actually "aperiodic", without exhibiting any ringing. The curve denoted $\zeta = 2$ in Figure 3.13 corresponds to the calculated result for $u(t) = 1.08 \exp(-0.27\omega_0 t) - 0.08 \exp(-3.73\omega_0 t)$.

3.2.7 Logarithmic Decrement

3.2.7.1 Definition of Logarithmic Decrement

When the damping factor ζ is small, a sinusoidal vibration continues, as shown in Figure 3.14. Since Eq. (3.82) can be rewritten as

$$u(t) = u_0 \exp(-\omega_0 t) \cos\left(\sqrt{1 - \zeta^2}\,\omega_0 t - \phi \right)$$

the resonance angular frequency is given by $\sqrt{1-\zeta^2}\,\omega_0$, not by ω_0, and the vibration peak points which satisfy

$$\cos\left(\sqrt{1-\zeta^2}\,\omega_0 t - \phi\right) = 1$$

are almost bounded by the envelope of $u(0)\exp\left(-\zeta\omega_0 t\right)$. Though the contact point on the envelope is slightly off from the amplitude maximum points, precisely speaking, the deviation is negligibly small. The "logarithmic decrement" δ is defined by the natural log of the ratio of two successive vibration amplitudes:

$$\delta = \ln\left(\frac{x_1}{x_2}\right) = \ln\frac{\exp(-\omega_0 t)}{\exp\left(-\omega_0(t + T_0)\right)} = \ln\left[\exp(\omega_0 T_0)\right] = \omega_0 \qquad (3.87)$$

Since the vibration resonance period T_0 is given by $T_0 = \dfrac{2\pi}{\sqrt{1-\zeta^2}\,\omega_0}$, the logarithmic decrement can be written as

$$\delta = \frac{2\pi\zeta}{\sqrt{1-\zeta^2}} \approx 2\pi\zeta \qquad (3.88)$$

3.2.7.2 Experimental Determination

Refer to Figure 3.14 for the notations. When Δx_2 is small, δ can be experimentally obtained as follows:

$$\delta = \ln\left(\frac{x_1}{x_2}\right) = \ln\left(\frac{x_2 + \Delta x_2}{x_2}\right) = \ln\left(1 + \frac{\Delta x_2}{x_2}\right) \cong \frac{\Delta x_2}{x_2}. \qquad (3.89)$$

So-called time constant τ determined from the envelope curve, $\exp(-t/\tau)$, experimentally is related with the damping factor ζ as $\tau = 1/\omega_0$.

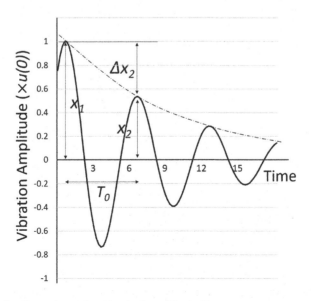

FIGURE 3.14 Damping factor determined by the logarithmic decrement.

The logarithmic decrement δ can also be determined from the energy loss. The total energy of this system is estimated as $\frac{1}{2}cx^2$. The energy loss per cycle ΔW can be estimated by

$$W - \Delta W = \frac{1}{2}c(x - \Delta x)^2$$

Thus,

$$\delta \cong \frac{\Delta x_2}{x_2} \cong \frac{\Delta W}{2W} \tag{3.90}$$

δ is obtained by a half of the ratio (lost energy/total energy).

3.3 BODE PLOT – FREQUENCY RESPONSE OF A SYSTEM

3.3.1 STEADY-STATE OSCILLATION

In Section 3.2, we considered "free" vibration of a mass–spring–damper system shown in Figure 3.15a [rewritten from Figure 3.6b]. Because of the damper, the system gradually loses energy, and the vibration amplitude was decreased. We consider now forced oscillation under a harmonic force $f(t) = f_0 \cdot \sin(\omega t)$ in this Section 3.3.1.

Supposing the viscous damper, the sum of spring force, damping force, and external force will generate the mass acceleration [see Figure 3.15b for the force direction]:

$$m\ddot{u} = -\xi \cdot \dot{u} - cu + f(t).$$

or

$$m\ddot{u} + \xi \cdot \dot{u} + cu = f(t). \tag{3.91}$$

In order to simplify the equation, we will adopt the following notations:
Resonance angular frequency for zero damping:

$$\omega_0^2 = c/m \quad \left[\omega_0 = \sqrt{c/m}\right] \tag{3.92}$$

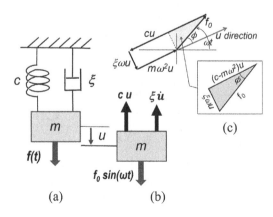

(a)   (b)

FIGURE 3.15 (a) Mass–spring–damper model and (b) mass under forced oscillation. (c) Force vector analysis under forced oscillation.

Damping ratio ζ and damping factor ξ:

$$2\zeta\omega_0 = \xi/m \quad \left[\zeta = \xi/2m\omega_0\right] \tag{3.93}$$

Equation (3.91) is transformed into

$$\ddot{u} + 2\zeta\omega_0 \cdot \dot{u} + \omega_0^2 u = \frac{1}{m}f(t). \tag{3.94}$$

Taking $L[u(t)] = U(s)$, $L[f(t)] = F(s)$ and theorems (3.40) and (3.41) with the initial conditions, $\dot{u}(t = 0) = 0$, $u(\text{t=0}) = u_0$, we obtain

$$\left[s^2 U - su_0\right] + 2\zeta\omega_0\left[sU - u_0\right] + \omega_0^2 U = (1/m)F(s) \tag{3.95}$$

Then,

$$\left[s^2 + \left(2\zeta\omega_0\right)s + \omega_0^2\right]U(s) = \left(s + 2\zeta\omega_0\right)u_0 + (1/m)F(s).$$

Thus, $U(s)$ is expressed by

$$U(s) = \frac{\left(s + 2\zeta\omega_0\right)u_0}{\left[s^2 + \left(2\zeta\omega_0\right)s + \omega_0^2\right]} + \frac{(1/m)F(s)}{\left[s^2 + \left(2\zeta\omega_0\right)s + \omega_0^2\right]} \tag{3.96}$$

The first term on the right-hand side is called the "transient solution" (or complementary function), which is the solution when we put the external force $f(t) = 0$. As we have discussed in Section 3.2, all solutions for different damping factors ζ include the time dependence of $\exp\left(-\zeta\omega_0 t\right)$. Thus, the transient solution will disappear with time lapse. To the contrary, the second term, will remain even after a long time lapse, is called "steady-state oscillation".

We consider only the second term (i.e., steady state oscillation) further below:

$$U(s) = G(s)F(s) \tag{3.97}$$

$$G(s) = \frac{1}{m\left[s^2 + \left(2\zeta\omega_0\right)s + \omega_0^2\right]} \tag{3.98}$$

The Laplace function $G(s)$ relates the input function $F(s)$ to the output function $U(s)$ and is called "transfer function". Because the denominator includes s^2, this function is called "second-order system". Knowing that when $F(s) = 1$, that is, $f(t) = \delta(t)$ (impulse function), $U(s) = G(s)$, the $G(s)$ can be obtained experimentally by taking Laplace transform of the output $u(t)$ under the "impulse input" (hitting the system by a hammer!). When we consider only the "steady-state oscillation" under a harmonic input, $f(t) = f_0 \sin(\omega t)$, since its Laplace transform is expressed by $\dfrac{\omega}{\left(s^2 + \omega^2\right)}$ and its "pole" exists at $s = j\omega$, we can discuss the frequency dependence of the displacement $u(t)$ from the "transfer function" by replacing s by "$j\omega$", which corresponds to the *Fourier transform*. By this replacement, Eq. (3.97) can be transformed as

$$U(j\omega) = \frac{\left(f_0/m\right)}{\left[-\omega^2 + j2\zeta\omega_0\omega + \omega_0^2\right]} \tag{3.99}$$

3.3.2   Steady State – Reconsideration

We considered the frequency dependency of the displacement in the previous section (Eq. (3.97)). Direct displacement in the time domain is considered here.[2] Supposing that the displacement $u(t)$ is excited with slight delay after the force $f(t)$, the steady-state vibration is described as

$$u(t) = u_0 \sin(\omega t + \phi) \quad (\phi < 0, \text{delay}) \tag{3.100}$$

We consider four forces balance: inertial force, damping force, spring force, and external force, the total sum as vectors should be zero. Refer to Figure 3.15c. Knowing the phase change by the time derivative, we obtain

$$\dot{u} = \omega u_0 \, \sin\left(\omega t + \phi + \frac{\pi}{2}\right)$$

$$\ddot{u} = \omega^2 u_0 \sin\left(\omega t + \phi + \pi\right) = -\omega^2 u_0 \sin\left(\omega t + \phi\right).$$

Now the original equation, $m\ddot{u} + \zeta \cdot \dot{u} + cu = f(t)$, is transformed to

$$m\omega^2 u_0 \sin\left(\omega t + \phi\right) - 2\zeta m\omega_0 \omega u_0 \sin\left(\omega t + \phi + \frac{\pi}{2}\right) - cu_0 \sin\left(\omega t + \phi\right) + f_0 \sin\left(\omega t\right) = 0$$

Figure 3.15c describes:

1. The output displacement $u(t)$ is delayed from the input force $f(t)$. The phase delay ϕ in the above equation is negative!
2. Spring force is opposite to the displacement.
3. Damping force is delayed 90° from the displacement and opposite to the velocity.
4. Inertial force is in phase with the displacement and opposite to the acceleration.
5. Four force vectors rotate at the angular velocity ω, by keeping the relative position fixed.

From the inserted triangle figure in Figure 3.15(c), the following relations can be derived:

$$u_0 = \frac{f_0}{\sqrt{\left(c - m\omega^2\right)^2 + \left(2m\zeta\omega_0\omega\right)^2}} \tag{3.101}$$

$$\tan\phi = -\frac{2m\zeta\omega_0\omega}{c - m\omega^2} \tag{3.102}$$

$$u(t) = \frac{f_0 \sin(\omega t + \phi)}{\sqrt{\left(c - m\omega^2\right)^2 + \left(2m\zeta\omega_0\omega\right)^2}} \tag{3.103}$$

Introducing non-dimensional normalized form, taking $u_0(\omega = 0)$, which is equal to f_0/c, and zero-damping resonance frequency $\omega_0 = \sqrt{c/m}$, the above equations are transformed as

$$\text{Magnification factor}: \frac{u_0}{u_0\,(\omega = 0)} = \frac{1}{\sqrt{\left[1 - \left(\dfrac{\omega}{\omega_0}\right)^2\right]^2 + \left(2\zeta\dfrac{\omega}{\omega_0}\right)^2}} \tag{3.104}$$

$$\text{Phase}: \tan\phi = -\frac{2\zeta\dfrac{\omega}{\omega_0}}{1-\left(\dfrac{\omega}{\omega_0}\right)^2} \tag{3.105}$$

Since we adopted $\sin(\omega t + \phi)$, negative ϕ is obtained for the phase delay.

Figure 3.16 illustrated the magnification factor $\left(\dfrac{u_0}{u_0(\omega=0)}\right)$ and phase as a function normalized frequency $\left(\dfrac{\omega}{\omega_0}\right)$ for various damping ratios ζ ($\zeta = 0.1, 0.2, 0.5, 1,$ and 2). When $(\omega/\omega_0) \ll 1.0$, the inertial and damping forces are small, leading to small phase deviation. When $(\omega/\omega_0) = 1.0$, the inertial force becomes large to balance with the spring force, and external force compensates the damping force, then phase becomes $-90°$. The Lissajous curve between force and displacement is a circle (or ellipse). When $(\omega/\omega_0) \gg 1.0$, the magnification factor becomes small, and the phase approaches to $-180°$. The force–displacement curve slope is opposite in comparison with the case $(\omega/\omega_0) \ll 1.0$.

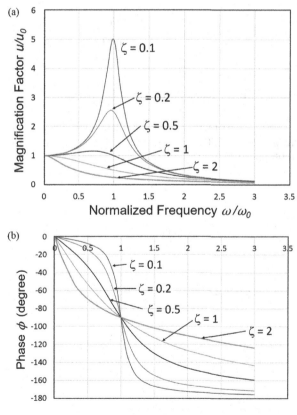

FIGURE 3.16 (a) Magnification factor $\left(\dfrac{u_0}{u_0(\omega=0)}\right)$ and (b) phase as a function normalized frequency $\left(\dfrac{\omega}{\omega_0}\right)$ for various damping ratio ζ.

3.3.3 BODE PLOT

Equation (3.97) can also be written using $T = 1/\omega_0$ (resonance period in radian scale):

$$G(j\omega) = (m/f_0)U(j\omega)$$

$$= \frac{1}{\left[(-\omega^2 T^2) + 2\zeta\, j\omega T + 1\right]} \tag{3.106}$$

The "Bode plot" is a representation of the transfer function (amplitude and phase) as a function of frequency on a *logarithmic scale*.

First, let us consider "asymptotic" straight curves for the "low"- and "high"-frequency regions.

- For $\omega \to 0$, $G(j\omega) \to 1$. Thus, gain $|G(j\omega)| = 1$, so that in decibels:

$$dB = 20 \cdot \log_{10}(1) = 0 \tag{3.107}$$

"0 dB/decade", that is, flat frequency dependence. Regarding the phase, the real number corresponds to 0°. Gain and phase Bode plots are left-hand side in Figure 3.17.

- Flor $\omega \to \infty$, $G(j\omega) \to 1/(-\omega^2 T^2)$

$$\text{Gain}: |G(j\omega)| = \left|\frac{1}{(-\omega^2 T^2)}\right| \tag{3.108}$$

so that in decibels:

$$dB = -20\log_{10}(\omega T)^2 = -40\log_{10}(\omega T) \tag{3.109}$$

"−40 dB/decade" (or −12 dB/octave) with frequency. Regarding the phase, negative real number corresponds to −180° as indicated by the gain and phase curves appearing right-hand side in Figure 3.17.

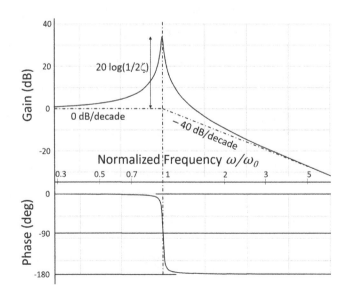

FIGURE 3.17 The Bode diagram for a standard second-order system.

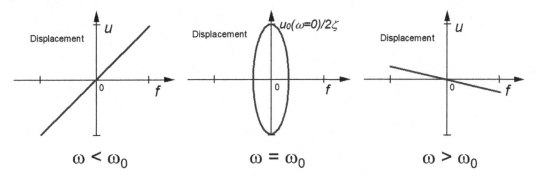

FIGURE 3.18 The Lissajous curves between force and displacement.

Second, we consider displacement around for the resonance ω_0 frequency.

- Resonance range: we will now consider the deviation from these two lines around the bend-point frequency, which is obtained from the relation $\omega T = 1$. Substituting $[\omega T = 1]$ in Eq. (3.106) yields:

$$G(j\omega) = \frac{1}{(2\zeta j\omega T)} = \frac{1}{(2\zeta j)} \qquad (3.110)$$

so that the gain and phase become $[1/(2\zeta)]$ and $-90°$, respectively. The constant ζ is the damping ratio. If $\zeta = 0$ (loss-free), an infinite amplitude will occur at the bend-point frequency (i.e., the resonance frequency). When ζ is large (>1), however, the resonance peak will disappear, and monotonous decrease in amplitude is observed.

The Lissajous curves between the force and displacement are illustrated in Figure 3.18: when $\left(\omega/\omega_0\right) \ll 1.0$, linear relation with a right-up positive slope, while for $\left(\omega/\omega_0\right) \gg 1.0$, linear relation with a right-down negative slope. At the resonance $\left(\omega/\omega_0\right) = 1.0$, an elliptical Lissajous relation (counter-clockwise rotation from $-90°$) with the magnification factor by $1/\zeta$. Note that this phase lag is due to the resonant vibration, but not directly related with the intrinsic mechanical loss ζ.

3.3.4 MECHANICAL QUALITY FACTOR

There are multiple definitions for the mechanical quality factor Q, which are approximately equivalent, but not exactly equivalent. One of these definitions is the frequency-to-bandwidth ratio of the mechanical resonator:

$$Q \overset{\text{def}}{=} \frac{f_r}{\Delta f} = \frac{\omega_r}{\Delta \omega} \qquad (3.111)$$

where f_r is the resonant frequency, Δf is the resonance width or *full width at half maximum* (FWHM), that is, the bandwidth over which the power of vibration is greater than half the power at the resonant frequency, $\omega_r = 2\pi f_r$ is the angular resonant frequency, and $\Delta \omega$ is the angular half-power bandwidth. From the Bode plot of the displacement in Figure 3.17, $\Delta \omega$ should be obtained from $1/\sqrt{2}$ of the maximum displacement (which corresponds to 1/2 of the maximum power) around the resonant frequency, which corresponds to 3 dB down level in a 20 dB-type gain plot.

Since $\dfrac{u_0(\omega_0)}{u_0(\omega=0)} = \dfrac{1}{2\zeta\dfrac{\omega}{\omega_0}}$, by putting $\dfrac{u_0}{u_0(\omega=0)} = \dfrac{1}{\sqrt{\left[1-\left(\dfrac{\omega}{\omega_0}\right)^2\right]^2 + \left(2\zeta\dfrac{\omega}{\omega_0}\right)^2}} = \dfrac{1}{\sqrt{2}}\dfrac{1}{2\zeta\dfrac{\omega}{\omega_0}}$,

we can obtain

$$1-\left(\frac{\omega}{\omega_0}\right)^2 = \pm 2\zeta\frac{\omega}{\omega_0}, \text{ depending on } \frac{\omega}{\omega_0} < 1 \text{ or } \frac{\omega}{\omega_0} > 1 \qquad (3.112)$$

We obtain $\dfrac{\omega}{\omega_0} = \sqrt{1+\zeta^2} - 1$ or $\dfrac{\omega}{\omega_0} = \sqrt{1+\zeta^2} + \zeta$ for $\dfrac{\omega}{\omega_0} < 1$ or $\dfrac{\omega}{\omega_0} > 1$, respectively. Thus, $\dfrac{\Delta\omega}{\omega_0} = 2\zeta$.

The mechanical quality factor Q_m is expressed by

$$Q_m = \frac{\omega_0}{\Delta\omega} = 1/2\zeta \qquad (3.113)$$

You can understand that the displacement amplification factor is proportional to Q_m.

3.3.5 COMPLEX ALGEBRA METHOD

3.3.5.1 Complex Displacement

Complex algebra method often facilitates solving differential equations, in the case of a forced oscillation under a harmonic external force, since the steady-state solution exhibits a harmonic oscillation with the same frequency as the input force. As explained in Figure 3.15c, the output displacement u is delayed from the input force f with phase lag ϕ, and keeping this phase lag, both vectors rotate at the angular frequency of ω. Supposing that these vectors are expressed by complex parameters as

$$f(t) = f_0\, e^{j\omega t} \qquad (3.114)$$

$$u(t) = u_0\, e^{j(\omega t-\phi)} = u_0\, e^{-j\phi}\, e^{j\omega t} \qquad (3.115)$$

f_0 and u_0 correspond to the absolute length of the force and displacement vectors, respectively. We write Eq. (3.115) by using the complex vibration amplitude as

$$u(t) = u_0^*\, e^{j\omega t}, \, u_0^* = u_0\, e^{-j\phi}$$

When we adopt Eqs. (3.114) and (3.115) into a mass–spring–damper model in Figure 3.15a,

$$m\ddot{u} + \xi \cdot \dot{u} + cu = f_0\, e^{j\omega t} \qquad (3.116)$$

Using $u(t) = u_0^*\, e^{j\omega t}$, we obtain

$$\left(-m\omega^2 + j\zeta\omega + c\right)u_0^* = f_0 \qquad (3.117)$$

Accordingly,

$$u_0^* = \frac{f_0}{\left(-m\omega^2 + j\zeta\omega + c\right)} = \frac{f_0 e^{-j\phi}}{\sqrt{\left(c-m\omega^2\right)^2 + (\xi\omega)^2}} \qquad (3.118)$$

The vibration amplitude and phase are obtained as:

$$u_0 = \frac{f_0}{\sqrt{\left(c - m\omega^2\right)^2 + (\xi\omega)^2}}$$

(3.119)

$$\phi = \tan^{-1}\left(\frac{\xi\omega}{c - m\omega}\right)$$

(3.120)

3.3.5.2 Complex Physical Parameter

From Eq. (3.116), $m\ddot{u} + \xi \cdot \dot{u} + cu = f_0\, e^{j\omega t}$, we assume harmonic vibration $u(t)$ with the same ω. Then

$$\left(-m\omega^2 + j\zeta\omega + c\right)u(t) = f_0\, e^{j\omega t}$$

(3.121)

We adopt a "complex spring constant" as

$$c^* = c\left(1 + j\frac{\zeta\omega}{c}\right)$$

(3.122)

Note that the viscoelastic damping-related imaginary part is frequency dependent. Chapter Problem 1.2 demonstrates the frequency independent case. Then, the dynamic equation is transformed as

$$\left(-m\omega^2 + c^*\right)u(t) = f_0\, e^{j\omega t}$$

(3.123)

We can understand that Eq. (3.123) corresponds to a simple mass–spring oscillator with a complex spring constant. Under an assumption $u(t) = u_0\, e^{j(\omega t - \phi)} = u_0\, e^{-j\phi}\, e^{j\omega t}$, remaining solution process is the same in Subsection 3.3.5.1, leading to the final results Eqs. (3.119) and (3.120).

CHAPTER ESSENTIALS

1. Electric loss (high frequency, Peta Hertz): EM wave is absorbed by the material's conducting loss. Absorption coefficient is frequency dependent.
2. Electric loss (low frequency): Inductor components are inevitably accompanied by resistive loss.
3. Dielectric loss: Though electronic polarization responds up to Peta Hertz, ionic and dipolar polarizations do have response limit, which provides the intrinsic dielectric loss. Small conductance in the material contributes as additional effective dielectric loss.
4. LCR circuit: $L\left(\dfrac{d^2Q}{dt^2}\right) + R\left(\dfrac{dQ}{dt}\right) + \dfrac{Q}{C} = V(t)$ or $L\left(\dfrac{d^2I}{dt^2}\right) + R\left(\dfrac{dI}{dt}\right) + \dfrac{I}{C} = \dfrac{dV}{dt}$

$$I(t) = \frac{V_0}{Z}\sin(\omega t - \phi); \quad Z = \sqrt{R^2 + \left(L\omega - \frac{1}{C\omega}\right)^2}, \quad \tan\phi = \frac{\left(L\omega - \dfrac{1}{C\omega}\right)}{R}$$

5. Three mechanical loss models:
   a. Solid damping – damping force $F = \zeta \cdot c u_{max}$
   b. Coulomb damping – damping force constant $\pm F$ – linear decay of the displacement amplitude
   c. Viscous damping – damping force proportional to velocity $F = \xi \cdot \dot{u}$ – exponential decay of the displacement amplitude

6. Steady-state oscillation for a mass–spring–dashpot model: $m\ddot{u} + \xi \cdot \dot{u} + cu = f(t)$. $\omega_0 = \sqrt{c/m}$ (ω_0: resonance angular frequency for zero damping) $\zeta = \xi/2m\omega_0$ (ξ, ζ: damping factor, ratio)

$$u(t) = \frac{f_0 \sin(\omega t + \phi)}{\sqrt{\left(c - m\omega^2\right)^2 + \left(2m\zeta\omega_0\omega\right)^2}}$$

$$u_0 = \frac{f_0}{\sqrt{\left(c - m\omega^2\right)^2 + \left(2m\zeta\omega_0\omega\right)^2}}$$

$$\tan\phi = -\frac{2m\zeta\omega_0\omega}{c - m\omega^2}$$

7. Bode plot: asymptotic curves – 0 dB/decade, –40 dB/decade.

$$\text{Resonance peak height} = 20\log_{10}\left(\frac{1}{2\zeta}\right)$$

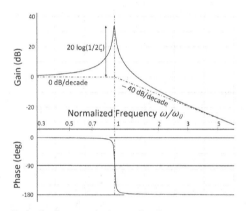

Bode plot for a standard second-order system.

8. Mechanical quality factor (amplification factor of the displacement):

$$Q_m = \frac{\omega_0}{\Delta\omega} = 1/2\zeta$$

9. Complex algebra method, including complex physical parameter such as complex elastic stiffness, is a useful tool for a harmonic steady-state oscillation to calculate the vibration amplitude and phase lag as a function of frequency. Bode diagram is a logarithmic plot of these parameters.

CHECK POINT

1. (T/F) When we add Coulomb damper to a mass–spring system under free vibration (no external force), vibration amplitude decreases exponentially. True or False?
2. What is the Laplace transform for the impulse $\delta(t)$ function?

3. What is the Laplace transform for the Heaviside Step function?
4. (T/F) The Laplace transform for sin (at) is $1/(s^2 + a^2)$. True or False?
5. (T/F) When $L[u(t)] = U(s)$ its first differentiation with respect to t is given by $L\left[\dfrac{du(t)}{dt}\right] = sU(s) - u(0)$. True or False?
6. (T/F) Mechanical quality factor and the damping ratio ζ are related as $Q_m = \dfrac{\omega_0}{\Delta\omega} = 1/\zeta$. True or False?
7. (T/F) Complex spring constant is equivalent to the viscoelastic damping model. True or False?
8. (T/F) The high-frequency portion of the Bode plot for the second-order system is approximated with an asymptotic straight line having a negative slope of 20 dB/decade. True or False?

CHAPTER PROBLEMS

1.1 a. In a mass–spring–damper model in free vibration condition, started from the initial displacement u_0, verify that the logarithmic decrement δ is given by the following equation, where u_n is the vibration amplitude after the n cycles:

$$\delta = \frac{1}{n}\ln\left(\frac{u_0}{u_n}\right).$$

b. The above δ is related with the damping ratio ζ as $\delta = 2\pi\zeta$ [Eq. (3.88)]. In free vibration condition, started from the initial u_0, calculate the required cycle number for reducing the displacement by a half (50%) as a function of the damping ratio ζ.

1.2 Let us consider a damper, the force of which is proportional to the displacement with a phase in the same as velocity. This damper is modeled as a complex spring constant $c(1 + j\gamma)$. Solve the dynamic equation in a mass–spring–damper system given as follows:

$$m\ddot{u} + c(1 + j\gamma)u = 0.$$

Hint: Suppose that $u(t) = u_0\, e^{(\alpha + j\beta)t}$, then determine α and β.

$$\alpha = \sqrt{\frac{c}{2m}}\sqrt{-1 + \sqrt{1+\gamma^2}} \cong \frac{\gamma}{2}\sqrt{\frac{c}{m}}, \beta = \pm\sqrt{\frac{c}{2m}}\sqrt{1 + \sqrt{1+\gamma^2}} \cong \pm\sqrt{\frac{c}{m}}; \delta = \ln\left(\frac{u_1}{u_2}\right) \cong \pi\gamma$$

1.3 Let us consider a mass–spring–dashpot system (Figure 3.15a). We apply a sinusoidal force $f_0\sin(\omega t)$ from the initial position $u(t = 0) = u_0$. Verify the following arguments:
   a. Transient response (vibration amplitude and phase) strongly depends on the initial condition.
   b. Steady-state oscillation (vibration amplitude and phase) is irrelevant to the initial condition.

1.4 Electrical LCR circuit and mechanical mass–spring–dashpot model are shown in the figure below. Provide the dynamic equations (under the input sinusoidal voltage or harmonic force) for both electric and mechanical systems. Then, discuss the equivalency for obtaining the steady-state solutions for both systems. This approach is the key to understand "equivalent circuit" concept introduced in Chapter 5.

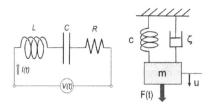

LCR circuit and mass-spring-dashpot model.

REFERENCES

1. R.B. Linsay, *Concepts and Methods of Theoretical Physics*, (D. Van Nostrand Co., New York, 1951).
2. W.T. Thomson, *Mechanical Vibrations*, (Prentice-Hall Inc., Englewood Cliffs, NJ, 1953).
3. R. Halmshaw, *Non-Destructive Testing*, 2nd edition, Chapter 2, (Edward Arnold, London, 1991), p. 273.
4. M. Guyot, V. Cagan, *Journal of Magnetism and Magnetic Materials*. **101**, 256 (1991).
5. M. Guyot, T. Merceron, V. Cagan, *Journal of Magnetism and Magnetic Materials*. **83**, 217 (1990).
6. G. A. Evans, M. Linzer, *Journal of the American Ceramic Society*. **56**(11), 575 (1973).
7. H. Aburatani, K. Uchino, *Proceedings of 9th International Symposium Application of Ferroelectrics*, IEEE, p. 871 (1996).
8. K. Uchino, *Journal of Nanotechnology & Material Science*. **1**(1), 1, Ommega Pub. (2014).
9. K. Uchino, T. Ishii, *Journal of Ceramic Society of Japan*. **96**, 863 (1988).
10. K. Uchino, Piezoelectric energy harvesting systems–Essentials to successful developments–, *Journal of Energy Technology*. **6**, 829–848 (2018). DOI: 10.1002/ente.201700785.
11. K. Uchino, *Micromechatronics*, (CRC Press, Boca Raton, FL, 2019).

4 Piezoelectric Loss Phenomenology

ABSTRACT

"Phenomenological Approach to Piezoelectric Losses", in which the loss phenomenology in piezoelectrics, including three losses, namely, dielectric, elastic, and piezoelectric losses, is introduced and interrelationships among these losses are discussed, based on the piezoelectric constitutive and dynamic equations. Most of the physical phenomena exhibit heat generation associated with losses, inevitably, because a sort of delay of the output reaction occurs from the input electric, magnetic, or mechanical force. We will describe the loss formulation to "piezoelectric" materials, including "coupling phenomenon losses" in particular, by introducing complex numbers for physical parameters in this chapter. Complex parameter usage limits the applicability only for small damping situation, such as the damping ratio ζ less than 0.1 (10%) from the thermodynamic theory viewpoint.

The content of this chapter overlaps somewhat intentionally with that in Chapter 2 for the reader who skips the previous fundamental chapters.

4.1 INTENSIVE AND EXTENSIVE LOSSES

4.1.1 ENERGY DESCRIPTION OF INTENSIVE AND EXTENSIVE PHYSICAL PARAMETERS

We need to obtain both "intensive" and "extensive" losses in the piezoelectric characterization, for realizing accurate analysis in piezo-device designing by using computer simulation tools such as *Finite Element Method*, because either loss will contribute in a different way depending on the realizing vibration mode in piezoelectrics. Also this loss distinction is essential in creating a microscopic loss model in piezo-materials [Chapter 9 introduces the microscopic and semi-microscopic loss models]; that is, in order to explain loss mechanisms theoretically, we usually utilize the "extensive" losses, which are calculated from "intensive" losses obtained experimentally easily.

According to IUPAC (the International Union of Pure and Applied Chemistry), an "extensive parameter" depends on the volume of the material, while an "intensive parameter" is the ratio of two extensive ones and, therefore, is independent on the volume of the material.[1] Mass, length, and displacement, for example, if we cut the object into a half, will become a half; that is, extensive properties; while temperature, force, and voltage do not change, even if we cut the object into a half; intensive properties. Consequently, stress (X, first derivative of force) and electric field (E, first derivative of voltage) are intensive parameters, which are externally controllable, while strain (x, first derivative of displacement) and dielectric displacement (D) (or polarization (P), as defined by the accumulated dipole moments per volume) are extensive parameters, which are internally determined in the material.[2,3]

We start with the Gibbs free energy, G, in terms of intensive parameters, in general differential form as:

$$dG = -x\,dX - D\,dE - S\,dT \tag{4.1}$$

where x and X are strain and stress, D and E are electric displacement and electric field, and S and T are entropy and temperature. Equation (4.1) is the energy expression in terms of the externally controllable (which is denoted as "intensive") physical parameters X, E, and T. If we assume the simplest linear phenomena for the elastic (Hooke's law), dielectric, and electromechanical coupling properties, we obtain the following Gibbs energy expression. Note that no higher-order terms are included, different from the case in Chapter 2:

$$G = -(1/2)s^E X^2 - d X E - (1/2)\varepsilon_0 \varepsilon^X E^2. \qquad (4.2)$$

The temperature dependence of the function is associated with the elastic compliance, s^E, the dielectric constant, ε^X, and the piezoelectric charge coefficient, d, and the direct entropy expression is omitted [refer to Example Problem 2.6]. It is also noteworthy that the electromechanical coupling term XE (not XE^2) indicates the crystal symmetry [piezoelectric class in Table 2.2] which allows the sign change of X or x simultaneously for the E direction change. The following two piezoelectric equations (i.e., *piezoelectric constitutive equations*) are derived from Eq. (4.2):

$$x = -\frac{\partial G}{\partial X} = s^E X + dE \qquad (4.3)$$

$$D = -\frac{\partial G}{\partial E} = dX + \varepsilon_0 \varepsilon^X E \qquad (4.4)$$

Note that the Gibbs energy function provides "intensive" physical parameters: E-constant elastic compliance s^E and X-constant permittivity ε^X.

On the other hand, when we consider the free energy in terms of the "extensive" (i.e., material-related) parameters of strain, x, and electric displacement, D, we start from the differential form of the Helmholtz free energy designated by A, such that:

$$dA = X \, dx + E \, dD - S \, dT, \qquad (4.5)$$

We now assume

$$A = (1/2)c^D x^2 - h \, x \, D + (1/2)\kappa_0 \kappa^x D^2. \quad \left[\kappa_0 = \left(\frac{1}{\varepsilon_0} \right) \right] \qquad (4.6)$$

We obtain from this energy function another pair of piezoelectric constitutive equations:

$$X = \frac{\partial A}{\partial x} = c^D x - hD \qquad (4.7)$$

$$E = \frac{\partial A}{\partial D} = -hx + \kappa_0 \kappa^x D \qquad (4.8)$$

where c^D is the elastic stiffness at constant electric displacement (open-circuit conditions), h is the inverse piezoelectric charge coefficient, and κ^x is the inverse dielectric constant at constant strain (mechanically clamped conditions). Note $\kappa_0 = 1/\varepsilon_0 = 1.129 \times 10^{11}$ m/F.

Equations (4.3), (4.4) and (4.7), (4.8) should be written in tensor or matrix notations, precisely speaking, and the reader can imagine that these pairs are mutually inverse tensor matrix relations. However, we obtain here the interrelations between the intensive and extensive physical parameters in the simplest 1D form.

4.1.2 Piezoelectric Constitutive Equations with Losses

4.1.2.1 Intensive Losses

Since the detailed mathematical treatment has been described in previous paper,[4] we summarize the essential results in this section. We start from the *piezoelectric constitutive equation*:

$$\begin{pmatrix} x \\ D \end{pmatrix} = \begin{pmatrix} s^E & d \\ d & \varepsilon_0 \varepsilon^X \end{pmatrix} \begin{pmatrix} X \\ E \end{pmatrix}$$ (4.9)

where, x is strain, X, stress, D, electric displacement, and E, electric field. Note that the original piezoelectric constitutive equations cannot yield a delay time-related loss, in general, without taking into account irreversible thermodynamic equations or dissipation functions. However, the "dissipation functions" are mathematically equivalent to the introduction of "complex physical constants" into the phenomenological equations, if the loss is small and can be treated as a perturbation (*dissipation factor tangent* $\ll 0.1$).

Therefore, we will introduce complex parameters ε^{X*}, s^{E*}, and d^*, using *, in order to consider the small hysteresis losses in dielectric, elastic, and piezoelectric constants (refer to Section 3.3.5):

$$\varepsilon^{X*} = \varepsilon^X (1 - j \tan \delta'),$$ (4.10)

$$s^{E*} = s^E (1 - j \tan \phi'),$$ (4.11)

$$d^* = d(1 - j \tan \theta').$$ (4.12)

θ' is the phase delay of the strain under an applied electric field or the phase delay of the electric displacement under an applied stress. Both delay phases should be exactly the same if we introduce the same complex piezoelectric constant d^* into two components of Eq. (4.9). δ' is the phase delay of the electric displacement to an applied electric field under a constant stress (e.g., zero stress) condition, and ϕ' is the phase delay of the strain to an applied stress under a constant electric field (e.g., short-circuit) condition. We will consider these phase delays as "intensive" losses, because these losses are related with the intensive physical parameters, ε^X, s^E, and d. Note that we take negative sign in front of loss tangent, supposing that the extensive parameters are induced after the intensive parameters are applied.

Since the areas on the D-E and x-X domains exhibit directly the electrical and mechanical energies, respectively (see Figure 4.1a and b), the stored energies (during a quarter cycle) and hysteresis

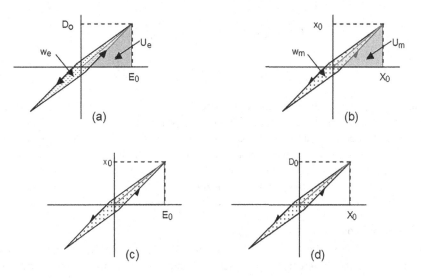

FIGURE 4.1   (a) D vs. E (stress free), (b) x vs. X (short-circuit), (c) x vs. E (stress free), and (d) D vs. X (short-circuit) curves with a slight hysteresis in each relation.

losses (during a full electric or stress cycle) for pure dielectric and elastic energies can be calculated as (refer to Example Problem 3.1 for the detailed calculation process):

$$U_e = (1/2)\varepsilon^X \varepsilon_0 E_0^2, \tag{4.13}$$

$$w_e = \pi \varepsilon^X \varepsilon_0 E_0^2 \tan \delta', \tag{4.14}$$

$$U_m = (1/2)s^E X_0^2, \tag{4.15}$$

$$w_m = \pi s^E X_0^2 \tan \phi'. \tag{4.16}$$

Here, U_e and U_m stand for electrical and mechanical energies stored during a quarter cycle, and w_e and w_m are the electrical and mechanical hysteresis losses, respectively. The dissipation factors, $\tan \delta'$ and $\tan \phi'$, can be experimentally obtained by measuring the dotted hysteresis area and the stored energy area; that is, $(1/2\pi)(w_e/U_e)$ and $(1/2\pi)(w_m/U_m)$, respectively. Note that the factor (2π) comes from integral per cycle.

The electromechanical hysteresis loss calculations, however, are more complicated, because the areas on the x-E and P-X domains do not directly provide energy. The areas on these domains can be calculated as follows, depending on the measuring ways; when measuring the induced strain under an electric field, the electromechanical conversion energy can be calculated as follows, by converting E to stress X:

$$U_{em} = \int x \, dX = \left(\frac{1}{s^E}\right) \int x \, dx = \left(d^2/s^E\right) \int_0^{E_0} E \, dE = (1/2)\left(d^2/s^E\right) E_0^2, \tag{4.17}$$

where $x = dE$ was used. Then, using Eqs. (4.11) and (4.12), and from the imaginary part, we obtain the loss during a full cycle as

$$w_{em} = \pi\left(d^2/s^E\right) E_0^2 (2 \tan \theta' - \tan \phi'). \tag{4.18}$$

Note that the area ratio in the strain vs. electric field measurement should provide the combination of piezoelectric loss $\tan \theta'$ and elastic loss $\tan \phi'$ (not $\tan \theta'$ directly!). When we measure the induced charge under stress, the stored energy U_{me} and the hysteresis loss w_{me} during a quarter and a full stress cycle, respectively, are obtained similar results as

$$U_{me} = \int P \, dE = (1/2)\left(d^2/\varepsilon_0 \varepsilon^X\right) X_0^2, \tag{4.19}$$

$$w_{me} = \pi\left(d^2/\varepsilon_0 \varepsilon^X\right) X_0^2 (2 \tan \theta' - \tan \delta'). \tag{4.20}$$

Now, the area ratio in the charge vs. stress measurement provides the combination of piezoelectric loss $\tan \theta'$ and dielectric loss $\tan \delta'$. Hence, from the measurements of D vs. E and x vs. X, we obtain $\tan \delta'$ and $\tan \phi'$, respectively, and either the piezoelectric (D vs. X) or converse piezoelectric measurement (x vs. E) provides $\tan \theta'$ through a numerical subtraction. The above equations provide a traditional loss measuring technique on piezoelectric actuators, an example of which is introduced in Section 7.2. You may recognize that the "piezoelectric" loss is a sort of "hidden" parameter, which cannot be directly measured, but obtained from the measurable combination loss!

4.1.2.2 Extensive Losses

So far, we have discussed the "intensive" dielectric, mechanical, and piezoelectric losses (with prime notation) in terms of "intensive" parameters X and E. In order to consider "physical" meanings of the losses (microscopic or semi-macroscopic model) in the material (e.g., domain dynamics), we will introduce the "extensive" losses[4] in terms of "extensive" parameters x and D. In practice, intensive losses are easily measurable, but extensive losses are not in the pseudo-DC measurement but obtainable from the intensive losses by using the *K-matrix* introduced later. When we start from the piezoelectric equations in terms of extensive physical parameters x and D,

$$\begin{pmatrix} X \\ E \end{pmatrix} = \begin{pmatrix} c^D & -h \\ -h & \kappa_0 \kappa^x \end{pmatrix} \begin{pmatrix} x \\ D \end{pmatrix} \tag{4.21}$$

where c^D is the elastic stiffness under D = constant condition (i.e., electrically open-circuit), κ^x is the inverse dielectric constant under x = constant condition (i.e., mechanically clamped), and h is the inverse piezoelectric constant d. We introduce the *extensive* dielectric, elastic, and piezoelectric losses as

$$\kappa^{x*} = \kappa^x (1 + j \tan \delta), \tag{4.22}$$

$$c^{D*} = c^D (1 + j \tan \phi), \tag{4.23}$$

$$h^* = h(1 + j \tan \theta). \tag{4.24}$$

The sign "+" in front of the imaginary "j" is taken by a general induction principle; that is "polarization induced after electric field application" and so on. However, in terms of the "extensive piezoelectric loss" $\tan \theta$, in the relation among electric displacement D (or polarization P) and strain x, we have no idea on which comes earlier either D or x. In other words, ferroelectricity-associated ferroelasticity or ferroelasticity-associated ferroelectricity? You will learn the "negative" piezo-loss $\tan \theta$ in practice in Section 9.3.

It is notable that the permittivity under a constant strain (e.g., zero strain or completely clamped) condition, ε^{x*}, and the elastic compliance under a constant electric displacement (e.g., open-circuit) condition, s^{D*}, can be provided as an inverse value of κ^{x*} and c^{D*}, respectively, in this simplest 1D expression. Thus, using exactly the same losses in Eqs. (4.22) and (4.23),

$$\varepsilon^{x*} = \varepsilon^x (1 - j \tan \delta), \tag{4.25}$$

$$s^{D*} = s^D (1 - j \tan \phi), \tag{4.26}$$

we will consider these phase delays again as "extensive" losses. Care should be taken in the case of a general 3D expression, where this part must be translated as "inverse" *matrix components* of κ^{x*} and c^{D*} tensors. In order to realize an x-constant status, we need to clamp 3D precisely; not just 1D, as introduced in the longitudinally clamped capacitance in the k_{31} mode plate specimen (see Section 4.3).

In order to obtain the interrelationship between the intensive and extensive losses, we remind the reader the physical property difference according to the boundary conditions: E constant and D constant, or X constant and x constant in the simplest 1D model. When an electric field is applied on a piezoelectric sample as illustrated in the **Top** part of Figure 4.2, this state will be equivalent to the superposition of the following two steps: first, the sample is completely clamped and the field E_0 is applied (pure electrical energy $(1/2)\,\varepsilon^x \varepsilon_0 \, E_0^2$ is stored); second, keeping the field at E_0, the mechanical constraint is released (additional mechanical energy $(1/2)\,(d^2/s^E)\,E_0^2$ is necessary). The total energy should correspond to the total input electrical energy $(1/2)\,\varepsilon^X \varepsilon_0 \, E_0^2$ under stress-free condition (left figure). That is, $(1/2)\varepsilon^X \varepsilon_0 E_0^2 = (1/2)\varepsilon^x \varepsilon_0 E_0^2 + (1/2)\left(d^2/s^E\right)E_0^2$. Similar energy calculation can be obtained from the **Bottom** part of Figure 4.2, leading to the following equations:

$$\varepsilon^x / \varepsilon^X = \left(1 - k^2\right), \tag{4.27}$$

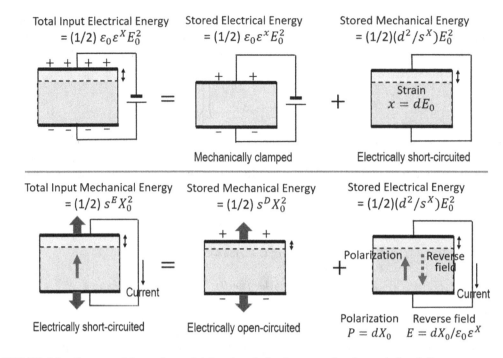

FIGURE 4.2 Conceptual figure for explaining the relation between ε^X and ε^x and s^E and s^D.

$$s^D/s^E = \left(1-k^2\right),\tag{4.28}$$

$$\kappa^X/\kappa^x = \left(1-k^2\right),\tag{4.29}$$

$$c^E/c^D = \left(1-k^2\right),\tag{4.30}$$

where

$$k^2 = d^2/\left(s^E\varepsilon_0\varepsilon^X\right) = h^2/\left(c^D\kappa_0\kappa^x\right).\quad\left[\kappa_0 = \left(\frac{1}{\varepsilon_0}\right)\right]\tag{4.31}$$

This k is called the *electromechanical coupling factor*, which is handled as a real number in most of the cases in this book.

4.1.2.3 *K*-Matrix in the Intensive and Extensive Losses

In order to obtain the relationships between the intensive and extensive losses, the following three equations derived from Eqs. (4.9) and (4.21) are essential:

$$s^E = \frac{1}{c^D}\frac{1}{\left(1-\dfrac{h^2}{\left(c^D\kappa_0\kappa^x\right)}\right)},\tag{4.32}$$

$$\varepsilon_0\varepsilon^X = \frac{1}{\left(\dfrac{1}{\kappa_0}\right)\kappa^x\left(1-\dfrac{h^2}{\left(c^D\kappa_0\kappa^x\right)}\right)},\tag{4.33}$$

$$d = \frac{1}{d} \frac{\dfrac{h^2}{\left(c^D \kappa_0 \kappa^x\right)}}{\left(1 - \dfrac{h^2}{\left(c^D \kappa_0 \kappa^x\right)}\right)}. \tag{4.34}$$

Replacing the parameters in Eqs. (4.32)–(4.34) by the complex parameters in Eqs. (4.10)–(4.12), (4.22)–(4.24), we obtain the relationships between the intensive and extensive losses:

$$\tan \delta' = \left(1/\left(1 - k^2\right)\right)\left[\tan \delta + k^2(\tan \phi - 2 \tan \theta)\right], \tag{4.35}$$

$$\tan \phi' = \left(1/\left(1 - k^2\right)\right)\left[\tan \phi + k^2(\tan \delta - 2 \tan \theta)\right], \tag{4.36}$$

$$\tan \theta' = \left(1/\left(1 - k^2\right)\right)\left[\tan \delta + \tan \phi - \left(1 + k^2\right)\tan \theta\right], \tag{4.37}$$

where k is the *electromechanical coupling factor* defined by Eq. (4.31) and here as a real number. It is important that the *intensive* dielectric, elastic, and piezoelectric losses (with prime) are mutually correlated with the *extensive* dielectric, elastic, and piezoelectric losses (non-prime) through the electromechanical coupling k^2, and that the denominator $(1 - k^2)$ comes basically from the ratios, $\varepsilon^x/\varepsilon^X = (1 - k^2)$ and $s^D/s^E = (1 - k^2)$, and this real part reflects to the dissipation factor when the imaginary part is divided by the real part.

Knowing the relationships between the intensive and extensive physical parameters (Eqs. (4.9) and (4.21)), and the electromechanical coupling factor k (Eq. (4.31)), we introduce so-called [K]-*matrix* to interrelate the intensive (prime) and extensive (non-prime) loss factors. [Try Example Problem 4.1 for your further understanding on the derivation process.]

$$\begin{bmatrix} \tan \delta' \\ \tan \phi' \\ \tan \theta' \end{bmatrix} = [K] \begin{bmatrix} \tan \delta \\ \tan \phi \\ \tan \theta \end{bmatrix}, \tag{4.38}$$

$$[K] = \frac{1}{1 - k^2} \begin{bmatrix} 1 & k^2 & -2k^2 \\ k^2 & 1 & -2k^2 \\ 1 & 1 & -1 - k^2 \end{bmatrix}, \quad k^2 = \frac{d^2}{s^E\left(\varepsilon^X \varepsilon_0\right)} = \frac{h^2}{c^D\left(\kappa^x \kappa_0\right)}. \tag{4.39}$$

The matrix $[K]$ is proven to be "invertible", that is, $K^2 = I$, or $K = K^{-1}$, where I is the identity matrix. Hence, the conversion relationship between the intensive (prime) and extensive (non-prime) exhibits full symmetry:

$$\begin{bmatrix} \tan \delta \\ \tan \phi \\ \tan \theta \end{bmatrix} = [K] \begin{bmatrix} \tan \delta' \\ \tan \phi' \\ \tan \theta' \end{bmatrix}. \tag{4.40}$$

The author emphasizes again that the extensive losses are more important for considering the physical micro/macroscopic models and can be obtained mathematically from a set of intensive losses, which are obtained directly from the experiments (in particular, pseudo-DC measurement).

Though we handle the electromechanical coupling factor k as a real parameter in the above, if we introduce a sort of "electromechanical coupling loss", it will be derived as follows from Eq. (4.31):

$$\left(\frac{k^{2''}}{k^{2'}}\right) = \tan\chi = -\left(2\tan\theta' - \tan\delta' - \tan\phi'\right) = \left(2\tan\theta - \tan\delta - \tan\phi\right) \tag{4.41}$$

Note that the electromechanical coupling loss is an eigen function of Eq. (4.39) with the $[K]$ *matrix*; that is, the unique constant irrelevant to the intensive or extensive description. The loss $\tan\chi$ can be either positive or negative, depending on the piezoelectric loss magnitude $\tan\theta'$, which contributes to the mechanical quality factor spectrum around the resonance and antiresonance frequencies, as discussed later.

Example Problem 4.1

1. Derive the $[K]$-matrix to interrelate the intensive and extensive losses:

$$[K] - \frac{1}{1-k^2}\begin{bmatrix} 1 & k^2 & -2k^2 \\ k^2 & 1 & -2k^2 \\ 1 & 1 & -1-k^2 \end{bmatrix} \tag{P4.1.1}$$

2. Verifying the following relationship first,

$$k^2 = \frac{d^2}{\left(s^E\varepsilon_0\varepsilon^X\right)} = \frac{h^2}{\left(c^D\kappa_0\kappa^x\right)}, \tag{P4.1.2}$$

then, verify electromechanical coupling loss relation:

$$\left(2\tan\theta' - \tan\delta' - \tan\phi'\right) = -\left(2\tan\theta - \tan\delta - \tan\phi\right). \tag{P4.1.3}$$

Hint
From the following pairs of equations, eliminate the parameters, X, E, x, and D.

$$x = s^E X + dE \tag{P4.1.4}$$

$$D = dX + \varepsilon_0\varepsilon^X E \tag{P4.1.5}$$

$$X = c^D x - hD \tag{P4.1.6}$$

$$E = -hx + \kappa_0\kappa^x D \tag{P4.1.7}$$

Solution

1. From Eqs. (4.9) and (4.21)

$$\begin{pmatrix} x \\ D \end{pmatrix} = \begin{pmatrix} s^E & d \\ d & \varepsilon_0\varepsilon^X \end{pmatrix}\begin{pmatrix} X \\ E \end{pmatrix} = \begin{pmatrix} s^E & d \\ d & \varepsilon_0\varepsilon^X \end{pmatrix}\begin{pmatrix} c^D & -h \\ -h & \kappa_0\kappa^x \end{pmatrix}\begin{pmatrix} x \\ D \end{pmatrix} \tag{P4.1.8}$$

Thus, the product of matrix should be the identity matrix:

$$\begin{pmatrix} s^E c^D - dh & -s^E h + d\kappa_0\kappa^x \\ dc^D - \varepsilon_0\varepsilon^X h & -dh + \varepsilon_0\varepsilon^X\kappa_0\kappa^x \end{pmatrix} = \begin{pmatrix} 1 & 0 \\ 0 & 1 \end{pmatrix} \tag{P4.1.9}$$

From these four each-component equations, we can derive the following equations easily:

$$\varepsilon^X \varepsilon_0 = \frac{1}{\kappa_0 \kappa^x \left[1 - \dfrac{h^2}{c^D \kappa_0 \kappa^x} \right]} = \frac{1}{\kappa_0 \kappa^x \left(1 - k^2 \right)} \tag{P4.1.10}$$

$$s^E = \frac{1}{c^D \left[1 - \dfrac{h^2}{c^D \kappa_0 \kappa^x} \right]} = \frac{1}{c^D \left(1 - k^2 \right)} \tag{P4.1.11}$$

$$d = \frac{\dfrac{h^2}{c^D \kappa_0 \kappa^x}}{h \left[1 - \dfrac{h^2}{c^D \kappa_0 \kappa^x} \right]} = \frac{k^2}{h \left(1 - k^2 \right)} \tag{P4.1.12}$$

Introducing the complex parameters, with prime losses on the left-hand side and non-prime losses on the right-hand side of the above three equations, we can obtain followings:

$$\tan \delta' = \left(1 / \left(1 - k^2 \right) \right) \left[\tan \delta + k^2 (\tan \phi - 2 \, \tan \theta) \right], \tag{P4.1.13}$$

$$\tan \phi' = \left(1 / \left(1 - k^2 \right) \right) \left[\tan \phi + k^2 (\tan \delta - 2 \, \tan \theta) \right], \tag{P4.1.14}$$

$$\tan \theta' = \left(1 / \left(1 - k^2 \right) \right) \left[\tan \delta + \tan \phi - \left(1 + k^2 \right) \tan \theta \right], \tag{P4.1.15}$$

[K]-matrix is automatically obtained from Eqs. (P4.1.13)–(P4.1.15).

2. Retrace Example Problem 2.9 for the first derivation process. Substituting X and E in Eqs. (P4.1.4) and (P4.1.5) by Eqs. (P4.1.6) and (P4.1.7), we obtain

$$x = s^E \left(c^D x - hD \right) + d \left(-hx + \kappa_0 \kappa^x D \right)$$

$$D = d \left(c^D x - hD \right) + \varepsilon_0 \varepsilon^X \left(-hx + \kappa_0 \kappa^x D \right).$$

Upon rearranging,

$$\left(1 + dh - s^E c^D \right) x = \left[d\kappa_0^x - hs^E \right] D$$

$$\left(dc^D - h\varepsilon_0 \varepsilon^X \right) x = \left[1 + dh - \kappa_0 \kappa^x \varepsilon_0 \varepsilon^X \right] D.$$

Finally,

$$\frac{d^2}{\left(s^E \varepsilon_0 \varepsilon^X \right)} = \frac{h^2}{\left(c^D \kappa_0 \kappa^x \right)}. \tag{P4.1.16}$$

Let us verify that Eq. (P4.1.16) corresponds to the electromechanical coupling factor, which is defined by k^2 = (stored electrical energy/input mechanical energy) or (stored mechanical energy/input electrical energy). We demonstrate the electric energy input case (try for the mechanical energy input case by yourself). See Figure 4.3a first, when we apply electric voltage between the top and bottom electrodes under stress-free condition ($X = 0$). Input electric energy must be equal to $(1/2)\varepsilon_0 \varepsilon^X E^2$ from Eq. (4.2), and the

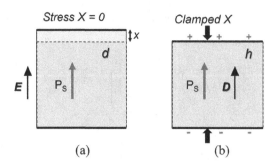

FIGURE 4.3 Calculation models of electromechanical coupling factor k for (a) electric field input and (b) electric displacement input.

strain generated by E should be dE from Eq. (P4.1.4). Since the converted/stored mechanical energy is obtained as $(1/2\, s^E)\, x^2$, we obtain:

$$k^2 = \left[(1/2)(d \cdot E)^2 / s^E\right] / \left[(1/2) \cdot \varepsilon_0 \varepsilon^X E^2\right]$$

$$= d^2 / \varepsilon_0 \varepsilon^X \cdot s^E. \tag{P4.1.17}$$

On the contrary, we consider now the extensive parameter description. See Figure 4.3b. When the specimen generates dielectric displacement D first along the spontaneous polarization direction under completely clamped condition ($x = 0$). Input electric energy must be equal to $(1/2)\kappa_0 \kappa^x D^2$ from Eq. (4.6). Since the blocking force (for clamping) is given by $X = -hD$ and the converted/stored mechanical energy is obtained as $\frac{1}{2}\left(\frac{1}{c^D}\right)X^2 = \frac{1}{2}\left(\frac{1}{c^D}\right)(hD)^2$, we obtain:

$$k^2 = \left[\frac{1}{2}\left(\frac{1}{c^D}\right)X^2\right] / \left[(1/2)\kappa_0 \kappa^x D^2\right].$$

$$= \frac{h^2}{\left(c^D \kappa_0 \kappa^x\right)} \tag{P4.1.18}$$

We can now understand that both $\dfrac{d^2}{\left(s^E \varepsilon_0 \varepsilon^X\right)}$ and $\dfrac{h^2}{\left(c^D \kappa_0 \kappa^x\right)}$ mean the electromechanical coupling factor, and the values are exactly the same.

Introducing the complex parameters, on both sides of Eq. (4.1.16), it is obvious to derive the electromechanical coupling loss equation:

$$\left(2\tan\theta' - \tan\delta' - \tan\phi'\right) = -\left(2\tan\theta - \tan\delta - \tan\phi\right). \tag{P4.1.19}$$

Example Problem 4.2

Verify the following $[K]$ matrix is *invertible*, that is, $K = K^{-1}$.

$$[K] = \frac{1}{1-k^2}\begin{bmatrix} 1 & k^2 & -2k^2 \\ k^2 & 1 & -2k^2 \\ 1 & 1 & -1-k^2 \end{bmatrix} \tag{P4.2.1}$$

Then, obtain the eigen function in terms of the vector component ($\tan\delta'$, $\tan\phi'$, $\tan\theta'$).

Solution

We calculate K^2.

$$[K] \times [K] = \frac{1}{1-k^2}\begin{bmatrix} 1 & k^2 & -2k^2 \\ k^2 & 1 & -2k^2 \\ 1 & 1 & -1-k^2 \end{bmatrix} \times \frac{1}{1-k^2}\begin{bmatrix} 1 & k^2 & -2k^2 \\ k^2 & 1 & -2k^2 \\ 1 & 1 & -1-k^2 \end{bmatrix}$$

$$= \frac{1}{(1-k^2)^2}\begin{bmatrix} 1+k^4-2k^2 & 0 & 0 \\ k^2+k^2-2k^2 & k^4+1-2k^2 & 0 \\ 0 & 0 & -2k^2-2k^2+(-1-k^2)^2 \end{bmatrix}$$

$$= \begin{bmatrix} 1 & 0 & 0 \\ 0 & 1 & 0 \\ 0 & 0 & 1 \end{bmatrix} = I$$

It is also interesting to obtain the eigen value and eigen vector of Eq. (P4.2.1). Taking the determinant

$$\begin{vmatrix} \dfrac{1}{1-k^2}-\lambda & \dfrac{k^2}{1-k^2} & \dfrac{-2k^2}{1-k^2} \\ \dfrac{k^2}{1-k^2} & \dfrac{1}{1-k^2}-\lambda & \dfrac{-2k^2}{1-k^2} \\ \dfrac{1}{1-k^2} & \dfrac{1}{1-k^2} & \dfrac{-1-k^2}{1-k^2}-\lambda \end{vmatrix} = 0, \tag{P4.2.2}$$

we obtain the eigen values of $\lambda_1 = \lambda_2 = 1$, $\lambda_3 = -1$, and the eigen vector for $\lambda_1 = \lambda_2 = 1$ as

$$x_1 = -x_2 + 2x_3$$
$$x_2 = x_2 \tag{P4.2.3}$$
$$x_3 = x_3$$

This results in that the value $(2\tan\theta' - \tan\delta' - \tan\phi') = -(2\tan\theta - \tan\delta - \tan\phi)$ is invariant, which corresponds to the loss for the electromechanical coupling factor $d^2/(s^E \varepsilon_0 \varepsilon^X) = h^2/(c^D \kappa^x \kappa_0)$. Note that negative sign in this loss equation comes from the definition difference of intensive and extensive losses in Eqs. (4.10)–(4.12) and (4.22)–(4.24). The loss for the *electromechanical coupling factor* appears in the latter sections.

4.2 CRYSTAL SYMMETRY AND LOSSES

As discussed in Section 2.6, there are 21×2 matrix components in elastic compliances and stiffnesses, 6×2 in dielectric permittivity and inverse permittivity components, and 18×2 components in piezoelectric constants and inverse piezoelectric constants, in general, in the lowest triclinic crystal symmetry *1*. We will handle mostly a polycrystalline ceramic (∞*mm*) such as lead zirconate titanate (PZT), which holds ten tensor components (five elastic compliances, three piezoelectric *d* constants, and two dielectric permittivity components), as given below [Note that $s_{66} = 2(s_{11} - s_{12})$ in the ∞ symmetry is not an independent parameter] in this textbook:

$$
\begin{bmatrix}
s_{11}^* & s_{12}^* & s_{13}^* & 0 & 0 & 0 \\
s_{12}^* & s_{11}^* & s_{13}^* & 0 & 0 & 0 \\
s_{13}^* & s_{13}^* & s_{33}^* & 0 & 0 & 0 \\
0 & 0 & 0 & s_{44}^* & 0 & 0 \\
0 & 0 & 0 & 0 & s_{44}^* & 0 \\
0 & 0 & 0 & 0 & 0 & s_{66}^*
\end{bmatrix},
\begin{bmatrix}
0 & 0 & 0 & 0 & d_{15}^* & 0 \\
0 & 0 & 0 & d_{15}^* & 0 & 0 \\
d_{31}^* & d_{31}^* & d_{33}^* & 0 & 0 & 0
\end{bmatrix}
\begin{bmatrix}
\varepsilon_{11}^* & 0 & 0 \\
0 & \varepsilon_{11}^* & 0 \\
0 & 0 & \varepsilon_{33}^*
\end{bmatrix},
$$

$$(4.42)$$

where * means a complex parameter, leading to 20 loss tensor components, taking into account both *intensive* and *extensive* losses. However, only ten components are independent, because of the relationship between the intensive and extensive losses, given by the K-matrix as Eqs. (4.38) and (4.39) in a 1D expression. This simplest K-matrix is not always applicable depending on the crystal symmetry.

4.3 RESONANCE AND ANTIRESONANCE

So far, we have considered the loss measurement from the hysteresis curves with a quasi-static or a pseudo-DC mode, as illustrated in Figure 4.1. We consider in this section a piezoelectric resonance method, that is, admittance/impedance spectrum measurement for determining dielectric, elastic, and piezoelectric three losses separately. As the reader already learned, the mechanical vibration resonance occurs in a finite material specimen with a principal length L at a frequency of $v/2L$, where v is the material's sound velocity. The key to understand is the fact that the sound velocity in a piezoelectric material exhibits some varieties, depending on the electric constraint condition; open-circuit or short-circuit, different from a pure elastic material.

4.3.1 RECTANGULAR PLATE k_{31} MODE

4.3.1.1 Preliminary Model Setting

Let us derive the necessary formulas for the longitudinal mechanical vibration of a piezo-ceramic plate through the transverse piezoelectric effect (d_{31}) as shown in Figure 4.4, which is occasionally used in this book in derivation examples. Length $L \gg$ width $w \gg$ thickness b is adopted. Assuming that the polarization is in the z direction and the x–y planes are the planes of the electrodes, the extensional vibration in the x direction is represented by the following dynamic equation:

$$
\rho\left(\partial^2 u/\partial t^2\right) = F = \left(\partial X_{11}/\partial x\right) + \left(\partial X_{12}/\partial y\right) + \left(\partial X_{13}/\partial z\right), \tag{4.43}
$$

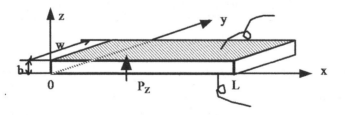

FIGURE 4.4 A rectangular piezo-ceramic plate ($L \gg w \gg b$) for a longitudinal mode through the transverse piezoelectric effect (d_{31}).

where u is the displacement in the x direction of a small volume element (ρ: mass density) in the ceramic plate. The relationship between the stress, the electric field (only E_z exists, because E_x and E_y should be zero due to the surface electrode), and the induced strain is described by the following set of equations (see Eq. (4.42) for polycrystalline specimen):

$$x_1 = s_{11}^E X_1 + s_{12}^E X_2 + s_{13}^E X_3 + d_{31}E_z$$

$$x_2 = s_{12}^E X_1 + s_{11}^E X_2 + s_{13}^E X_3 + d_{31}E_z$$

$$x_3 = s_{13}^E X_1 + s_{13}^E X_2 + s_{33}^E X_3 + d_{33}E_z$$

$$x_4 = s_{44}^E X_4$$ (4.44)

$$x_5 = s_{44}^E X_5$$

$$x_6 = 2\left(s_{11}^E - s_{12}^E\right)X_6$$

Let us consider the mechanical resonance intuitively first. When a frequency-sweeping AC electric field (i.e., constant voltage condition) is applied to this piezoelectric plate, length, width, and thickness, extensional resonance vibrations are successively excited. Supposing a typical PZT plate with dimensions 100 mm × 10 mm × 1 mm, these resonance frequencies correspond roughly to 10 kHz, 100 kHz, and 1 MHz, respectively. If we consider here only the fundamental length extensional modes for this configuration (neglecting higher order harmonics), when the frequency of the applied field is well below 10 kHz, the induced displacement follows the AC field cycle, and the displacement magnitude is given by $d_{31}E_3L$ (no significant phase lag, or phase = 0°). As we approach to the fundamental resonance frequency, a delay in the length displacement with respect to the applied field begins to develop, and the amplitude of the displacement becomes enhanced. At the exact resonance frequency, the phase difference will be −90°, and displacement magnification will reach $(8/\pi^2)Q_m \cdot d_{31}E_3L$ (detailed derivation process is in Section 4.3.2). Then, above 10 kHz, the length displacement no longer follows the applied field (phase lag approaches −180°), and the amplitude of the displacement is significantly reduced. This corresponds to the "length-1D-clamped" condition. Refer to Section 3.3.3 Bode plot for understanding the amplitude and phase change with frequency. With increasing the drive frequency higher around 100 kHz, the width vibration is amplified now by the factor of Q_m, by clamping the length vibration (almost zero strain) and keeping the thickness vibration strain $d_{33}E_3$ (no amplification). Further increase in the drive frequency around 1 MHz, the thickness vibration will be enhanced by clamping the length and width vibrations. Note that this intuitive thought does not take into account higher-order harmonic vibration modes tentatively, but is useful to understand the concept of "damped" capacitance.

4.3.1.2   Dynamic Equations for the k_{31} Mode[5]

When a very long, thin plate is driven by E_z in the vicinity of this fundamental length-mode resonance, X_2 and X_3 may be set equal to zero throughout the plate (i.e., the uniform strain distribution along y and z directions due to much lower drive frequency than their width and thickness resonance mode). Since shear stress will not be generated by the applied electric field E_z (because the electric field is parallel to the spontaneous polarization) only the following single equation (first equation of Eq. (4.44)) applies:

$$X_1 = x_1/s_{11}^E - \left(d_{31}/s_{11}^E\right)E_z$$ (4.45)

Substituting Eq. (4.45) into Eq. (4.43) and assuming that $x_1 = \partial u/\partial x$ (strain definition along the x axis) and $\partial E_z/\partial x = 0$ (since each top or bottom electrode is maintained at the same potentials, the internal electric field should be the same), we obtain the following dynamic equation:

$$\rho\left(\partial^2 u/\partial t^2\right) = \left(\partial X_{11}/\partial x\right) = \frac{1}{s_{11}^E}\left(\partial x_1/\partial x\right) - \left(d_{31}/s_{11}^E\right)\left(\partial E_z/\partial x\right) = \frac{1}{s_{11}^E}\left(\partial^2 u/\partial x^2\right) \tag{4.46}$$

Assuming $E_z = E_z\, e^{j\omega t}$, $u = u\, e^{j(\omega t - \delta)}$ (displacement is delayed by δ from the applied field in general), and the sound velocity v_{11} in the piezo-ceramic along the x (or 1) direction given by

$$v_{11}^E = 1/\sqrt{\rho\, s_{11}^E} \tag{4.47}$$

we obtain

$$-\omega^2 u = \left(v_{11}^E\right)^2\left(\partial^2 u/\partial x^2\right). \tag{4.48}$$

Superscript "E" on the elastic compliance and sound velocity comes from the E-constant condition. Thus, general solution is described as

$$u = A\sin\left(\omega x/v_{11}^E\right) + B\cos\left(\omega x/v_{11}^E\right),\ \text{and}\ x_1 = \frac{\omega}{v_{11}^E}A\cos\left(\omega x/v_{11}^E\right) - \frac{\omega}{v_{11}^E}B\sin\left(\omega x/v_{11}^E\right),$$

Taking into account the boundary conditions, $X_1 = 0$ at $x = 0$ and L (due to the mechanically free condition at the plate ends), A and B can be determined (A and B include the above phase lag δ information in general). Then, the following solutions are obtained:

$$\textbf{(Displacement)}\quad u(x) = \left(\frac{v_{11}^E}{\omega}\right)d_{31}E_z\,\frac{\sin\left[\dfrac{\omega(2x-L)}{2v_{11}^E}\right]}{\cos\left(\dfrac{\omega L}{2v_{11}^E}\right)} \tag{4.49}$$

$$\textbf{(Strain)}\quad \partial u/\partial x = x_1 = d_{31}E_z\left(\frac{\cos\left[\dfrac{\omega(2x-L)}{2v_{11}^E}\right]}{\cos\left(\dfrac{\omega L}{2v_{11}^E}\right)}\right) \tag{4.50}$$

4.3.1.3 Admittance/Impedance Calculation for the k_{31} Mode (Zero Loss Case)

When the specimen is utilized as an electrical component such as a filter or a vibrator, the electrical admittance [(induced current/applied voltage) ratio] plays an important role. We will establish the formula for characterizing the admittance spectrum obtained from the k_{31}-type piezo-plate specimen. Now essential is the electric displacement constitutive equation (Eq. (4.9)). Since the electrodes are on the top and bottom of the piezo-plate, as shown in Figure 4.4, the required equation is only D_3:

$$D_3 = d_{31}X_1 + \varepsilon_{33}^X\varepsilon_0 E_3 \tag{4.51}$$

As adopted in Section 4.3.1.1, only E_3 and X_1 are active ($X_2 = X_3 = 0$) in the above equation. Since the current flow into the specimen is described by the surface charge increment, $\partial D_3/\partial t$ (note that D_3 is position dependent (i.e., free-charge is distributing sinusoidally on the electrode), though $\partial E_3/\partial x = 0$, since the stress is sinusoidally distributed in the specimen, which generates polarization via direct piezoelectric effect), and the total current is given by:

$$i = j\omega w \int_0^L D_3 \, dx = j\omega w \int_0^L \left(d_{31} X_1 + \varepsilon_{33}^X \varepsilon_0 E_z \right) dx$$

$$= j\omega w \int_0^L \left[d_{31} \left\{ x_1 / s_{11}^E - \left(d_{31} / s_{11}^E \right) E_z \right\} + \varepsilon_{33}^X \varepsilon_0 E_z \right] dx. \tag{4.52}$$

Using Eq. (4.49), the admittance for the mechanically free sample is calculated to be:

$$Y = (-i/V) = \left(i/E_z b \right)$$

$$= \left(j\omega w L/E_z b \right) \left[\int_0^L \left(d_{31}^2 / s_{11}^E \right) \left(\frac{\cos\left[\frac{\omega(L-2x)}{2v_{11}^E} \right]}{\cos\left(\frac{\omega L}{2v_{11}^E} \right)} \right) E_z + \varepsilon_{33}^X \varepsilon_0 - \left(d_{31}^2 / s_{11}^E \right) E_z \right] dx$$

$$= (j\omega w L/b) \varepsilon_0 \varepsilon_{33}^{LC} \left[1 + \left(d_{31}^2 / \varepsilon_0 \varepsilon_{33}^{LC} \cdot s_{11}^E \right) \left(\tan\left(\omega L/2v_{11}^E \right) / \left(\omega L/2v_{11}^E \right) \right) \right],$$

$$= j\omega C_d \left[1 + \frac{k_{31}^2}{1 - k_{31}^2} \frac{\tan(\Omega_{11})}{\Omega_{11}} \right] \quad \left[\Omega_{11} = \left(\omega L/2v_{11}^E \right) \right]$$

$$= j\omega C_0 \left[\left(1 - k_{31}^2 \right) + k_{31}^2 \frac{\tan(\Omega_{11})}{\Omega_{11}} \right] \quad \left[k_{31}^2 = d_{31}^2 / \varepsilon_0 \varepsilon_{33}^X s_{11}^E \right] \tag{4.53}$$

where w is the width, L the length, b the thickness of the rectangular piezo-sample, and V is the applied voltage. Note that $E_z = -grad(V)$ (i.e., $V = -E_z b$), and the current direction measured externally should be taken oppositely to the internal "displacement current" flow direction, which gives $Y = -i/V$. ε_{33}^{LC} is the permittivity in a *longitudinally clamped* sample, which is given by

$$\varepsilon_{33}^{LC} = \varepsilon_{33}^X - \left(\frac{d_{31}^2}{\varepsilon_0 s_{11}^E} \right) = \varepsilon_0 \varepsilon_{33}^X \left(1 - k_{31}^2 \right). \quad \left[k_{31}^2 = d_{31}^2 / \varepsilon_0 \varepsilon_{33}^X s_{11}^E \right] \tag{4.54}$$

Note, however, that this is not three-dimensionally clamped permittivity ε_{33}^x, precisely speaking, which also reflects to the dielectric loss later. Accordingly, Eq. (4.53) can be understood as follows under a constant input voltage/electric field: the first term $(j\omega w L/b)\varepsilon_0 \varepsilon_{33}^{LC} = j\omega C_d$ is called "damped" (longitudinally clamped) capacitance, which is directly proportional to ω, while the second term $j\omega C_0 k_{31}^2 \frac{\tan(\Omega_{11})}{\Omega_{11}}$ is called "motional" capacitance, which is originated from the resonator's size (length) change via the mechanical vibration and strongly dependent on ω like $\tan(wL/2v_{11}^E)$. When ω is small, $\frac{\tan(\Omega_{11})}{\Omega_{11}} \to 1$, then the motional admittance becomes $j\omega C_0 k_{31}^2$. The total input energy will split into the motional (mechanical) and damped (electric) energy with the ratio k_{31}^2 vs. $\left(1 - k_{31}^2 \right)$, respectively. However, as ω approaches to the resonance frequency, motional admittance (or capacitance) increases dramatically like $\tan(wL/2v_{11}^E)$, which we can understand the accumulation/amplification of energy with respect to "time". Figure 4.5 shows an example admittance magnitude and phase spectra for a rectangular piezo-ceramic plate ($L = 20$) for a fundamental longitudinal

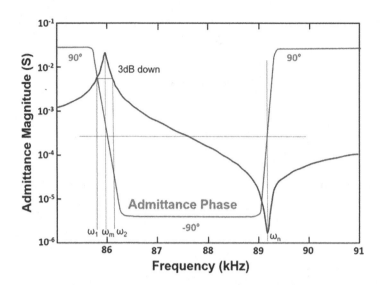

FIGURE 4.5 Admittance magnitude and admittance phase spectra for a rectangular piezo-ceramic plate for a fundamental longitudinal mode (k_{31}) through the transverse piezoelectric effect (d_{31}).

mode (k_{31}) through the transverse piezoelectric effect (d_{31}), on the basis of Eq. (4.53). Note that the shown data include losses (refer to Eq. (4.72) with losses), and the 3 dB down method to obtain mechanical quality factor Q_m is also inserted in advance.

The piezoelectric resonance is achieved where the admittance becomes infinite or the impedance is zero (by neglecting the loss). The resonance frequency f_R is calculated from Eq. (4.53) ($\tan(wL/2v_{11}{}^E) = \infty$ by putting $wL/2v_{11}{}^E = \pi/2$), and the fundamental frequency is given by

$$f_R = \omega_R/2\pi = v_{11}^E/2L = 1\big/\left(2L\sqrt{\rho s_{11}^E}\right). \tag{4.55}$$

This resonance mode corresponds to the fundamental standing wave with the velocity $v_{11}{}^E$ on a rod with length L (i.e., $f_R = v_{11}{}^E/2L$). On the other hand, the antiresonance state is generated for zero admittance or infinite impedance:

$$\left(\omega_A L/2v_{11}^E\right)\cot\left(\omega_A L/2v_{11}^E\right) = -d_{31}^2\big/\varepsilon_0\varepsilon_{33}^{LC}\cdot s_{11}^E = -k_{31}^2\big/\left(1-k_{31}^2\right). \tag{4.56}$$

The final transformation is provided by the definition,

$$k_{31} = d_{31}\big/\sqrt{s_{11}^E \cdot \varepsilon_{33}^X\varepsilon_0}. \tag{4.57}$$

4.3.1.4　Strain Distribution on the k_{31} Plate

The position dependence of x_1 is transformed to the following expression from Eq. (4.49):

$$x_1(x) = d_{31}E_z\left[\frac{\cos\left[\omega(L-2x)/2v_{11}^E\right]}{\cos\left(\omega L/2v_{11}^E\right)}\right] \tag{4.58}$$

The strain distribution is basically sinusoidal with the maximum at the center of plate ($x = L/2$) (see the numerator). When ω is close to ω_R, ($\omega_R L/2v_{11}{}^E$) = $\pi/2$, leading to the denominator $\cos(\omega_R L/2v_{11}{}^E) \to 0$. Significant strain magnification is obtained. It is worth noting that the stress X_1 is zero at the plate ends ($x = 0$ and L), but the strain x_1 is not zero but is equal to $d_{31}E_z$.

4.3.1.5 Resonance and Antiresonance Modes[6]

The resonance and antiresonance states are both mechanical resonance states with amplified strain/displacement states, but they are very different from the driving viewpoints. The mode difference is described by the following intuitive model. In a hypothetically high electromechanical coupling material with k almost equal to 1, the resonance or antiresonance states appear for $\tan(\omega L/2v) = \infty$ or 0 [i.e., $\omega L/2v = (m - 1/2)\pi$ or $m\pi$ (m: integer)], respectively, by neglecting the damped capacitance, since $\left(1 - k_{31}^2\right)$ is close to zero. The strain amplitude x_1 distribution for each state [calculated using Eq. (4.58)] is illustrated in Figure 4.6. In the resonance state, large strain amplitudes and large capacitance changes (called *motional capacitance*) are induced (Figure 4.6a), and under a constant applied voltage, the current can easily flow into the device (i.e., admittance Y is infinite). In contrast, at the antiresonance, the strain induced in the device compensates completely (i.e., a half extends and a half shrinks in Figure 4.6c), resulting in no motional capacitance change, and the current cannot flow easily into the sample (i.e., Y is zero). In other words, both plate ends become the nodal line, leading to no total displacement change (the antinode lines inside the sample displace largely at the antiresonance mode). Thus, for a high k material, the first antiresonance frequency f_A should be twice as large as the first resonance frequency f_R.[6]

In a typical case like a PZT, where $k_{31} = 0.3$, the antiresonance state varies from the previously mentioned double-f_R mode and becomes closer to the resonance mode (Figure 4.6b). The low-coupling material exhibits an antiresonance mode where the capacitance change due to the size change (*motional capacitance*) is significantly compensated by the current required to charge up the static capacitance (called *damped capacitance*). Thus, the antiresonance frequency f_A will approach the resonance frequency f_R, and the plate ends are not the nodal lines anymore but vibrating largely similar to the resonance mode. Only the difference between the drive schemes for the resonance or antiresonance modes appears "low voltage/high current" or "high voltage/low current", due to the admittance/impedance difference.

4.3.2 Rod Specimen for the k_{33} Mode[5]

4.3.2.1 Dynamic Equations for the k_{33} Mode

Let us consider now the longitudinal vibration k_{33} mode in comparison with the k_{31} mode. When the resonator is long in the z direction and the electrodes are deposited on each end of the rod, as shown in Table 4.1 and Figure 4.7, the following conditions are satisfied:

$$X_1 = X_2 = X_4 = X_5 = X_6 = 0 \text{ and } X_3 \neq 0. \tag{4.59}$$

Thus, the necessary constitutive equations are

$$X_3 = \left(x_3 - d_{33}E_z\right)/s_{33}^E \tag{4.60}$$

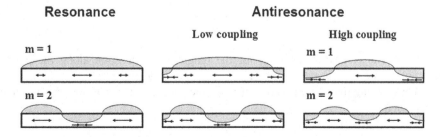

FIGURE 4.6 (a–c) Strain distribution in the resonance and antiresonance states. Longitudinal vibration through the transverse piezoelectric effect (d_{31}) in a rectangular plate.[6]

TABLE 4.1

The Characteristics of Various Piezoelectric Resonators with Different Shapes and Sizes

	Factor	Boundary Conditions	Resonator Shape	Definition
a	k_{31}	$X_1 \neq 0, X_2 = X_3 = 0$ $x_1 \neq 0, x_2 \neq 0, x_3 \neq 0$		$\dfrac{d_{31}}{\sqrt{s_{11}^E \varepsilon_0 \varepsilon_{33}^X}}$
b	k_{33}	$X_1 = X_2 = 0, X_3 \neq 0$ $x_1 = x_2 \neq 0, x_3 \neq 0$		$\dfrac{d_{33}}{\sqrt{s_{33}^E \varepsilon_0 \varepsilon_{33}^X}}$
c	k_p	$X_1 = X_2 \neq 0, X_3 = 0$ $x_1 = x_2 \neq 0, x_3 \neq 0$		$k_{31}\sqrt{\dfrac{2}{1-\sigma}}$
d	k_t	$X_1 = X_2 \neq 0, X_3 \neq 0$ $x_1 = x_2 = 0, x_3 \neq 0$		$k_{33}\sqrt{\dfrac{\varepsilon_0 \varepsilon_{33}^x}{c_{33}^D}}$
e	$k_{24} = k_{15}$	$X_1 = X_2 = X_3 = 0, X_4 \neq 0$ $x_1 = x_2 = x_3 = 0, x_5 \neq 0$		$\dfrac{d_{15}}{\sqrt{s_{55}^E \varepsilon_0 \varepsilon_{11}^X}}$

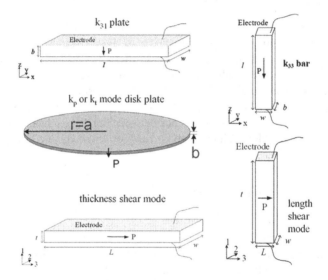

FIGURE 4.7 Sketches of the sample geometries for five required vibration modes.

$$D_3 = \varepsilon_{33}^X E_z + d_{33} X_3 \tag{4.61}$$

for this configuration. Assuming a local displacement u in the z direction, the dynamic equation is given by:

$$\rho \frac{\partial^2 u}{\partial t^2} = \frac{1}{s_{33}^E}\left[\frac{\partial^2 u}{\partial z^2} - d_{33}\frac{\partial E_z}{\partial z}\right] \tag{4.62}$$

from Eq. (4.60). The electrical condition for the longitudinal vibration along the polarization direction is not $\partial E_z/\partial z = 0$ but rather $\partial D_z/\partial z = 0$ because no electrode covers along the vibration direction z. Thus, taking the first derivative of both sides of Eq. (4.61) in terms of z, we obtain the following equation:

$$\varepsilon_0 \varepsilon_{33}^X \frac{\partial E_z}{\partial z} + \frac{d_{33}}{s_{33}^E}\left[\left(\frac{\partial^2 u}{\partial z^2}\right) - d_{33}\left(\frac{\partial E_z}{\partial z}\right)\right] = 0$$

Then,

$$\varepsilon_0 \varepsilon_{33}^X \left(1 - k_{33}^2\right) \left(\frac{\partial E_z}{\partial z}\right) = -\frac{d_{33}}{s_{33}^E} \left(\frac{\partial^2 u}{\partial z^2}\right)$$

Thus, taking into account

$$k_{33}^2 = \frac{d_{33}^2}{s_{33}^E \varepsilon_0 \varepsilon_{33}^X} \tag{4.63}$$

and

$$s_{33}^D = \left(1 - k_{33}^2\right) s_{33}^E \tag{4.64}$$

we finally obtain the dynamic equation,

$$\rho \frac{\partial^2 u}{\partial t^2} = \frac{1}{s_{33}^D} \frac{\partial^2 u}{\partial z^2} \tag{4.65}$$

Compared with Eq. (4.48) $\left(v = 1/\sqrt{\rho s_{11}^E}\right)$ in the surface electroded (E-constant) sample along the vibration direction, non-side-electroded (D-constant) k_{33} sample exhibits $\left(v_{33}^D = 1/\sqrt{\rho s_{33}^D}\right)$, which is faster (elastically stiffened) than that in E-constant condition. Taking a similar calculation process to the k_{31} mode, we obtain

$$[\textbf{Displacement}] \quad u_3 = \frac{d_{33}}{\varepsilon_0 \varepsilon_{33}^X} \frac{v_{33}^D}{\omega} D_3 \left[\sin\left(\frac{\omega}{2v_{33}^D}(2z - L)\right) \middle/ \cos\left(\frac{\omega L}{2v_{33}^D}\right)\right], \tag{4.66}$$

$$[\textbf{Strain}] \quad x_3 = \frac{d_{33}}{\varepsilon_0 \varepsilon_{33}^X} D_3 \left[\cos\left(\frac{\omega}{2v_{33}^D}(2z - L)\right) \middle/ \cos\left(\frac{\omega L}{2v_{33}^D}\right)\right] \tag{4.67}$$

In comparison with the resonance/antiresonance strain distribution status in the k_{31} mode in Figure 4.6, Figure 4.8 illustrates the strain distribution status in the k_{33} mode. Because k_{31} and k_{33}

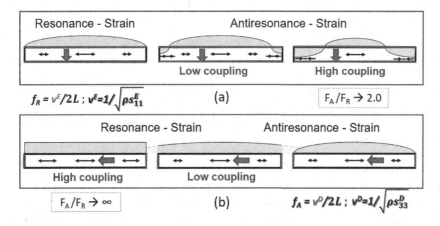

FIGURE 4.8 Strain distribution in the resonance and antiresonance states. Longitudinal vibration through the transverse d_{31} (a) and longitudinal d_{33} (b) piezo-effect in a rectangular plate.

modes possess E-constant and D-constant constraints, respectively, in k_{31}, the resonance frequency is directly related with $v_{11}{}^E$ or $s_{11}{}^E$, while in k_{33}, the antiresonance frequency is directly related with $v_{33}{}^D$ or $s_{33}{}^D$, $c_{33}{}^D$. The antiresonance in k_{31} and the resonance in k_{33} are subsidiary, originated from the electromechanical coupling factors. It is also worth noting that with increasing the k value toward 1, the ratio f_A/f_R approaches 2 in k_{31}, while it can reach ∞ in k_{33}, and that the strain distribution becomes almost flat or uniform in k_{33}, though the stress distributes sinusoidally with zero at the plate ends.

Admittance can also be calculated as

$$Y = \frac{i}{-V} = \frac{j\omega\varepsilon_0\varepsilon_{33}^{LC}\left(\dfrac{wb}{L}\right)}{\left[1-k_{33}^2\left\{\dfrac{\tan\left(\dfrac{\omega L}{2v_{33}^D}\right)}{\left(\dfrac{\omega L}{2v_{33}^D}\right)}\right\}\right]} = j\omega C_d + \frac{j\omega C_d}{\left[-1+1/k_{33}^2\left\{\dfrac{\tan(\Omega_{33})}{(\Omega_{33})}\right\}\right]}$$

$$= j\omega C_d + \frac{1}{\left[-\dfrac{1}{j\omega C_d}+1/j\omega C_d k_{33}^2\left\{\dfrac{\tan(\Omega_{33})}{(\Omega_{33})}\right\}\right]} \tag{4.68}$$

Here we used $\Omega_{33} = \left(\dfrac{\omega L}{2v_{33}^D}\right)$, $\varepsilon_{33}^{LC} = \varepsilon_{33}^X\left(1-k_{33}^2\right)$, $s_{33}^D = s_{33}^E\left(1-k_{33}^2\right)$, $k_{33}^2 = \dfrac{d_{33}^2}{\varepsilon_0\varepsilon_{33}^X s_{33}^E}$, $v_{33}^D = 1/\sqrt{\rho s_{33}^D}$,

and $C_d = \varepsilon_0\varepsilon_{33}^{LC}\left(\dfrac{wb}{L}\right)$. The second expression is to show the *damped admittance* and the *motional admittance*, separately, and the final expression is explicitly to reveal that the *motional admittance* branch should include the *negative capacitance* (with exactly the same damped capacitance value) in series with the pure vibration-related contribution proportional to $\tan(\Omega_{33})/(\Omega_{33})$.

The resonance frequency is obtained from $Y = \infty$, that is,

$$\left(\frac{\omega L}{2v_{33}^D}\right) = k_{33}^2 \tan\left(\frac{\omega L}{2v_{33}^D}\right) \tag{4.69}$$

To the contrary, the antiresonance frquency is obtained from $Y = 0$, that is,

$$\tan\left(\frac{\omega L}{2v_{33}^D}\right) = \infty, \text{ or } \frac{\omega L}{2v_{33}^D} = \frac{\pi}{2}, \text{ leading to } f_A = \frac{v_{33}^D}{2L} \tag{4.70}$$

Different from the k_{31} case, the k_{33} mode exhibits the antiresonance as a natural mechanical resonance frequency with a half-wave length exactly on the rod length under the sound velocity of v_{33}^D (i.e., stiffened vibration), and the resonance is a subsidiary vibration mode associated with the electromechanical coupling.

4.3.2.2   Boundary Condition: *E*-Constant vs. *D*-Constant

Both dielectric permittivity ε and elastic compliance s exhibit significant difference in terms of electromechanical coupling factor k under different boundary conditions: mechanical stress-free or clamped; electric short-circuit or open-circuit, as described in Eqs. (4.27) and (4.28). We also discussed the k_{31} mode vibrator with E-constant s_{11}^E and k_{33} mode with D-constant s_{33}^D for analyzing the dynamic equations. We consider here the relation between these status differences.

The key principle is from *Gauss law* (static electromagnetic Maxwell relation):

$$div\ \boldsymbol{D} = \rho;\text{or } div\ \boldsymbol{E} = \frac{1}{\varepsilon_0\varepsilon_r}\left(\rho - div\ \boldsymbol{P}\right) \tag{4.71}$$

If there exist free charges ρ, *div* $\boldsymbol{E}$ can be equal to zero by compensating $\boldsymbol{P}$ with ρ, leading to E-constant with respect to space/coordinate. In contrast, if no charges, *div* $D = 0$ (D-constant), by generating so-called *depolarization field* (reverse electric field) $E = -\left(\dfrac{\boldsymbol{P}}{\varepsilon_0\varepsilon_r}\right)$. Figure 4.9 visualizes the boundary condition models for E-constant and D-constant conditions under dynamic waves (higher than 1 kHz, not particularly limiting to the resonance mode). A mechanical/sound wave generates space-distributed polarization modulation (in addition to the spontaneous polarization P_S) in a piezoelectric plate, as $D = dX$. When the surface has electrodes as in the k_{31} mode (Figure 4.9a), charges can easily be supplied through the electrodes, as illustrated. Thus, E-constant (zero in the short-circuit or close to zero under the voltage supply connected) between the top and bottom electrodes or $\left(\dfrac{\partial E_z}{\partial x}\right) = 0$ is derived when the field is applied between the electrodes. On the contrary, when the surface does not have electrode, no charge is supplied (Figure 4.9b), the depolarization field is induced to maintain $\left(\dfrac{\partial D_z}{\partial x}\right) = 0$, leading to D-constant condition and the depolarization field (i.e., a large voltage distribution should be observed on the bare piezo-ceramic surface). In the case of k_{33} mode (Figure 4.9c), though the edges are electroded, there is no electrode along the wave propagation z axis. Thus, no charge is available to compensate the polarization modulation along the z direction, and the sound velocity along the polarization direction should be a D-constant sound velocity with s_{33}^{D}. It is worth to note that situation is different when the driving frequency is pseudo-DC. Even in a k_{33} rod specimen, when the operating frequency is low, the depolarization field attracts "stray" charges in the specimen or on the specimen surface from the surrounding atmosphere, and it is cancelled out (i.e., "screening"); that is, approaching to the E-constant condition, rather than D-constant status. The effective elasticity seems to change from the original s_{33}^{D} to s_{33}^{E} with decreasing the operating frequency from 1 kHz down to 0.001 Hz, in practice.

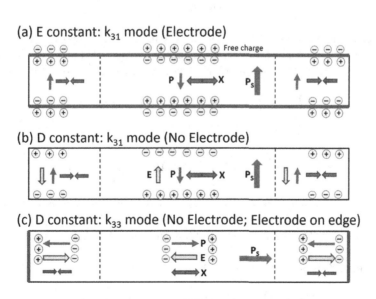

FIGURE 4.9 (a–c) Boundary conditions: E-constant vs. D-constant under dynamic waves.

4.3.3 Resonance/Antiresonance Dynamic Equations with Losses

4.3.3.1 Loss and Mechanical Quality Factor in k_{31} Mode[5]

a. Resonance Q_A:

Now, we introduce the complex parameters into the admittance formula Eq. (4.53) around the resonance frequency[5]: $\varepsilon_{33}^{X*} = \varepsilon_3^X\left(1 - j\tan\delta_{33}'\right)$, $s_{11}^{E*} = s_{11}^E\left(1 - j\tan\phi_{11}'\right)$, and $d_{31}^* = d_{31}\left(1 - j\tan\theta_{31}'\right)$.

$$Y = Y_d + Y_m$$

$$= j\omega C_d\left(1 - j\tan\delta_{33}'''\right) + j\omega C_d K_{31}^2\left[1 - j\left(2\tan\theta_{31}' - \tan\phi_{11}'\right)\right]\left[\tan\left(\omega L/2v_{11}^{E*}\right)/\left(\omega L/2v_{11}^{E*}\right)\right], \quad (4.72)$$

where

$$C_0 = \left(wL/t\right)\varepsilon_0\varepsilon_{33}^X, \left(\text{free electrostatic capacitance, real number}\right) \quad (4.73)$$

$$C_d = \left(1 - k_{31}^2\right)C_0. \left(\text{damped/clamped capacitance, real number}\right) \quad (4.74)$$

$$K_{31}^2 = \frac{k_{31}^2}{1 - k_{31}^2} \quad (4.75)$$

Note that the loss for the first term ("damped/clamped" admittance) is represented by the dielectric loss $\tan\delta'''$:

$$\tan\delta_{33}''' = \left[1/\left(1 - k_{31}^2\right)\right]\left[\tan\delta_{33}' + k_{31}^2\left(\tan\phi_{11}' - 2\tan\theta_{31}'\right)\right] \quad (4.76)$$

Though the formula is identical to Eq. (4.35), $\tan\delta_{33}'''$ is not exactly to the "extensive" non-prime loss $\tan\delta$, because the extensive loss should be under three-dimensionally clamped condition, not under just 1D longitudinally clamped. Taking into account

$$v_{11}^{E*} = \frac{1}{\sqrt{\rho s_{11}^E\left(1 - j\tan\phi_{11}'\right)}} \approx v_{11}^E\left(1 + j\frac{\tan\phi_{11}'}{2}\right) \quad (4.77)$$

we further calculate $1/[\tan(\omega L/2v^*)]$ with an expansion-series approximation around the A-type resonance frequency $(\omega_A L/2v) = \pi/2$, taking into account that the resonance state is defined in this case for the minimum impedance point. Using new frequency parameters,

$$\Omega_A = \omega_A L/2v_{11}^E = \pi/2, \Delta\Omega_A = \Omega - \pi/2 \left(\ll 1\right), \quad (4.78)$$

and

$$\frac{1}{\tan\Omega^*} = \cot\left(\frac{\pi}{2} + \Delta\Omega_A - j\frac{\pi}{4}\tan\phi_{11}'\right) = \Delta\Omega_A - j\frac{\pi}{4}\tan\phi_{11}'$$

the *motional admittance* Y_m is approximated around the first resonance frequency ω_A by

$$Y_m = j\left(8/\pi^2\right)\omega_A C_d K_{31}^2\left[1 - j\left(2\tan\theta_{31}' - \tan\phi_{11}'\right)\right]/\left[\left(4/\pi\right)\Delta\Omega_A - j\tan\phi_{11}'\right]. \quad (4.79)$$

The maximum Y_m is obtained at $\Delta\Omega_A = 0$:

$$Y_m^{max} = \left(8/\pi^2\right)\omega_A C_d K_{31}^2\left(\tan\phi_{11}'\right)^{-1} = \left(8/\pi^2\right)\omega_A C_0 k_{31}^2 Q_A, \quad (4.80)$$

The mechanical quality factor for A-type resonance $Q_A = (\tan \phi_{11}')^{-1}$ can be proved as follows: Q_A is defined by $Q_A = \omega_A/2\Delta\omega$, where $2\Delta\omega$ is a full width of the 3 dB down (i.e., $1/\sqrt{2}$, because $20 \log_{10}(1/\sqrt{2}) = -3.01$) of the maximum value $Y_m{}^{max}$ at $\omega = \omega_A$. Since $|Y| = |Y|^{max}/\sqrt{2}$ can be obtained when the "conductance = susceptance"; $\Delta\Omega_A = (\pi/4)\tan \phi_{11}'$ [see the denominator of Eq. (4.79)],

$$Q_A = \Omega_A/2\Delta\Omega_A = (\pi/2)/2(\pi/4)\tan\phi_{11}' = \left(\tan\phi_{11}'\right)^{-1}. \tag{4.81}$$

Similarly, the maximum displacement u^{max} is obtained at $\Delta\Omega = 0$:

$$U^{max} = \left(8/\pi^2\right)d_{31}E_Z L \; Q_A. \tag{4.82}$$

The maximum displacement at the resonance frequency is $(8/\pi^2)Q_A$ times larger than that at a non-resonance frequency, $d_{31} E_Z L$. Under the constant voltage/field drive, the displacement is amplified at the resonance frequency, while under the constant current drive, the displacement U and the impedance Z are amplified at the antiresonance frequency by the factor of $(8/\pi^2)Q_B$, as explained in Example Problem 4.3.

Example Problem 4.3

Under pseudo-DC operation, the input electric energy is split into the converted mechanical energy by k^2 and the stored electric energy by $(1 - k^2)$, leading to the damped and motional capacitance ratio $(1 - k^2)$ vs. k^2. However, under the resonance drive, though the damped admittance is provided by $\omega_A C_0(1 - k_{31}^2)$, the maximum of the motional admittance for the fundamental resonance frequency is describe by

$$Y_m{}^{max} = \left(8/\pi^2\right)w_A C_0 k_{31}^2 Q_A. \tag{P4.3.1}$$

The calibration factor $(8/\pi^2)$ (≈ 0.81) is required for the fundamental resonance frequency, rather than just 1. Explain why this calibration factor is required for the fundamental resonance condition.

Hint
Calculate the motional admittance for higher-order resonance harmonics. Only one fundamental resonance mode does not spend all mechanically convertible energy. You may also use the relation:

$$\sum \left[\frac{1}{(2m-1)^2} \right] = \left(\frac{\pi^2}{8} \right) \tag{P4.3.2}$$

Solution
We start from Eq. (4.67):

$$Y_m = j\omega C_d K_{31}^2 \left[1 - j\left(2\tan\theta_{31}' - \tan\phi_{11}'\right)\right]\left[\tan\left(\omega L/2v_{11}^{E*}\right)/\left(\omega L/2v_{11}^{E*}\right)\right], \tag{P4.3.3}$$

Note that the A-type resonance is obtained at $\left(\dfrac{\omega_{A,n}L}{2v_{11}^E}\right) = n\left(\dfrac{\pi}{2}\right)$, where $n = 1, 3, 5, \ldots$ (the nth higher-order harmonics) and $v_{11}^{E^2} = 1/\rho s_{11}^E$. Then, taking into account the complex elastic compliance

$$v_{11}^{E*} = \frac{1}{\sqrt{\rho s_{11}^E\left(1 - j\tan\phi_{11}'\right)}} = v_{11}^E\left(1 + j\frac{\tan\phi_{11}'}{2}\right), \tag{P4.3.4}$$

we further calculate $1/[\tan(\omega L/2v^*)]$ with an expansion-series approximation around the A-type resonance frequency $(\omega_A L/2v_{11}^E) = n(\pi/2)$, taking into account that the resonance state is defined in this case for the minimum impedance (maximum admittance) point. Using new frequency parameters,

$$\frac{1}{\tan\Omega^*} = \cot\left(n\frac{\pi}{2} + \Delta\Omega_A - j\frac{1}{2}\left(n\frac{\pi}{2}\right)\tan\phi'_{11}\right) = \Delta\Omega_A - jn\frac{\pi}{4}\tan\phi'_{11} \qquad (P4.3.5)$$

the *motional admittance* Y_m is approximated around the n-th resonance frequency $\omega_{A,n}$ by

$$Y_{m,n} = j\left(8/\pi^2 n^2\right)\omega_A C_0 k_{31}^2\Big/\left[(4/n\pi)\Delta\Omega_A - j\tan\phi'_{11}\right]. \qquad (P4.3.6)$$

The maximum Y_m is obtained at $\Delta\Omega_A = 0$:

$$Y_{m,n}^{\max} = \left(8/\pi^2 n^2\right)\omega_A C_0 k_{31}^2\left(\tan\phi'_{11}\right)^{-1} = \left(8/\pi^2 n^2\right)\omega_A C_0 k_{31}^2 Q_m \qquad (P4.3.7)$$

Supposing that the intensive elastic loss $\tan\phi_{11}'$ or the mechanical quality factor Q_m is insensitive to the frequency difference among the higher-order harmonic resonance frequencies, we can understand that each harmonic mode is originated from the effective motional capacitance equal to $(8/\pi^2 n^2)\, C_0 k_{31}^2$, and the admittance is enhanced by the factor of Q_m. Under the resonance, the input cyclic electric energy will excite the mechanical vibration and motional capacitance synchronously by a factor of Q_m by spending the cyclic excitation number proportional to Q_m. The motional capacitance is proportional to $(1/n^2)$ for the n-th order harmonic resonance mode. Knowing a general relationship $\sum\left[\dfrac{1}{(2m-1)^2}\right] = \left(\dfrac{\pi^2}{8}\right)$ (m: positive integer), when we add motional capacitances for all harmonic resonance modes:

$$\left(8/\pi^2\right)C_0\, k_{31}^2 \sum_{n=1,3,5,\ldots}\left(\frac{1}{n^2}\right) = C_0\, k_{31}^2 \qquad (P4.3.8)$$

Since the total motional capacitance for all harmonic resonance modes corresponds exactly to the free capacitance minus damped capacitance, the calibration factor $(8/\pi^2 n^2)$ can be understood as the distribution ratio of the mechanical energy to all n-th harmonic modes. The above concept on the higher-order harmonic modes will be used in the Equivalent Circuit Model explained in the next Chapter 5.

b. **Antiresonance Q_B:**

On the other hand, a higher quality factor at the antiresonance is usually observed in comparison with that at the resonance point,[7,8] the reason of which was interpreted by Mezheritsky from the combination of three loss factors.[8] In this subsection, we provide an alternative and, more importantly, a user-friendly formula to determine piezoelectric losses by analyzing the admittance/impedance spectra at resonance and antiresonance.[9] The antiresonance corresponds to the minimum admittance of Eq. (4.53):

$$Y = (j\omega wL/b)\varepsilon_0\varepsilon_{33}^X\left[\left(1-k^2\right) + k^2 \tan\left(\omega L/2v_{11}^E\right)\Big/\left(\omega L/2v_{11}^E\right)\right], \left(v_{11}^E = 1\Big/\sqrt{\rho s_{11}^E}\right)$$

$$= j\omega C_0\left[\left(1-k_{31}^2\right) + k_{31}^2\frac{\tan(\Omega_{11})}{\Omega_{11}}\right].$$

Here, $v_{11}^E = 1/\sqrt{\rho s_{11}^E}$, $\Omega_{11} = \left(\omega L/2v_{11}^E\right)$, and $k_{31}^2 = d_{31}^2/\varepsilon_0\varepsilon_{33}^X s_{11}^E$.

In the resonance discussion, we neglected the damped admittance, because the motional admittance is significantly large due to $\tan(\omega L/2v_{11}^{E*}) \nearrow \infty$. On the contrary, in the antiresonance discussion, we consider basically the subtraction between the damped and motional admittances; that is, the admittance should be exactly to zero when the loss is not included or is only the minimum when we consider the losses (i.e., complex parameters) in Eq. (4.72).

We introduce the normalized admittance Y' for further calculation:

$$Y' = 1 - k_{31}^2 + k_{31}^2 \frac{\tan\left(\omega l / 2 v_{11}^E\right)}{\omega l / 2 v_{11}^E} = 1 - k_{31}^2 + k_{31}^2 \frac{\tan(\Omega)}{\Omega}. \tag{4.83}$$

Since the expansion series of $\tan\Omega$ is convergent in this case, taking into account

$$1 / v_{11}^{E*} = \sqrt{\rho s_{11}^E \left(1 - j \tan\phi_{11}'\right)} = v_{11}^E \left(1 - j \frac{\tan\phi_{11}'}{2}\right)$$

we can apply the following expansion approximation in terms of $\tan\phi_{11}'$:

$$\tan\left(\Omega^*\right) = \tan\left(\Omega - j \frac{\Omega \tan\phi_{11}'}{2}\right) = \tan\Omega - j \frac{\Omega \tan\phi_{11}'}{2\cos^2\Omega}$$

Introducing losses for the parameters in Eq. (4.83) leads to

$$Y' = 1 - k_{31}^2 \left[1 - j\left(2\tan\theta_{31}' - \tan\delta_{33}' - \tan\phi_{11}'\right)\right] + k_{31}^2 \left[1 - j\left(2\tan\theta_{31}' - \tan\delta_{33}' - \tan\phi_{11}'\right)\right] \frac{\tan\Omega^*}{\Omega^*}. \tag{4.84}$$

Note that the "electromechanical coupling loss" $\left(2\tan\theta_{31}' - \tan\delta_{33}' - \tan\phi_{11}'\right)$ contributes significantly in this antiresonance discussion. We separate Y' into conductance G (real part) and susceptance B (imaginary part) as $Y' = G + jB$:

$$G = 1 - k_{31}^2 + k_{31}^2 \frac{\tan\Omega}{\Omega}. \tag{4.85}$$

$$B = \left(k_{31}^2 - k_{31}^2 \frac{\tan\Omega}{\Omega}\right)\left(2\tan\theta_{31}' - \tan\delta_{33}' - \tan\phi_{11}'\right) - \frac{k_{31}^2}{2}\left(\frac{1}{\cos^2\Omega} - \frac{\tan\Omega}{\Omega}\right)\tan\phi_{11}'. \tag{4.86}$$

The antiresonance frequency Ω_b should satisfy $G = 0$ [refer to Section 7.2.5.1], and

$$1 - k_{31}^2 + k_{31}^2 \frac{\tan\Omega_B}{\Omega_B} = 0. \tag{4.87}$$

Using new parameters,

$$\Omega = \Omega_B + \Delta\Omega_B, \tag{4.88}$$

similar to $\Delta\Omega_A$ for the resonance, $\Delta\Omega_B$ is also a small number, and the first-order approximation can be utilized.

$$\frac{\tan\Omega}{\Omega} = \frac{\tan\Omega_B}{\Omega_B} + \frac{1}{\Omega_B}\left(\frac{1}{\cos^2\Omega_B} - \frac{\tan\Omega_B}{\Omega_B}\right)\Delta\Omega_B$$

Neglecting high-order term which has two or more small factors (loss factor or $\Delta\Omega_B$).

$$G = \frac{k_{31}^2}{\Omega_B}\left(\frac{1}{\cos^2 \Omega_B} - \frac{\tan \Omega_B}{\Omega_B}\right)\Delta\Omega_B. \tag{4.89}$$

$$B = \left(2\tan\theta_{31}' - \tan\delta_{33}' - \tan\phi_{11}'\right) - \frac{k_{31}^2}{2}\left(\frac{1}{\cos^2 \Omega_B} - \frac{\tan \Omega_B}{\Omega_B}\right)\tan\phi_{11}'. \tag{4.90}$$

Consequently, the minimum absolute value of admittance can be achieved when $\Delta\Omega_B$ is 0. The antiresonance frequency Ω_B is determined by Eq. (4.89). In order to find the 3dB-up point, let $G = B$, where $\sqrt{G^2 + B^2} = \sqrt{2}B$ is satisfied:

$$\frac{k_{31}^2}{\Omega_B}\left(\frac{1}{\cos^2 \Omega_B} - \frac{\tan \Omega_B}{\Omega_B}\right)\Delta\Omega_B = \left(2\tan\theta_{31}' - \tan\delta_{33}' - \tan\phi_{11}'\right) - \frac{k_{31}^2}{2}\left(\frac{1}{\cos^2 \Omega} - \frac{\tan \Omega}{\Omega}\right)\tan\phi_{11}'. \tag{4.91}$$

Further, since the antiresonance quality factor is given by

$$Q_{B,31} = \frac{\Omega_B}{2|\Delta\Omega_B|}, \tag{4.92}$$

Equation (4.91) can be represented as

$$\frac{k_{31}^2}{2Q_{B,31}}\left(\frac{1}{\cos^2 \Omega_B} - \frac{\tan \Omega_B}{\Omega_B}\right) = -\left(2\tan\theta_{31}' - \tan\delta_{33}' - \tan\phi_{11}'\right) + \frac{k_{31}^2}{2}\left(\frac{1}{\cos^2 \Omega_B} - \frac{\tan \Omega_B}{\Omega_B}\right)\tan\phi_{11}'.$$

We can now obtain the result as

$$\frac{1}{Q_{B,31}} = \tan\phi_{11}' - \frac{2}{k_{31}^2}\left(2\tan\theta_{31}' - \tan\delta_{33}' - \tan\phi_{11}'\right)\Bigg/\left(\frac{1}{\cos^2 \Omega_B} - \frac{\tan \Omega_B}{\Omega_B}\right).$$

or, as a final formula:

$$\frac{1}{Q_{B,31}} = \frac{1}{Q_{A,31}} - \frac{2}{1 + \left(\frac{1}{k_{31}} - k_{31}\right)\Omega_B^2}\left(2\tan\theta_{31}' - \tan\delta_{33}' - \tan\phi_{11}'\right) \tag{4.93}$$

Note the following relation:

$$\frac{1}{\cos^2 \Omega_B} - \frac{\tan \Omega_B}{\Omega_B} = \frac{\left(1 - k_{31}^2\right)^2 \Omega_B^2 + k_{31}^2}{k_{31}^4} \tag{4.94}$$

You may understand that k_{31} mode, where the wave propagation direction with the electrode is perpendicular to the spontaneous polarization direction, the primary mechanical resonance (a half-wave length vibration of the plate length) corresponds to the "resonance" mode with the sound velocity s_{11}^E and the "antiresonance" mode corresponds to the subsidiary mode via the electromechanical coupling. Also from the Eq. (4.96), we can understand that when $\left(2\tan\theta_{31}' - \tan\delta_{33}' - \tan\phi_{11}'\right) > 0$, $Q_{B,31} > Q_{A,31}$; while $\left(2\tan\theta_{31}' - \tan\delta_{33}' - \tan\phi_{11}'\right) < 0$, $Q_{B,31} < Q_{A,31}$.

4.3.3.2   Loss and Mechanical Quality Factor in k_{33} Mode[10]

The length extensional mode is shown in Figure 4.7, where $L \gg w, b$. Taking into account the dynamic equation given by Eq. (4.68), we can derive the impedance expression of the k_{33} mode bar in a similar fashion to k_{31} mode.[11]

$$Z(\omega) = \frac{1}{j\omega C_d} \left(1 - k_{33}^2 \frac{\tan \Omega}{\Omega} \right), \tag{4.95}$$

where

$$C_d = \frac{wb}{L} \varepsilon_0 \varepsilon_{33}^X \left(1 - k_{33}^2 \right), \tag{4.96}$$

$$\Omega = \frac{\omega L}{2} \sqrt{\rho s_{33}^D} = \frac{\omega L}{2} \sqrt{\rho s_{33}^E \left(1 - k_{33}^2 \right)}, \tag{4.97}$$

$$k_{33}^2 = \frac{d_{33}^2}{\varepsilon_0 \varepsilon_{33}^X s_{33}^E}. \tag{4.98}$$

By introducing the complex parameters,

$$\left(k_{33}^2 \right)^* = k_{33}^2 \left(1 - j\chi_{33} \right), \tag{4.99}$$

$$\chi_{33} = 2\tan \theta_{33}' - \tan \delta_{33}' - \tan \phi_{33}' \text{ (“electromechanical coupling loss”)}, \tag{4.100}$$

$$C_d^* = C_d \left(1 - j\tan \delta_{33}''' \right) \text{ (damped capacitance loss)}, \tag{4.101}$$

$$\tan \delta_{33}''' = \frac{1}{1 - k_{33}^2} \left[\tan \delta_{33}' - k_{33}^2 \left(2\tan \theta_{33}' - \tan \phi_{33}' \right) \right], \tag{4.102}$$

$$\Omega^* = \Omega \sqrt{1 - j\tan \phi_{33}'''}, \tag{4.103}$$

$$\tan \phi_{33}''' = \frac{1}{1 - k_{33}^2} \left[\tan \phi_{33}' - k_{33}^2 \left(2\tan \theta_{33}' - \tan \delta_{33}' \right) \right]. \tag{4.104}$$

Note again that the parameters in k_{33} mode have similar forms as k_{31} mode, and the difference is that the loss factors by $-\chi_{33}$, $\tan \delta_{33}'''$, and $\tan \phi_{33}'''$, which show identical forms to the "extensive" loss parameters in terms of "intensive" losses but are not the same, strictly speaking. The difference between the extensive (non-prime) losses and these triple-prime losses comes from the 3D or 1D mechanically clamped conditions. Refer to the k_t mode description in the next subsection, where the elastic loss factors are purely "extensive losses" since the elastic stiffness c_{33}^D is the primary parameter, and the mechanical 3D clamp is practically satisfied. Therefore, a similar derivation process to the k_{31} mode can be applied, and the results are given by

$$Q_{B,33} = \frac{1}{\tan \phi_{33}'''} = \frac{1 - k_{33}^2}{\tan \phi_{33}' - k_{33}^2 \left(2\tan \theta_{33}' - \tan \delta_{33}' \right)}, \tag{4.105}$$

$$\frac{1}{Q_{A,33}} = \frac{1}{Q_{B,33}} + \frac{2}{k_{33}^2 - 1 + \Omega_A^2 / k_{33}^2} \left(2\tan \theta_{33}' - \tan \delta_{33}' - \tan \phi_{33}' \right). \tag{4.106}$$

Different from the k_{31} mode, you may approximately understand that k_{33} and k_t modes, where the wave propagation direction is parallel to the spontaneous polarization direction, the primary mechanical resonance (a half-wave length vibration of the length or thickness) corresponds to the "antiresonance" mode with the sound velocity s_{33}^D, and the "resonance" state is the subsidiary mode via the piezoelectric coupling. Also from the Eq. (4.106), we can understand that when $\left(2 \tan \theta_{33}' - \tan \delta_{33}' - \tan \phi_{33}'\right) > 0$, $Q_{B,33} > Q_{A,33}$; while $\left(2 \tan \theta_{33}' - \tan \delta_{33}' - \tan \phi_{33}'\right) < 0$, $Q_{B,31} < Q_{A,31}$. From Eq. (4.95) and setting $Z = 0$, the resonance frequency is provided simply by

$$\Omega_A = k_{33}^2 \tan \Omega_A. \tag{4.107}$$

Note again that a half-wave length of the vibration is longer than the rod length at the resonance frequency (more uniform strain distribution than the sinusoid); a half-wave length is realized at the antiresonance frequency.

4.3.3.3 Loss and Mechanical Quality Factor in Other Modes

To obtain the loss factor matrix, five vibration modes need to be characterized in PZT ceramics with ∞mm crystallographic symmetry (ten independent dielectric, elastic, and piezoelectric loss components for intensive and extensive parameters), as summarized in Table 4.1 and Figure 4.7. The methodology is based on the equations of quality factors Q_A (resonance) and Q_B (antiresonance) in various modes with regard to loss factors and other properties.[10,11] We can measure Q_A and Q_B for each mode by using the 3 dB-up/down method (or quadrantal frequency method) in the impedance/admittance spectra (see Figure 4.5). The experimental techniques to determine the mechanical quality factors are described in Chapter 7. In addition to some derivations based on fundamental relations of the material properties, all the 20 loss factors can be obtained for piezoelectric ceramics. We derived the relationships between mechanical quality factors Q_A (resonance) and Q_B (antiresonance) in all required five modes shown in Table 4.1. The results are summarized below:[10]

   a. k_{31} mode: (intensive elastic loss)

$$Q_{A,31} = \frac{1}{\tan \phi_{11}'},$$

$$\frac{1}{Q_{B,31}} = \frac{1}{Q_{A,31}} - \frac{2}{1 + \left(\frac{1}{k_{31}} - k_{31}\right)^2 \Omega_{B,31}^2}\left(2 \tan \theta_{31}' - \tan \delta_{33}' - \tan \phi_{11}'\right)$$

$$\Omega_{A,31} = \frac{\omega_a l}{2 v_{11}^E} = \frac{\pi}{2}\ \left[v_{11}^E = 1/\sqrt{\rho s_{11}^E}\right], \Omega_{B,31} = \frac{\omega_b l}{2 v_{11}^E}, 1 - k_{31}^2 + k_{31}^2 \frac{\tan \Omega_B}{\Omega_B} = 0$$

   b. k_t mode: (extensive elastic loss)

$$Q_{B,t} = \frac{1}{\tan \phi_{33}},$$

$$\frac{1}{Q_{A,t}} = \frac{1}{Q_{B,t}} - \frac{2}{k_t^2 - 1 + \Omega_{A,t}^2/k_t^2}\left(2 \tan \theta_{33} - \tan \delta_{33} - \tan \phi_{33}\right)$$

$$\Omega_{B,t} = \frac{\omega_b l}{2 v_{33}^D} = \frac{\pi}{2}\ \left[v_{33}^D = 1/\sqrt{\rho/c_{33}^D}\right], \Omega_{A,t} = \frac{\omega_a l}{2 v_{33}^D}, \Omega_{A,t} = k_t^2 \tan \Omega_{A,t}$$

c. k_{33} mode:

$$Q_{B,33} = \frac{1}{\tan\phi_{33}'''} = \frac{1-k_{33}^2}{\tan\phi_{33}' - k_{33}^2\left(2\tan\theta_{33}' - \tan\delta_{33}'\right)}$$

$$\frac{1}{Q_{A,33}} = \frac{1}{Q_{B,33}} + \frac{2}{k_{33}^2 - 1 + \Omega_A^2/k_{33}^2}\left(2\tan\theta_{33}' - \tan\delta_{33}' - \tan\phi_{33}'\right)$$

$$\Omega_{B,33} = \frac{\omega_b l}{2v_{33}^D} = \frac{\pi}{2}\ \left[v_{33}^D = 1\big/\sqrt{\rho s_{33}^D}\right], \Omega_{A,33} = \frac{\omega_a l}{2v_{33}^D}, \Omega_{A,33} = k_{33}^2\tan\Omega_{A,33}$$

d. k_{15} mode (constant E – length shear mode): (intensive elastic loss)

$$Q_{A,15}^E = \frac{1}{\tan\phi_{55}'},$$

$$\frac{1}{Q_{B,15}^E} = \frac{1}{Q_{A,15}^E} - \frac{2}{1+\left(\dfrac{1}{k_{15}}-k_{15}\right)^2\Omega_B^2}\left(2\tan\theta_{15}' - \tan\delta_{11}' - \tan\phi_{55}'\right)$$

$$\Omega_B = \frac{\omega_b L}{2v_{55}^E} = \frac{\omega_b L}{2}\sqrt{\rho s_{55}^E}, 1 - k_{15}^2 + k_{15}^2\frac{\tan\Omega_B}{\Omega_B} = 0$$

e. k_{15} mode (constant D – thickness shear mode): (extensive elastic loss)

$$Q_{B,15}^D = \frac{1}{\tan\phi_{55}}$$

$$\frac{1}{Q_{A,15}^D} = \frac{1}{Q_{B,15}^D} - \frac{2}{k_{15}^2 - 1 + \Omega_A^2/k_{15}^2}\left(2\tan\theta_{15} - \tan\delta_{11} - \tan\phi_{55}\right)$$

$$\Omega_A = \frac{\omega_a t}{2v_{55}^D} = \frac{\omega_a t}{2}\sqrt{\frac{\rho}{c_{55}^D}}, \Omega_A = k_{15}^2\tan\Omega_A$$

Note again that because k_{31} and k_{33}/k_t modes possess E-constant and D-constant constraints, respectively, in k_{31}, the resonance frequency is directly related with $v_{11}{}^E$ or $s_{11}{}^E$ as $f_A = \dfrac{v_{11}^E}{2L} = 1/2L\sqrt{\rho s_{11}^E}$; while in k_{33}/k_t, the antiresonance frequency is directly related with $v_{33}{}^D$ or $s_{33}{}^D$, $c_{33}{}^D$ as $f_B = \dfrac{v_{33}^D}{2L} = 1/2L\sqrt{\rho s_{33}^D}$ or $1/2b\sqrt{\rho/c_{33}^D}$. It is important to distinguish k_{33} ($X_1 = X_2 = 0$, $x_1 = x_2 \neq 0$) from k_t ($X_1 = X_2 \neq 0$, $x_1 = x_2 = 0$) from the boundary conditions. Note the relations: $s_{33}^E = s_{33}^E\left(1 - k_{33}^2\right)$ and $c_{33}^E = c_{33}^D\left(1 - k_t^2\right)$, and $k_{33} > k_t$, in general. The pure "extensive" loss $\tan\phi_{33}$ is obtained from the loss relating with c_{33}^D from the definition, that is, in the k_t mode. When the length of a rod k_{33} is not very long, the mode approaches the k_t, and $c_{33}{}^D \approx 1/s_{33}{}^D$. The antiresonance in k_{31} and the resonance in k_{33}/k_t are subsidiary, originated from the electromechanical coupling factors. We also remind the reader the relation for the *electromechanical coupling factor losses* from Eq. (4.41):

$$\left(2\tan\theta' - \tan\delta' - \tan\phi'\right) = -\left(2\tan\theta - \tan\delta - \tan\phi\right). \tag{4.108}$$

Since the side is not clamped ($x_1 = x_2 \neq 0$) in the k_{33} mode (different from the k_t mode), the triple prime losses in the previous subsection is not exactly equal to non-prime extensive losses. 3D-clamped k_t mode exhibits the purely "extensive" non-prime losses, though the Q_m formulas for the k_{33} mode seem to be rather close to the extensive losses transformed from the intensive losses.

4.3.4 Q_A AND Q_B IN THE IEEE STANDARD

It is also important to discuss the assumption in the IEEE Standard,[12] where the difference of the mechanical quality factors among the resonance and antiresonance modes is neglected: that is, $Q_A = Q_B$. This is historically originated from the neglection of the coupling loss (i.e., piezoelectric loss) and the assumption of $\tan \phi' \gg \tan \delta'$ around the resonance region, leading to only one loss factor, that is, the intensive elastic loss. However, if we adopt our three-loss model, this situation ($Q_A = Q_B$) occurs only when $(2\tan \theta' - \tan \delta' - \tan \phi') = 0$ or $\tan \theta' = (\tan \delta' + \tan \phi')/2$. The IEEE Standard corresponds to only when the piezoelectric loss is equal to the average value of the dielectric and elastic losses, which exhibits a serious contradiction to the well-known PZT experimental results, that is, $Q_A < Q_B$. As we can realize in Figure 4.5 from the peak sharpness, the PZTs exhibit Q_A (resonance) $< Q_B$ (antiresonance), irrelevant to the vibration mode (Figure 4.5 is an example of the k_{31} mode). This concludes that $(\tan \delta_{33}' + \tan \phi_{11}' - 2\tan \theta_{31}') < 0$ or $(\tan \delta_{33}' + \tan \phi_{11}')/2 < \tan \theta_{31}'$ for k_{31}, and $(\tan \delta_{33} + \tan \phi_{33} - 2\tan \theta_{33}) > 0$ or $(\tan \delta_{33} + \tan \phi_{33})/2 > \tan \theta_{33}$ for k_t. It is worth noting that the intensive piezoelectric loss is larger than the average of the dielectric and elastic intensive losses in Pb-contained piezo-ceramics. We introduce in Chapter 9 that in Pb-free piezoelectric ceramics, the piezoelectric loss contribution is not significant, and $(\tan \delta_{33}' + \tan \phi_{11}')/2 > \tan \theta_{31}'$, which may suggest different loss mechanisms depending on different piezo-ceramic materials.

CHAPTER ESSENTIALS

1. Piezoelectric constitutive equations:

$$\text{Intensive parameter description} \quad x = -\frac{\partial G}{\partial X} = s^E X + dE$$

$$D = -\frac{\partial G}{\partial E} = dX + \varepsilon_0 \varepsilon^X E$$

s^E – elastic compliance under constant field, ε^X – dielectric constant under constant stress, and d – piezoelectric charge coefficient

$$\text{Extensive parameter description} \quad X = \frac{\partial A}{\partial x} = c^D x - hD$$

$$E = \frac{\partial A}{\partial D} = -hx + \kappa_0 \kappa^x D \quad \left[\kappa_0 = \frac{1}{\varepsilon_0} \right]$$

c^D – elastic stiffness under constant electric displacement, κ^x – inverse dielectric constant under constant strain, and h – inverse piezoelectric charge coefficient.

2. Interrelationship between intensive and extensive parameters:

$$c^D = \frac{1}{s^E} \frac{1}{\left(1 - \frac{d^2}{(s^E \varepsilon_0 \varepsilon^X)}\right)}, \kappa_0 \kappa^x = \frac{1}{\varepsilon_0 \varepsilon^X} \frac{1}{\left(1 - \frac{d^2}{(s^E \varepsilon_0 \varepsilon^X)}\right)}, h = \frac{1}{d} \frac{\frac{d^2}{(s^E \varepsilon_0 \varepsilon^X)}}{\left(1 - \frac{d^2}{(s^E \varepsilon_0 \varepsilon^X)}\right)}$$

3. Electromechanical coupling factor

$$k^2 = \frac{d^2}{\left(s^E \varepsilon_0 \varepsilon^X\right)} = \frac{h^2}{\left(c^D \kappa_0 \kappa^x\right)}.$$

4. Constraint dependence of permittivity and elastic compliance:

$$\varepsilon^x/\varepsilon^X = \left(1-k^2\right), \ s^D/s^E = \left(1-k^2\right), \ \kappa^X/\kappa^x = \left(1-k^2\right), \ c^E/c^D = \left(1-k^2\right),$$

5. Intensive and extensive loss definitions:

$$\varepsilon^{X*} = \varepsilon^X (1 - j\tan\delta') \quad k^{x*} = k^x(1 + j\tan\delta)$$

$$s^{E*} = s^E(1 - j\tan\phi') \quad c^{D*} = c^D(1 + j\tan\phi)$$

$$d^* = d(1 - j\tan\theta') \quad h^* = h\,(1 + j\tan\theta)$$

6. Intensive and extensive loss interrelation (1D expression):

$$\begin{bmatrix} \tan\delta' \\ \tan\phi' \\ \tan\theta' \end{bmatrix} = [K] \begin{bmatrix} \tan\delta \\ \tan\phi \\ \tan\theta \end{bmatrix}, \text{or} \begin{bmatrix} \tan\delta \\ \tan\phi \\ \tan\theta \end{bmatrix} = [K] \begin{bmatrix} \tan\delta' \\ \tan\phi' \\ \tan\theta' \end{bmatrix},$$

$$\text{where } [K] = \frac{1}{1-k^2} \begin{bmatrix} 1 & k^2 & -2k^2 \\ k^2 & 1 & -2k^2 \\ 1 & 1 & -1-k^2 \end{bmatrix}$$

7. Strain distribution in the resonance and antiresonance states (see figure below). (a) k_{31}: resonance – a half-wave length; and (b) k_{33}: antiresonance – a half-wave length.

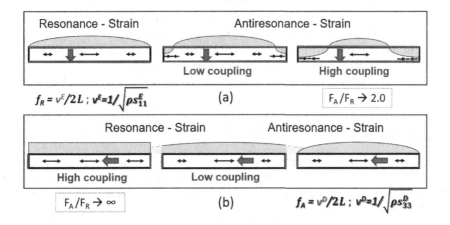

8. When the vibration orientation is in parallel to the P_S (k_{33} and k_t cases), the induced polarization is cancelled by the depolarization field, leading to D-constant.

9. Mechanical quality factors at resonance and antiresonance frequencies:

   a. k_{31} mode: intensive elastic loss

$$\frac{1}{Q_{A,31}} = \frac{1}{\tan\phi'_{11}}, \frac{1}{Q_{B,31}} = \frac{1}{Q_{A,31}} - \frac{2}{1+\left(\frac{1}{k_{31}}-k_{31}\right)^2 \Omega_{B,31}^2}\left(2\tan\theta'_{31} - \tan\delta'_{33} - \tan\phi'_{11}\right)$$

   b. k_t mode: extensive elastic loss

$$Q_{B,t} = \frac{1}{\tan\phi_{33}}, \frac{1}{Q_{A,t}} = \frac{1}{Q_{B,t}} - \frac{2}{k_t^2 - 1 + \Omega_{A,t}^2/k_t^2}\left(2\tan\theta_{33} - \tan\delta_{33} - \tan\phi_{33}\right)$$

CHECK POINT

1. (T/F) Because the polarization is induced after the electric field is applied (time delay), the P vs. E hysteresis loop should show the clockwise rotation. True or False?

2. (T/F) The hysteresis area of the strain x vs. electric field E corresponds directly to the piezoelectric loss factor $\tan\theta'$. True or False?

3. (T/F) The permittivity under mechanically clamped condition is smaller than that under mechanically free condition. True or False?

4. (T/F) The elastic compliance under open-circuit condition is larger than that that under short-circuit condition. True or False?

5. (T/F) There is a highly resistive (no electric carrier/impurity in a crystal) piezoelectric single crystal (spontaneous polarization P_S) with a monodomain state without surface electrode in vacuum. The "depolarization electric field" in the crystal is given by $E = \left(\dfrac{P_S}{\varepsilon_0\varepsilon}\right)$ (i.e., electric field is along the P_S). True or False?

6. (T/F) The piezoelectric resonance is only the mechanical resonance mode, and the antiresonance is not the mechanical resonance. True or False?

7. (T/F) The fundamental resonance mode of the k_{33} mode has an exact half-wave length vibration on the rod specimen. True or False?

8. (T/F) The fundamental resonance mode of the k_{31} mode has an exact half-wave length vibration on the plate specimen. True or False?

9. Provide the relationship between the mechanical quality factor Q_M at the resonance frequency and the intensive elastic loss in the k_{31}-type specimen.

10. Provide the relationship between the mechanical quality factor Q_M at the antiresonance frequency and the extensive elastic loss in the k_t-type specimen.

11. (T/F) The strain distribution in a high k_{33} rod specimen is more uniform at the antiresonance mode than that at the resonance mode. True or False?

12. When $(\tan\delta'_{33} + \tan\phi'_{11})/2 < \tan\theta'_{31}$ is satisfied, which is larger Q_A or Q_B for the k_{31}-type specimen?

CHAPTER PROBLEMS

4.1 When we neglect the piezoelectric loss $\tan\theta'$ (i.e., $\tan\theta' = 0$) among three losses, $\tan\delta'$, $\tan\phi'$, and $\tan\theta'$, discuss the relation of the mechanical quality factors at the resonance Q_A and antiresonance frequencies Q_B. Which is larger Q_A or Q_B, in the case of $\tan\delta' > 0$, $\tan\phi' > 0$?

4.2 A "Hard" PZT shows the following performances:

$$s_{33}^E = 14.6 \times 10^{-12} \ \mathrm{m^2/N}, \ k_t = 0.52, \ k_{33} = 0.64.$$

k_t and k_{33} modes generate 3D and 1D-clamped conditions, respectively. When we calculate the D-constant elastic compliances for both modes, we obtain

$$s_{33}^D = s_{33}^E \left(1 - k_t^2\right) = 10.6 \times 10^{-12} \ \mathrm{m^2/N}, \ s_{33}^D = s_{33}^E \left(1 - k_{33}^2\right) = 8.6 \times 10^{-12} \ \mathrm{m^2/N}.$$

Consider the physical reason why 1D-clamped condition exhibits stiffer elasticity.

REFERENCES

1. D.R. Tobergte, S. Curtis, IUPAC. Compendium of Chemical Terminology, (the "Gold Book") 53 (2013).
2. K. Aizu, *Journal of the Physical Society of Japan.* **38**, 1592 (1975).
3. C. Kittel, *Introduction to Solid State Physics,* 6th Edition, Chap.13, (John Wiley & Sons, New York, 1986).
4. K. Uchino, S. Hirose, Loss mechanisms in piezoelectrics: How to measure different losses separately, *IEEE Transactions on Ultrasonics, Ferroelectrics, and Frequency Control.* **48**, 307–321 (2001).
5. K. Uchino, *Micromechatronics,* 2nd Edition, (CRC Press, Boca Raton, FL, 2019). ISBN-13: 978-0-367-20231-6.
6. K. Uchino, *Ferroelectric Devices,* 2nd Edition, (CRC Press, Boca Raton, FL, 2010).
7. S. Hirose, M. Aoyagi, Y. Tomikawa, S. Takahashi, K. Uchino, *Ultrasonics.* 34, 213 (1996).
8. A.V. Mezheritsky, *IEEE Transactions on Ultrasonics, Ferroelectrics, and Frequency Control.* 49 (2002) 484.
9. Y. Zhuang, S.O. Ural, A. Rajapurkar, S. Tuncdemir, A. Amin, K. Uchino, Derivation of piezoelectric losses from admittance spectra, *Japanese Journal of Applied Physics.* **48**, 041401 (2009).
10. Y. Zhuang, S.O. Ural, S. Tuncdemir, A. Amin, K. Uchino, Analysis on loss anisotropy of piezoelectrics with ∞mm crystal symmetry, *Japanese Journal of Applied Physics.* 49, 021503 (2010).
11. S. Zhang, R. Xia, L. Lebrun, D. Anderson, T.R. Shrout, Piezoelectric materials for high power, high temperature applications, *Materials Letters.* **59**(27), 3471–3475 (2005).
12. ANSI/IEEE Std 176-1987, *IEEE Standard on Piezoelectricity,* (The Institute of Electrical and Electronics Engineers, New York, 1987), p. 56.

5 Equivalent Circuits with Piezo Losses

ABSTRACT

"Equivalent Circuits with Piezo Losses" expands the equivalent circuit approach in order to facilitate the experimental analysis easier. Mechanical and electrical systems are occasionally equivalent from the mathematical formula's viewpoint. Therefore, an electrician tries to understand a mechanical system behavior from a more familiar LCR electrical circuit analysis. However, two important notes must be taken into account: (1) mechanical loss is handled as "viscous damping" and (2) equivalent circuit approach is almost successful, as long as we consider a steady sinusoidal (harmonic) vibration. When we consider a transient response, such as a pulse drive of a mechanical system with finite specimen size, the equivalent circuit analysis generates a significant discrepancy. The equivalent (electric) circuit (EC) is a widely used tool which greatly simplifies the process of design and analysis of the piezoelectric devices, in which the circuit can only graphically characterize the mechanical loss by applying a resistor (and dielectric loss sometimes).[1] Different from a pure mechanical system, a piezoelectric vibration exhibits an "antiresonance" mode in addition to a "resonance" mode, due to the existence of the damped capacitance (i.e., only the partial of the input electric energy is transduced into the mechanical energy). Without introducing the *piezoelectric loss*, it is difficult to explain the difference of the mechanical quality factors at the resonance and antiresonance modes. We consider new equivalent circuits of piezoelectric devices with these three losses in this chapter.

As discussed in the Chapter 4, without introducing the *piezoelectric loss*, it is difficult to explain the difference of the mechanical quality factors at the resonance and antiresonance modes. Damjanovic, therefore, introduced an additional branch into the standard circuit, which is used to present the influence of the piezoelectric loss.[2] However, concise and decoupled formulas of three (dielectric, elastic, and piezoelectric) losses have not been derived, which can be used for the measurements of losses in piezoelectric material as a user-friendly method.

5.1 EQUIVALENCY BETWEEN MECHANICAL AND ELECTRICAL SYSTEMS

There are two classifications of an LCR electrical circuit: series connection and parallel connection. Though both circuits are equivalent, in general, focused usage is different.

5.1.1 LCR SERIES CONNECTION EQUIVALENT CIRCUIT

The dynamic equation for a pure mechanical system composed of a mass, a spring, and a damper illustrated in Table 5.1a is expressed by

$$M\left(d^2u/dt^2\right) + \zeta\left(du/dt\right) + cu = F(t), \text{ or} \tag{5.1a}$$

$$M\left(dv/dt\right) + \zeta v + c\int_0^t v dt = F(t) \tag{5.1b}$$

where u is the displacement of a mass M, v is the velocity (= du/dt), c spring constant, ζ damping constant of the dashpot, and F is the external force. Note that when a continuum elastic material is considered, the actual damping may be "solid damping" (as we discussed in Chapter 3), but we

TABLE 5.1

Equivalency between Mechanical and Electrical Systems, Composed of *M* (Mass), *c* (Spring Constant), *ζ* (Viscous Damper), *L* (Inductance), *C* (Capacitance), and *R* (Resistance)

Mechanical	Electrical (F – V)	Electrical (F – I)
Force F(t)	Voltage V(t)	Current I(t)
Velocity v/ů	Current I	Voltage V
Displacement u	Charge q	–
Mass M	Inductance L	Capacitance C
Spring Compliance 1/c	Capacitance C	Inductance L
Damping z	Resistance R	Conductance G

(a) Mechanical system, (b) LCR series connection, and (c) LCR parallel connection.

consider or approximate here the *viscous damping*, where the damping force is described in proportion to the velocity, from a mathematical simplicity viewpoint.

On the other hand, the dynamic equation for an electrical circuit composed of a series connection of an inductance L, a capacitance C, and a resistance R illustrated in Table 5.1b is expressed by

$$L\left(d^2q/dt^2\right) + R\left(dq/dt\right) + \left(1/C\right)q = V(t), \text{ or} \tag{5.2a}$$

$$L\left(dI/dt\right) + RI + \left(1/C\right)\int_0^t I\,dt = V(t) \tag{5.2b}$$

where q is charge, I is the current ($= dq/dt$), and V is the external voltage. Taking into account the equation similarity, the engineer introduces an equivalent circuit; that is, consider a mechanical system with using an equivalent electrical circuit, which is intuitively simpler for an electrical engineer. In contrast, consider an electrical circuit with using an equivalent mechanical system, which is intuitively simpler for a mechanical engineer. Equivalency between these two systems is summarized in the center column in Table 5.1.

When we consider steady sinusoidal vibrations of the system at the frequency $\omega\left(V(t) = V_0 e^{j\omega t}, I(t) = I_0 e^{j\omega t - \delta}\right)$, Eq. (5.2) can be transformed into

$$[j\omega L + R + (1/j\omega C)]I = V, \text{ or} \tag{5.3a}$$

$$Y = I/V = [j\omega L + R + (1/j\omega C)]^{-1}. \tag{5.3b}$$

Under a certain constant voltage (such as 1 V), the current (A) behavior provides the frequency dependence of the *circuit admittance*. Thus, this series connection equivalent circuit (EC) is useful to discuss the piezoelectric resonance mode with the admittance maximum peak under a constant voltage condition.

We consider the *Bode plot* of Eq. (5.3b). The admittance $|Y|$ *gain* is plotted in Figure 5.1a as a function of frequency ω in both logarithmic scale. The steady-state oscillation plot exhibits:

1. 20 dB/decade ($\propto \omega C$) asymptotic curve with 90° phase in the low-frequency region.
2. The peak at ω_0, resonance angular frequency for zero damping, given by $\omega_0 = 1/\sqrt{LC}$, with the peak height $|Y|_{max} = (1/R)$, and $Q = \sqrt{L/C}/R$, which corresponds to the quality factor.
3. −20 dB/decade ($\propto 1/\omega L$) asymptotic curve with −90° phase in the high-frequency region.

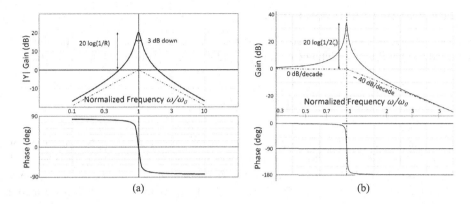

FIGURE 5.1 Bode diagram for a series LCR circuit: (a) admittance and (b) charge (second-order system).

Let us calculate the quality factor Q in the LCR circuit defined by $\omega_R/2\Delta\omega_R$, where $\Delta\omega_R$ is the half width of the admittance frequency spectrum to provide the $1/\sqrt{2}$ (3 dB down) of the maximum admittance ($1/R$) at the resonance frequency ω_R. Since these cut-off frequencies are provided by

$$\frac{1}{\sqrt{2}} = \frac{1}{\sqrt{\left[\left(\frac{\omega_c L}{R}\right) - \left(\frac{1}{\omega_c RC}\right)\right]^2 + 1}}$$

Then, two roots for the cut-off frequency ω_C are given by

$$\omega_{c1,c2} = \mp\frac{R}{2L} + \sqrt{\left(\frac{R}{2L}\right)^2 + \left(\frac{1}{LC}\right)},$$

Since $2\Delta\omega_R = \omega_{c2} - \omega_{c1}$ and $\omega_R = 1/\sqrt{LC}$, the quality factor is expressed by

$$Q = \omega_R/2\Delta\omega_R = \left(1/\sqrt{LC}\right)/(R/L) = \sqrt{L/C}/R \qquad (5.4)$$

On the contrary, when we consider the charge q, rather than current I under the voltage [Eq. (5.2a)]:

$$L\left(d^2q/dt^2\right) + R\left(dq/dt\right) + (1/C)q = V(t)$$

Taking the harmonic oscillation, the above equation is transformed to

$$\left[-\omega^2 L + j\omega R + (1/C)\right]q = V, \text{ or} \qquad (5.5a)$$

$$q/V = [-\omega^2 L + j\omega R + (1/C)]^{-1} \qquad (5.5b)$$

The reader is reminded of this formula of the second-order system, and the *Bode plot* of Eq. (3.106), already discussed in Section 3.3.3. Refer to Figure 5.1b, where the *gain* of charge q is plotted as a function of frequency ω in both logarithmic scales. The steady-state oscillation plot exhibits:

1. 0 dB/decade asymptotic curve in the low-frequency region.
2. The peak at ω_0, resonance angular frequency for zero damping, given by $\omega_0 = 1/\sqrt{LC}$, with the peak height $(1/2\zeta) = \sqrt{L/C}/R = Q$, which corresponds to the quality factor.
3. −40 dB/decade asymptotic curve in the high-frequency region.

5.1.2   LCR PARALLEL CONNECTION EQUIVALENT CIRCUIT

Let us now consider the dynamic equation for an electrical circuit composed of a parallel connection of an inductance L_B, a capacitance C_B, and a conductance G_B illustrated in Table 5.1c:

$$C_B (dV/dt) + G_B V + (1/L_B) \int_0^t V \, dt = I(t) \tag{5.6}$$

where I is the current from the current supply, and V is the same voltage applied on three components. In comparison with Eq. (5.1b), equivalency between the two mechanical and electrical systems is summarized in the last column in Table 5.1. When we consider steady sinusoidal vibrations of the system at frequency $\omega \left(I(t) = I_0 e^{j\omega t}, V(t) = V_0 e^{j\omega t - \delta} \right)$, Eq. (5.6) can now be transformed into

$$[j\omega C_B + G_B + (1/j\omega L_B)]V = I, \text{ or} \tag{5.7a}$$

$$Z = V/I = [j\omega C_B + G_B + (1/j\omega L_B)]^{-1} \tag{5.7b}$$

Under a certain constant current (such as 1 A), the voltage (V) behavior provides the frequency dependence of the *circuit impedance*. Thus, this parallel connection EC is preferred to discuss the piezoelectric "antiresonance mode" (i.e., B-type resonance) with the impedance maximum peak.

Example Problem 5.1

Two equivalent circuits in Table 5.1b, c are modeled for the same mechanical system in Table 5.1a. Therefore, we can expect the mutual relationships between these inductance, capacitance, and resistance/conductance values. Obtain the mutual relationships.

Hint
Since the voltage–current behavior should be equivalent in these series and parallel connection circuits, the admittance in Eq. (5.3b) should be an inverse of the impedance in Eq. (5.7b).

Solution

Equating Z with $1/Y$:

$$Z = \left[j\omega C_B + G_B + (1/j\omega L_B) \right]^{-1} = 1/Y = \left[j\omega L + R + (1/j\omega C) \right], \tag{P5.1.1}$$

we obtain the following equation:

$$C_B \left(\frac{1}{C} - \omega^2 L \right) + \frac{1}{L_B} \left(L - \frac{1}{\omega^2 C} \right) + RG_B + j \left[G_B \left(\omega L - \frac{1}{\omega C} \right) + R \left(\omega C_B - \frac{1}{\omega L_B} \right) \right] = 1 \tag{P5.1.2}$$

In order to keep the same resonance frequency in both circuits,

$$\omega^2 = \frac{1}{LC} = \frac{1}{L_B C_B} \tag{P5.1.3}$$

should be maintained. Thus, Eq. (P5.1.2) indicates another equation,

$$RG_B = 1 \tag{P5.1.4}$$

Though we have the circuit component flexibility, as long as $LC = L_B C_B$, the simplest solution is the utilization of the same L, C, and R for L_B, C_B, and $1/G_B$.

5.2 EQUIVALENT CIRCUIT (LOSS-FREE) OF THE K_{31} MODE

We introduce the EC of piezoelectric devices, a widely used tool which greatly simplifies the process of designing the devices. The key in a piezoelectric is illustrated schematically in Figure 5.2, where the input electric energy is partially converted to the output mechanical energy by the factor or k^2, while the remaining energy $(1 - k^2)$ is stored in a capacitor (so-called damped capacitance). The loss observed as heat generation is usually small (around a couple of %), which is proportional to the loss tangent/dissipation factor. Different from the previous section of a simple LCR series connection, where only the resonance mode shows up, when we include the damped capacitance, the antiresonance mode appears, where the damped and motional capacitances are basically cancelled out. In other words, the existence of the damped capacitance is essential to generate the antiresonance mode.

We consider first the simplest equivalent circuit (loss-free) for the k_{31} mode piezo-plate, as shown in Figure 5.3, on which we already analyzed the resonance/antiresonance modes in detail in Section 4.3.2. You are reminded of the admittance equation:

$$
\begin{aligned}
Y &= (j\omega wL/b)\varepsilon_0\varepsilon_{33}^{LC}\left[1+\left(d_{31}^2/\varepsilon_0\varepsilon_{33}^{LC}s_{11}^E\right)\left(\tan\left(\omega L/2v_{11}^E\right)/\left(\omega L/2v_{11}^E\right)\right)\right] \\
&= (j\omega wL/b)\varepsilon_0\varepsilon_{33}^X\left[(1-k_{31})+k_{31}^2\left(\tan\left(\omega L/2v_{11}^E\right)/\left(\omega L/2v_{11}^E\right)\right)\right] \\
&= j\omega C_d\left[1+\frac{k_{31}^2}{1-k_{31}^2}\frac{\tan(\Omega_{11})}{\Omega_{11}}\right] \\
&= j\omega C_0\left[\left(1-k_{31}^2\right)+k_{31}^2\frac{\tan(\Omega_{11})}{\Omega_{11}}\right]
\end{aligned}
\tag{5.8}
$$

where w is the width, L the length, and b the thickness of the rectangular piezo-sample. We adopt the following notations for make the formulas simpler:

Input electrical energy 100%

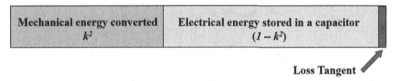

| Mechanical energy converted k^2 | Electrical energy stored in a capacitor $(1 - k^2)$ |

Loss Tangent

FIGURE 5.2   Energy conversion in a piezoelectric.

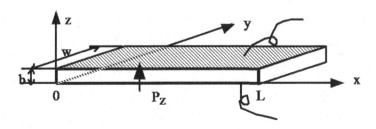

FIGURE 5.3   A rectangular piezo-ceramic plate $(L \gg w \gg b)$ for a longitudinal mode through the transverse.

$$k_{31}^2 = d_{31}^2 / \varepsilon_0 \varepsilon_{33}^X s_{11}^E$$

$$\varepsilon_0 \varepsilon_{33}^{LC} = \varepsilon_0 \varepsilon_{33}^X \left(1 - k_{31}^2\right)$$

$$C_0 = \varepsilon_0 \varepsilon_{33}^X \frac{Lw}{b} \quad \text{(Free capacitance)}$$

$$C_d = \varepsilon_0 \varepsilon_{33}^{LC} \frac{Lw}{b} \quad \text{(Damped capacitance)}$$

$$\Omega_{11} = \left(\omega L / 2 v_{11}^E\right)$$

Equation (5.8) indicates that the first term is originated from the *longitudinally clamped capacitance* [proportional to $(1 - k^2)$], and the second term is the *motional capacitance* associated with the mechanical vibration [proportional to k^2]. By splitting Y into the damped admittance Y_d and the motional part Y_m,

$$Y_d = j\omega C_d \tag{5.9}$$

$$Y_m = j\omega C_d \left[\frac{k_{31}^2}{1 - k_{31}^2} \frac{\tan(\Omega_{11})}{\Omega_{11}}\right] \tag{5.10}$$

The damped branch can be represented by a capacitor with *damped capacitance C_d* in Figure 5.4a. We connect the motional branch (mechanical vibration) in "parallel" to this damped capacitance, because Eq. (5.8) indicates the summation of these two admittances.

5.2.1 Resonance Mode

Because the maximum Y corresponds to the resonance mode, we can analyze merely the motional Y_m, which is much larger than Y_d. Since $Y_m = j\omega C_d \left[\dfrac{k_{31}^2}{1 - k_{31}^2} \dfrac{\tan(\Omega_{11})}{\Omega_{11}}\right]$ is infinite (∞) around the A-type resonance frequency, that is, $\omega_A L / 2 v_{11}^E = \Omega_{11,A,n} = n\pi/2 \ (n = 1, 3, 5, \ldots)$, taking *Mittag-Leffler's theorem* of $\dfrac{\tan(\Omega_{11})}{\Omega_{11}}$ around $\omega_{A,n}$, we get

$$\frac{\tan(\Omega_{11})}{\Omega_{11}} = \sum_{n:\text{odd}}^{\infty} \left(\frac{8}{n^2 \pi^2}\right) \bigg/ \left\{1 - \left(\frac{\Omega_{11}}{\Omega_{11,A,n}}\right)^2\right\} \tag{5.11}$$

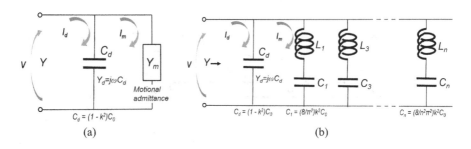

(a)   (b)

FIGURE 5.4 Equivalent circuit for the k_{31} mode (loss-free): (a) conceptual EC and (b) LC series EC.

When we use an LC series connection EC model in Table 5.1b on the motional branch, we convert the mass contribution to L and elastic compliance to C and create L_n and C_n series connections as shown in Figure 5.4b. Note here that a similar approach can be made for the parallel connection EC, because of the equivalency among the series and parallel ECs [refer to Example Problem 5.1]. Each pair of (L_1,C_1), (L_3,C_3), ... (L_n,C_n) contributes to the fundamental, the second, and the n-th resonance vibration mode, respectively. Remember that n is only for the odd number, or even number n does not show up in the piezoelectric resonance, which corresponds basically to the antiresonance mode in a hypothetically high k material. Though each branch is activated only at its own n-th resonance frequency, the capacitance's contribution remains even at an inactive frequency range, in particular, at a low-frequency range. (Note a series connection of the impedance of capacitance, $1/j\omega C$, and inductance, $j\omega L$ in this EC. Under a low-frequency region, the inductance contribution will disappear.) Neglecting the damped admittance, the motional impedance of this LC circuit around the resonance $\omega_{A,n}$ is approximated by

$$1/Y_{m,n} = j\omega L_n + 1/j\omega C_n \approx j\left(L_n + 1/\omega_{A,n}^2 C_n\right)\left(\omega - \omega_{A,n}\right) \tag{5.12}$$

where $\omega_{A,n}^2 = 1/L_n C_n$. Using Eqs. (5.11) and (5.12), we can obtain the following equation:

$$Y_m = j\omega C_d\left[\frac{k_{31}^2}{1-k_{31}^2}\sum_{n:odd}^{\infty}\left(\frac{8}{n^2\pi^2}\right)\middle/\left\{1-\left(\frac{\Omega_{11}}{\Omega_{11,A,n}}\right)^2\right\}\right] = \sum_{n,odd}^{\infty}[j/\left(L_n + 1/\omega_{A,n}^2 C_n\right)\left(\omega_{A,n} - \omega\right)] \tag{5.13}$$

Taking into account further approximation, $\dfrac{1}{\left\{1-\left(\dfrac{\Omega_{11}}{\Omega_{11,A,n}}\right)^2\right\}} \approx \dfrac{\omega_{11,A,n}}{2\left(\omega_{11,A,n} - \omega_{11}\right)}$ for each n-th

branch, we can obtain the following two equations which express the L_n, C_n in terms of the transducer's physical parameters:

$$L_n = \left(bLs_{11}^E/4v_{11}^{E2}wd_{31}^2\right)/2 = (\rho/8)(Lb/w)\left(s_{11}^{E2}/d_{31}^2\right) \tag{5.14}$$

$$C_n = 1/\omega_{A,n}^2 L_n = \left(L/n\pi v_{11}^E\right)^2 (8/\rho)(w/Lb)\left(d_{31}^2/s_{11}^{E2}\right)$$
$$= \left(8/n^2\pi^2\right)(Lw/b)\left(d_{31}^2/s_{11}^{E2}\right)s_{11}^E \tag{5.15}$$

$$\omega_{A,n} = 1/\sqrt{L_n C_n} = n\pi/L\sqrt{\rho s_{11}^E} \tag{5.16}$$

Note initially that L_n is a constant, irrelevant to n. All harmonics have the same L, which is originated from the ceramic density ρ or the specimen mass M. C_n is proportional to $1/n^2$ and the elastic compliance s_{11}^E. Note that the parameter $\left(d_{31}/s_{11}^E\right)$ is distinguished in both Eqs. (5.14) and (5.15), which will be explained in Section 5.3 as a *force factor* $\Phi = 2wd_{31}/s_{11}^E$. The total motional capacitance $\sum_n C_n$ is calculated as follows, using an important relation $\sum\left[\dfrac{1}{(2m-1)^2}\right] = \left(\dfrac{\pi^2}{8}\right)$:

$$\sum_n C_n = \sum_n \frac{1}{n^2}\left(\frac{8}{\pi^2}\right)\left(\frac{Lw}{b}\right)\left(\frac{d_{31}^2}{s_{11}^E}\right) = \left(\frac{Lw}{b}\right)\left(\frac{d_{31}^2}{s_{11}^E}\right) = k_{31}^2 C_0 \tag{5.17}$$

Therefore, we can understand that the total capacitance $C_0 = (wL/b)\varepsilon_0\varepsilon_{33}^X$ is split into the damped capacitance $C_d = \left(1-k_{31}^2\right)C_0$ and the total motional capacitance $k_{31}^2 C_0$, which is reasonable from the energy conservation viewpoint.

5.2.2 ANTIRESONANCE MODE

We consider next the antiresonance mode at the n-th mode, where the total admittance $Y = 0$ in Eqs. (5.8) and (5.12):

$$Y = j\omega C_d \left[1 + \frac{k_{31}^2}{1 - k_{31}^2} \frac{\tan(\Omega_{11})}{\Omega_{11}} \right] = j\omega C_d + \frac{1}{\dfrac{1}{j\omega C_n} + j\omega L_n} = 0 \tag{5.18}$$

This admittance corresponds to the closed-circuit admittance under externally open-circuit condition (i.e., smallest admittance condition). Accordingly, we obtain the antiresonance (B-type) frequency ω_B in terms of the EC parameters:

$$\omega_{B,n} = \sqrt{\left(1 + \frac{C_n}{C_d}\right)\Big/ L_n C_n} = \sqrt{\left(\frac{1}{C_n} + \frac{1}{C_d}\right)\Big/ L} \tag{5.19}$$

5.3 EQUIVALENT CIRCUIT OF THE K_{31} MODE WITH LOSSES

5.3.1 IEEE STANDARD EQUIVALENT CIRCUIT

Figure 5.5 shows the IEEE Standard EC for the k_{31} mode with only one elastic loss $(\tan\phi')$.[3] This elastic loss introduction in the mechanical branch is based on the assumption that the elastic loss in a piezoelectric material follows a "viscous damping" model, merely from the mathematical simplicity viewpoint. The dielectric or piezoelectric losses are neglected. In this EC model merely for the fundamental resonance mode, in addition to Eqs. (5.9), (5.14), and (5.15) in the loss-free EC, the circuit analysis provides the following R and Q (electrical quality factor, which corresponds to the mechanical quality factor in the piezo-plate) relation:

$$Q = \sqrt{L_A/C_A}\Big/ R_A \tag{5.20}$$

This R introduction means the inclusion of merely the elastic loss (i.e., $\tan\phi_{11}'$ in the k_{31} mode), leading automatically to the relation Q_A (resonance) $= Q_B$ (antiresonance). In order to demonstrate the usefulness of the equivalent circuit model for the piezoelectric device analysis, a simulation tool is introduced. PSpice is a popular circuit analysis software code for simulating the performance of electric circuits, which is widely distributed for students in the university Electrical

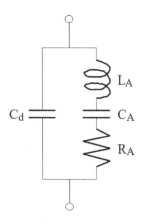

FIGURE 5.5 Equivalent circuit for the k_{31} mode (IEEE).

Engineering department. EMA Design Automation, Inc., in the United States <http://www.ema-eda.com> is distributing a free-download "OrCAD Capture", for this circuit design solution software. The reader can access to the download site, if you want:

http://www.orcad.com/products/orcad-liteoverview?gclid=COaXitWJp9ECFcxKDQodCGMB0w

Figure 5.6 shows the PSpice simulation process of the IEEE Standard k_{31} type. Figure 5.6a shows an equivalent circuit for the k_{31} mode. L, C, and R values were calculated for PZT4 with $40 \times 6 \times 1\,\text{mm}^3$ [Eqs. (5.14), (5.15), and (5.20)], and Figure 5.6b plots the simulation results on the current

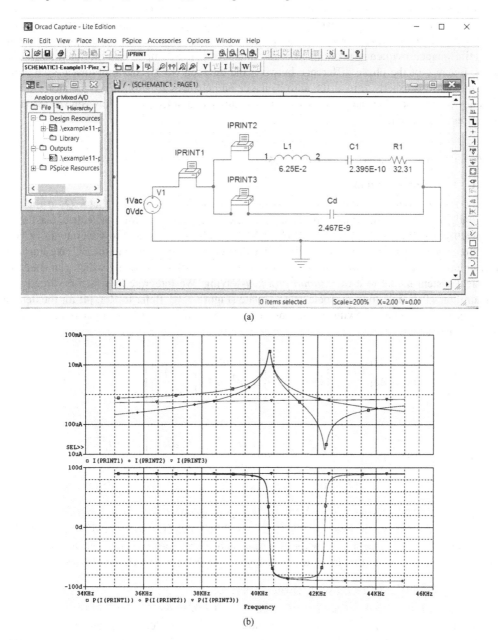

(a)

(b)

FIGURE 5.6 PSpice simulation of the IEEE-type k_{31} mode. (a) Equivalent circuit for the k_{31} mode. L, C, and R values were calculated for PZT4 with $40 \times 6 \times 1\,\text{mm}^3$. (b) Simulation results on admittance magnitude and phase spectra.

under 1 V_{ac}, that is, admittance magnitude and phase spectra. IPRINT1 (current measurement), IPRINT2, and IPRINT3 are the measure of the total admittance ($\square$ line), motional admittance ($\bigcirc$ line), and damped admittance (∇ line), respectively. First, the damped admittance shows a slight increase with the frequency ($j\omega C_d$) with +90° phase in a full frequency range. Second, the motional admittance shows a peak at the resonance frequency, where the phase changes from +90° (i.e., capacitive) to −90° (i.e., inductive). In other words, the phase is exactly zero at the resonance frequency. The admittance magnitude decreases above the resonance frequency with a rate of −20 dB down in a *Bode plot*. Third, by adding the above two, the total admittance is obtained. The admittance magnitude shows two peaks, maximum and minimum, which correspond roughly to the resonance and antiresonance points, respectively. You can find that the peak sharpness (i.e., the mechanical quality factor) is the same for both peaks, because only one loss is included in the equivalent circuit. The antiresonance frequency is obtained at the intersect of the damped and motional admittance curves. Because of the phase difference between the damped (+90°) and motional (−90°) admittance, the phase is exactly zero at the antiresonance and changes to +90° above the antiresonance frequency. Remember that the phase is −90° (i.e., inductive) at a frequency between the resonance and antiresonance frequencies.

5.3.2 EQUIVALENT CIRCUIT WITH THREE LOSSES

5.3.2.1 Hamilton's Principle

In Section 4.3.1, we derived the difference between the mechanical quality factor Q_A at the resonance and Q_B at the antiresonance in the k_{31} specimen, based on the three dielectric, elastic, and piezoelectric loss factors, $\tan\delta'$, $\tan\phi'$, and $\tan\theta'$. We now consider how to generate an EC with these three losses in order to realize the difference between Q_A and Q_B even in the EC. We start from the *Hamilton's Principle*, a powerful tool for "mechanics" problem-solving, which can transform a physical system model to a *variational problem*-solving. We integrate loss factors directly into the Hamilton's Principle for a piezoelectric k_{31} plate (Figure 5.3).[4] Refer to Ref. [4] for the detailed derivation process. The following admittance expression can be derived, which is equivalent to Eq. (4.72) in Chapter 4:

$$Y^* = j\omega \cdot \frac{Lw}{b} \cdot \left(\varepsilon_0 \varepsilon_{33}^{X'} - \text{Re}\left[\frac{\left(d_{31}^*\right)^2}{s_{11}^{E*}} \right] \right) + \omega \cdot \frac{Lw}{b} \cdot \left(\varepsilon_0 \, \varepsilon_{33}^{X''} + \text{Im}\left[\frac{\left(d_{31}^*\right)^2}{s_{11}^{E*}} \right] \right)$$

$$+ j\omega \cdot \frac{8Lw}{b\pi^2} \cdot \text{Re}\left[\frac{\left(d_{31}^*\right)^2}{s_{11}^{E*}} \right] \cdot \frac{\frac{\pi^2}{L^2 \rho s_{11}^{E*}}}{\frac{\pi^2}{L^2 \rho s_{11}^{E*}} - \omega^2} + j\omega \cdot \frac{8Lw}{b\pi^2} \cdot \left(j\text{Im}\left[\frac{\left(d_{31}^*\right)^2}{s_{11}^{E*}} \right] \right) \cdot \frac{\frac{\pi^2}{L^2 \rho s_{11}^{E*}}}{\frac{\pi^2}{L^2 \rho s_{11}^{E*}} - \omega^2} \quad (5.21)$$

Among the above four terms in Eq. (5.21), the first and second terms correspond to the damped capacitance and its dielectric loss (i.e., the "extensive"-like dielectric loss ($\tan\delta'''$) in Eq. (4.72) in the previous chapter), respectively, while the third and fourth terms correspond to the motional capacitance and the losses combining with "intensive" elastic and piezoelectric losses.

5.3.2.2 k_{31} Equivalent Circuit with Three Losses

Damjanovic[2] introduced a motional branch to describe the third term in Eq. (5.21), which contains a motional resistor, a motional capacitor, and a motional inductor. Meanwhile, an additional branch is also injected into the classical circuit[1] to pictorially express the last term in Eq. (5.21) to present the influence of the piezoelectric loss, where the new resistance, capacitance, and inductance are all proportional to corresponding motional elements with the proportionality constant being

$$jIm\left[\frac{\left(d_{31}^*\right)^2}{s_{11}^{E*}}\right]\Bigg/Re\left[\frac{\left(d_{31}^*\right)^2}{s_{11}^{E*}}\right].$$

Shi et al. proposed a concise EC shown in Figure 5.7a with three losses.[4] Compared with the IEEE Standard EC with only one elastic loss or the Damjanovic's EC with a full set of L, C, and R, only one additional electrical element G_m' is introduced into the classical circuit.[4] The new coupling conductance can reflect the coupling effect between the elastic and the piezoelectric losses. The admittance of this new EC can be mathematically expressed as:

$$Y^* = G_d + j\omega C_d + \frac{G_m' + j\omega C_m}{\left(1 + G_m'/G_m - \omega^2 L_m C_m\right) + j\left(\omega C_m/G_m + \omega L_m G_m'\right)} \tag{5.22}$$

The parameters of the new EC can, therefore, be obtained by comparing Eq. (5.21) with Eq. (5.22) as new expressions of three "intensive" loss factors:

$$\tan\phi' = \omega C_m/G_m \quad \left[G_m = 1/R_m\right] \tag{5.23a}$$

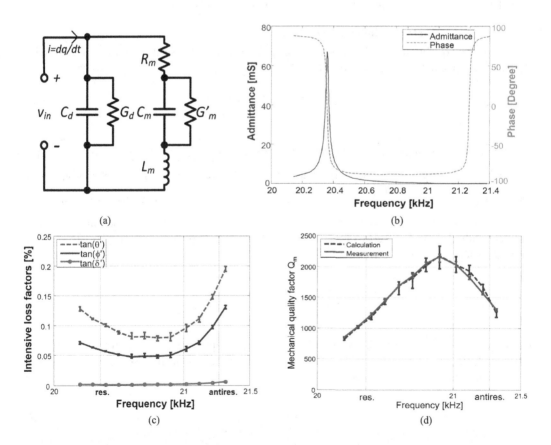

(a)   (b)

(c)   (d)

FIGURE 5.7 (a) Equivalent circuit proposed with three intensive loss factors; (b) admittance spectrum to be used in the simulation; (c) frequency spectra of intensive loss dielectric, elastic, and piezoelectric loss factors obtained from the admittance spectrum (b) fitting; and (d) calculated mechanical quality factor Q_m as a function frequency around the resonance and antiresonance frequencies.

$$\tan\theta' = \tan\left(\phi' - \beta'\right) \tag{5.23b}$$

$$\tan\delta' = k_{31}^2 \tan\left(2\theta' - \phi'\right) + \frac{G_d}{\omega C_d} \tag{5.23c}$$

where the phase delay $\tan\beta' = \frac{\omega C_m}{G_m'} - \sqrt{\left(\frac{\omega C_m}{G_m'}\right)^2 + 1}$ denotes the disparity between the piezoelectric and elastic components. From Eq. (4.18), we learned that the piezoelectric loss and elastic loss are always coupled in the E excitation measurement. The value of β' generally holds negative or approaches to zero (when $G_m' \to 0$), which implies that the piezoelectric loss is persistently larger or equal to the elastic component. β' means also the negative of the "force factor" loss [refer to Section 5.5.1.2]. The significance of the piezoelectric loss has been therefore verified in theory from the equivalent circuit viewpoint.

Using the experimental data in Figure 5.7b, almost frequency-independent circuit parameters as $C_d = 3.2\,\text{nF}$, $C_m = 0.29\,\text{nF}$, $L_m = 210\,\text{mH}$, and $G_d = 0$ (extensive-like (longitudinally clamped) dielectric loss $\tan\delta'''$ is small) can be obtained, and the frequency-dependent parameters (G_m and G_m').[4] By manipulating Eqs. (5.24a)–(5.23c), we determined intensive dielectric, elastic, and piezoelectric losses as a function of frequency, as shown in Figure 5.7c.

5.3.2.3   Quality Factor in the Equivalent Circuit

The mechanical quality factor, Q_m, is always applied to evaluate the effect of losses. When arriving at steady state, it can be expressed by:

$$Q_m = 2\pi \cdot \frac{\text{energy stored/cycle}}{\text{energy lost/cycle}} \tag{5.24}$$

The denominator is supposed to compensate the dissipation, w_{loss}; that is, $\int_V w_{\text{loss}}\, dV = \frac{\pi}{2}\left|v_3^*\right|\left|q_3^*\right|$ $\cos\varphi$, where v_3, q_3 are voltage and charge, and the phase difference between current and input voltage, φ, ranges within $\left[-\frac{\pi}{2}, \frac{\pi}{2}\right]$. Meanwhile, the reactive portion of the input energy returns to the amplifier and is neither used nor dissipated. Furthermore, the maximum stored and kinetic energies also get equilibrium in an electric cycle. With definitions of energy items and appropriate substitutions, Q_m can be calculated as:[5]

$$Q_m = \frac{\omega_a^2 - \omega_r^2}{\cos\varphi} \cdot \frac{\omega^2}{\left|\omega^2 - \left(\omega_r^*\right)^2\right|\left|\omega^2 - \left(\omega_a^*\right)^2\right|} \tag{5.25}$$

As ω^2 approaches to ω_r^2 or ω_a^2, the phase difference will approach zero. Therefore, for low k_{31}^2 materials, with substituting Eq. (5.25), mechanical quality factors at the resonance and antiresonance frequencies can be calculated as:

$$Q_A = \frac{1}{\tan\phi'} = \frac{1}{R_m}\sqrt{\frac{L_m}{C_m}} \tag{5.26}$$

$$Q_B = \frac{1}{\tan\phi' + \dfrac{8K_{31}^2}{\pi^2}\left[\tan\phi' + \tan\delta' - 2\tan\theta'\right]} \qquad \left[K_{31}^2 = k_{31}^2 / \left(1 - k_{31}^2\right)\right] \tag{5.27}$$

Equations (5.26) and (5.27) obtained from a new equivalent circuit are basically the same (for a small k_{31} case) as we derived analytically in Chapter 4 Eqs. (4.81) and (4.93). Hence, the calculation of Q_m at these special frequencies has been verified by the well-accepted conclusion. Not only at these frequencies, Eq. (5.25) also infers an advanced calculation method of Q_m for a wide bandwidth. Figure 5.7d shows the frequency spectrum of the mechanical quality factor Q_m calculated from Eq. (5.25). You can clearly find that (1) the Q_B at antiresonance is larger than Q_A at resonance, and (2) the maximum Q_m (i.e., the highest efficiency) can be obtained at a frequency between the resonance and antiresonance frequencies. This frequency can theoretically be obtained by taking the first derivative of Eq. (5.25) in terms of ω to be equal to zero, which suggests the best operating frequency of the transducer to realize the maximum efficiency. Though the physical origin/mechanism has not been clarified yet, this frequency drive corresponds to the natural mechanical resonance frequency under a suitable impedance connected condition (not short- or open-circuit condition). Practical applications of this operating frequency will be demonstrated in Chapter 8.

5.4 EQUIVALENT CIRCUIT OF THE K_{33} MODE WITH LOSSES

Remember that the k_{33} mode is governed by the sound velocity v^D, not by v^E, and that the antiresonance is the primary mechanical resonance given by $f = v^D/2L$, and the resonance is the subsidiary mode originated from the electromechanical coupling factor k_{33}. The difference is primarily originated from the *depolarization field* created in the "longitudinal" piezoelectric effect oscillator (k_{33}, k_t), in comparison with the "transversal" piezoelectric effect oscillator (k_{31}). Let us consider to formulate the EC for the k_{33} mode, as shown in Figure 5.8a.

5.4.1 RESONANCE/ANTIRESONANCE OF THE K_{33} MODE

Referring to the derivation process introduced in Section 4.3.3.2, we just summarize the key formulas first.

- The constitutive equations

$$X_3 = \left(x_3 - d_{33}E_z\right)/s_{33}^E \tag{5.28}$$

$$D_3 = \varepsilon_0\varepsilon_{33}^X E_z + d_{33}X_3 \tag{5.29}$$

- Dynamic equation

$$\rho\frac{\partial^2 u}{\partial t^2} = \frac{1}{s_{33}^D}\frac{\partial^2 u}{\partial z^2} \quad \left(s_{33}^D = (1-k_{33}^3)\,s_{33}^E\right) \tag{5.30}$$

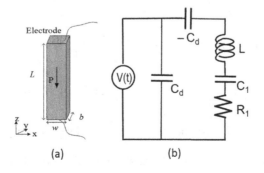

(a)   (b)

FIGURE 5.8 (a) k_{33} mode piezo-ceramic rod and (b) equivalent circuit for the k_{33} mode.

- Admittance is expressed as

$$Y = \frac{j\omega\varepsilon_0\varepsilon_{33}^{LC}\left(\frac{wb}{L}\right)}{\left[1 - k_{33}^2\left\{\frac{\tan\left(\frac{\omega L}{2v_{33}^D}\right)}{\left(\frac{\omega L}{2v_{33}^D}\right)}\right\}\right]} = j\omega C_d + \frac{j\omega C_d}{\left[-1 + 1/k_{33}^2\left\{\frac{\tan(\Omega_{33})}{\Omega_{33}}\right\}\right]} \tag{5.31}$$

Here we used $\Omega_{33} = \left(\frac{\omega L}{2v_{33}^D}\right)$, $\varepsilon_{33}^{LC} = \varepsilon_{33}^X\left(1 - k_{33}^2\right)$, $s_{33}^D = s_{33}^E\left(1 - k_{33}^2\right)$, $k_{33}^2 = \frac{d_{33}^2}{\varepsilon_0\varepsilon_{33}^X s_{33}^E}$, $v_{33}^D = 1/\sqrt{\rho s_{33}^D}$,

and $C_d = \varepsilon_0\varepsilon_{33}^{LC}\left(\frac{wb}{L}\right)$. The second expression is to show the *damped admittance* and the *motional admittance*, separately.

5.4.2 Resonance/Antiresonance of the k_{33} Mode

When we consider the resonance condition, $Y = \infty$, the resonance frequency is obtained from Eq. (5.31) (the denominator of the motional admittance = 0) as

$$\left(\frac{\omega L}{2v_{33}^D}\right)\cot\left(\frac{\omega L}{2v_{33}^D}\right) = k_{33}^2 \quad \left[v_{33}^D = 1/\sqrt{\rho s_{33}^D}\right] \tag{5.32}$$

Since the resonance is the subsidiary mode, the resonance frequency of the k_{33} mode depends strongly on the electromechanical coupling factor k_{33} value.

To the contrary, the antiresonance mode is obtained by putting $Y = 0$ (the denominator of the total admittance = infinite), which provides the condition, $\tan\left(\frac{\omega L}{2v_{33}^D}\right) = \infty$. Thus, the antiresonance frequency is determined by $n\left(v_{33}^D/2L\right)$ ($n = 1,3,5, ...$), and the vibration mode shows an exact half-wave length on the specimen with length L under sound velocity v_{33}^D; while the resonance is the subsidiary vibration mode as discussed above. This provides intriguing contrast to the k_{31} mode, where the resonance mode is the primary vibration with a half-wave length of the specimen of L, and the antiresonance is the subsidiary vibration mode.

5.4.3 Equivalent Circuit of the k_{33} Mode

We can rewrite the Eq. (5.31), as follows:

$$Y = j\omega C_d + \frac{1}{-\dfrac{1}{j\omega C_d} + \dfrac{1}{j\tan\left(\dfrac{\omega L}{2v_{33}^D}\right)\dfrac{2bwd_{33}^2}{\rho v_{33}^D L^2 s_{33}^{E\,2}}}}, \tag{5.33}$$

where $C_d = \left(\frac{\varepsilon_0\varepsilon_{33}^{x3} bw}{L}\right)$. $\varepsilon_0\varepsilon_{33}^{x3}$ is the same as the longitudinally clamped permittivity, $\varepsilon_{33}^{LC} = \varepsilon_{33}^X\left(1 - k_{33}^2\right)$. From Eq. (5.33), we can understand that the equivalent circuit of the k_{33} mode is composed of the first term damped admittance with a "damped capacitance" and the second term "motional admittance". Also, the motional branch is obtained by a series connection of so-called "negative capacitance – C_d" (exactly the same absolute value of the damped capacitance in the electric branch) and the pure motional admittance, $j \tan\left(\dfrac{\omega L}{2 v_{33}^D}\right) \dfrac{2 b w d_{33}^2}{\rho v_{33}^D L^2 s_{33}^{E\,2}}$. Figure 5.8b illustrates the fundamental mode EC by translating the motional admittance with only a pair of L and C. The IEEE Standard model includes only one resistance R_1, which corresponds to the elastic loss $\tan\phi'''$ in the material's parameter. The admittance should be the minimum at the antiresonance frequency, where the pure mechanical resonance status is realized, because the damped capacitance should be compensated by this negative capacitance – C_d in the closed-loop circuit. On the contrary at the resonance, the admittance should be the maximum, and the effective motional capacitance in the motional branch is provided by $1\left/\left(\dfrac{1}{C_1} + \dfrac{1}{-C_d}\right)\right.$, which provides $s_{33}^D \left(= s_{33}^E (1 - k_{33}^2)\right)$, rather than s_{33}^E (i.e., origin of C_1). The reader can understand intuitively that the negative capacitance comes from the "depolarization field", or the D-constant status of the k_{33} vibration mode, different from the k_{31} E-constant mode. Figure 5.8b integrated a resistance R_1 in series with L_1 and C_1 in the pure mechanical branch. In comparison with Eqs. (5.14), (5.15), and (5.20) in the k_{31} mode, the EC components, L, C, and R of the k_{33} mode can be denoted as:

$$L_n = \left(b L s_{33}^D / 4 v_{33}^{D\,2} w d_{33}^2\right)/2 = (\rho/8)(Lb/w)\left(s_{33}^{D\,2}/d_{33}^2\right) \tag{5.34}$$

$$C_n = 1/\omega_{A,n}^2 L_n = (L/n\pi v_{33}^D)^2 (8/\rho)(w/Lb)\left(d_{33}^2/s_{33}^{D\,2}\right)$$
$$= (8/n^2\pi^2)(Lw/b)\left(d_{33}^2/s_{33}^{D\,2}\right)s_{33}^D \tag{5.35}$$

$$R_n = \sqrt{L_n/C_n}\,/Q \tag{5.36}$$

Here, $s_{33}^D = s_{33}^E\left(1 - k_{33}^2\right)$, $k_{33}^2 = \dfrac{d_{33}^2}{\varepsilon_0 \varepsilon_{33}^X s_{33}^E}$ and $Q = \tan\phi'''$ as the material's constants.

We may adopt three resistances/conductances into Figure 5.8(b), similar to the k_{31} case, to analyze the Q_A and Q_B difference.

5.5 FOUR- AND SIX-TERMINAL EQUIVALENT CIRCUITS (EC) – K_{31} CASE

Though the new two-terminal EC with three dielectric, elastic, and piezoelectric losses is useful for the basic no-load piezoelectric samples, we need to extend it to four- and six-terminal EC models in order to consider the load effect for practical transducer/actuator applications with composite structures such as Langevin transducers.

5.5.1 Four-Terminal Equivalent Circuit

5.5.1.1 Four-Terminal Equivalent Circuit (Zero Loss)

We consider again the k_{31}-type piezoelectric plate, whose admittance is described by Eq. (5.8). When we consider the damped electric branch and the motional mechanical branch together, we can generate a two-terminal EC as shown in Figure 5.9a. However, since the electric and mechanical branches are physically different, it is more reasonable to discuss these branches separately, which intuitively create a four-terminal (or two-port) equivalent circuit, as exemplified in Figure 5.9b. The electric branch (left-hand side) is separated from the mechanical branch (right-hand side) by a *transformer*, which transforms voltage and current (electrical energy) to force and vibration velocity (mechanical energy), respectively, with the transformer ratio of Φ or $1/\Phi$, in order to change the unit from the electric to mechanical parameters. This Φ is called *force factor*. In this case, the port

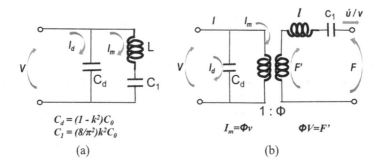

$C_d = (1 - k^2)C_0$
$C_1 = (8/\pi^2)k^2C_0$

(a)

$I_m = \Phi v$ $\Phi V = F'$

(b)

FIGURE 5.9 (a) Two- and (b) four-terminal ECs for k_{31} mode (zero loss).

on the mechanical branch can be mechanically loaded (symmetrically in the four-terminal model), depending on the piezoelectric composite structure.

Let us formulate electric component parameters of the four-terminal EC of the k_{31} plate in Figure 5.3. The motional current I_m is given by

$$I_m = E_z bY_m = E_z bj\omega C_d \left(\frac{k_{31}^2}{1 - k_{31}^2}\right)\frac{\tan(\omega L/2v_{11})}{(\omega L/2v_{11})} \tag{5.37}$$

while the vibration velocity $\dot{u}$ at the plate edge is described from Eq. (4.49) as

$$(\partial u/\partial t)_{x=L} = jd_{31}E_z v_{11}\tan(\omega L/2v_{11}). \tag{5.38}$$

Taking into account the definition of the transformer ratio Φ in terms of current and vibration velocity by

$$I_m = \Phi\dot{u} = \Phi(\partial u/\partial t)_{x=L}, \tag{5.39}$$

Φ can be obtained as

$$\Phi = \frac{2wd_{31}}{s_{11}^E} \tag{5.40}$$

Note the general relations: $F' = \Phi V$ and $I_m = \Phi\dot{u}$. Using the former in terms of voltage V and force F' by

$$\Phi V = F', \tag{5.41}$$

$$\Phi E_z b = \left(j\omega l + \frac{1}{j\omega c_1}\right)(\partial u/\partial t)_{x=L} \tag{5.42}$$

Note here that mechanical force F' is obtained by the product of vibration velocity $(\partial u/\partial t)_{x=L}$ and the mechanical impedance $\left(j\omega l + \frac{1}{j\omega c_1}\right)$. Since the voltage is given by the product of motional current I_m and the impedance in the two-terminal model,

$$V = E_z b = \left(j\omega L + \frac{1}{j\omega C_1}\right)I_m \tag{5.43}$$

we obtain the relationship between the L, C_1 in the one-port model and l, c_1 in the two-port model:

$$\left(j\omega l + \frac{1}{j\omega c_1} \right) = \left(j\omega L + \frac{1}{j\omega C_1} \right)\Phi^2 \tag{5.44}$$

We finally obtain the following relations in terms of the force factor Φ:

$$\Phi^2 L = l, \ C_1/\Phi^2 = c_1 \tag{5.45}$$

The force factor $\Phi = 2wd_{31}/s_{11}^E$ has a practical value around 0.1 for PZTs.

Example Problem 5.2

Figure 5.10 shows a composite piezoelectric oscillator, which is composed of the k_{31}-type piezoelectric plate and two metal plates bonded on both ends of the piezo-plate. Supposing the piezo-plate length L and the metal length is $L/2$ symmetrically on both ends, consider the equivalent circuit of this composite oscillator to analyze the vibration mode.

Hint
Four-terminal (two-port) EC for the k_{31} mode is given by Figure 5.11a. Consider the elastic material's equivalent circuit.

Solution

Figure 5.11a shows a two-port equivalent circuit for the k_{31} mode, on which mechanical load can be applied. The resistance connected in series corresponds to the mechanical loss. We consider some load application cases.

Recall the parameters in the two-terminal model:

$$L_n = (\rho/8)(Lb/w)\left(s_{11}^{E2}/d_{31}^2 \right)$$

$$C_n = (8/n^2\pi^2)(Lw/b)\left(d_{31}^2/s_{11}^{E2} \right)s_{11}^E$$

$$R_n = \sqrt{L_n/C_n}/Q$$

The force factor, inductance, capacitance, and resistance on the mechanical branch can be obtained as:

$$\Phi = \frac{2wd_{31}}{s_{11}^E} \tag{P5.1.1}$$

$$l_n = \Phi^2 L_n = (\rho/2)(Lbw)$$

$$c_n = C_n/\Phi^2 = (2/n^2\pi^2)(L/wb)s_{11}^E \tag{P5.1.2}$$

$$r_n = \Phi^2 R_n = (l_n/c_n)^{1/2}/Q$$

Note the difference from the two-terminal model: the L, C components in the two-terminal EC include the piezoelectric constant explicitly, but the l, c components in the four-terminal EC above

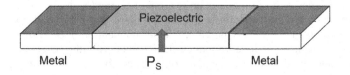

FIGURE 5.10 Composite piezoelectric oscillator.

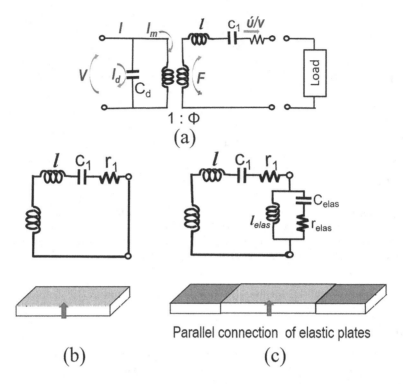

FIGURE 5.11 (a) Four-terminal equivalent circuit for the k_{31} mode. (b) No load (short-circuit) condition. (c) Elastic plates attached in parallel.

do not, because the electromechanical coupling is defined in the "force factor" of the transformer, and the mechanical branch parameters should be only pure elastic parameters.

When the two terminals on the mechanical branch are short-circuited (Figure 5.11b), i.e., mechanically free condition (the force F on the piezo-plate ends is zero), this will be transformed to the two-terminal EC. On the contrary, when the two terminals on the mechanical branch are open-circuited, this condition corresponds to completely clamped (strain-free) on the both plate ends.

Figure 5.11c shows the model where the metal plate with the same width, thickness, and length L is bonded symmetrically by cutting a half on the both ends of the piezoelectric plate. The load is modeled by the LC "parallel" connection in this case with the parameters, l_{elast}, c_{elast}, and r_{elast}, where ρ, s_{metal}, and Q are the metal's density, elastic compliance, and the inverse elastic loss, respectively:

$$l_{\text{elast}} = (\rho)(Lbw)$$

$$c_{\text{elast}} = \left(1/n^2\pi^2\right)(L/wb)\, s_{\text{metal}} \qquad\qquad (P5.1.3)$$

$$r_{\text{elast}} = \left(l_{\text{elast}}/c_{\text{elast}}\right)^{1/2}\big/Q$$

For the reader's reference, if the metal plate is bonded symmetrically by cutting a half of the thickness on the both top and bottom surfaces of the piezo-plate, the load is modeled by the above LC components in "series" connection. A series connection of LC components does not decrease the mechanical resonance frequency significantly, as in the parallel connection. You may understand this situation by taking into account the mechanical impedance series or parallel connection.

5.5.1.2 Four-Terminal Equivalent Circuit with Three Losses

Uchino proposes a *four-terminal equivalent circuit* for a k_{31} mode plate, including elastic, dielectric, and piezoelectric losses (Figure 5.12a), which can handle symmetrical external mechanical losses.[6] The four-terminal EC includes an ideal transformer with a voltage step-up ratio Φ to connect the electric (damped capacitance) and the mechanical (motional capacitance) branches, where $\Phi = 2wd_{31}/s_{11}^E$, called *force factor* in this *electromechanical transformer*. New capacitance l, c_1, and r_1 are related with L, C_1, and R_1 in the two-terminal EC given in Eqs. (5.14), (5.15), and (5.20):

$$l = \Phi^2 L; \; c_1 = C_1/\Phi^2 ; \; r_1 = \Phi^2 R_1, \tag{5.46}$$

where Φ is so-called *force factor*. Regarding the three losses, as shown in Figure 5.12a in a k_{31} piezo-plate, in addition to the IEEE standard "elastic" loss r_1 and "dielectric" loss R_d, we introduce the *coupling loss* r_{cpl} in the force factor $\left(\Phi = 2wd_{31}/s_{11}^E\right)$ as inversely proportional to $(\tan\phi_{11}' - \tan\theta_{31}')$, which can be either positive or negative, depending on the $\tan\theta_{31}'$ magnitude. Figure 5.12b shows the PSpice software simulation results for three values of r_{cpl}. We can find that (1) the resonance Q_A does not change with changing r_{cpl}, (2) when $r_{cpl} = 100$ kΩ (i.e., $\tan\theta_{31}' \approx 0$), $Q_A > Q_B$, (3) when $r_{cpl} = 1$ GΩ (i.e., $\tan\phi_{11}' - \tan\theta_{31}' \approx 0$), $Q_A = Q_B$, and (4) when $r_{cpl} = -100$ kΩ (i.e., $\tan\phi_{11}' - \tan\theta_{31}' < 0$), $Q_A < Q_B$. Taking into account a typical PZT's case, where $\tan\theta' > (1/2)(\tan\delta' + \tan\phi')$, the well-known experimental result $Q_A < Q_B$ can be expected from the negative r_{cpl} (i.e., negative "force factor loss"). Thus, the large piezoelectric loss $\tan\theta_{31}'$ in PZTs is the key to exhibit the negative force factor loss, which leads to the factor $Q_A < Q_B$.

5.5.2 Six-Terminal Equivalent Circuit

5.5.2.1 Mason's Equivalent Circuit

Mason introduced a famous six-terminal (three-port) EC model, relating with *distributed element model*.[7] As illustrated in Figure 5.13, two ports in the mechanical branch of the six-terminal (three-port) EC for the k_{31} piezoelectric plate correspond to the two edges of the plate, on which different mechanical loads can individually be applied, exemplified by a Langevin transducer with difference in head and tail masses sandwiching the center piezoelectric disk. Also Mason's EC does not include any approximation, such as L and C component combination which limits the usage only for a particular resonance mode and can be applied to any frequency.

Let us determine the electronic components, Z_1, Z_2 and the *force factor* Φ' in the six-terminal EC model. We denote the displacement along the length of a plate specimen (Figure 5.3), u, force and vibration velocity on the edge, F_1 and $\dot{u}_1$ at $x = 0$, F_2 and $\dot{u}_2$ at $x = L$, in addition to the

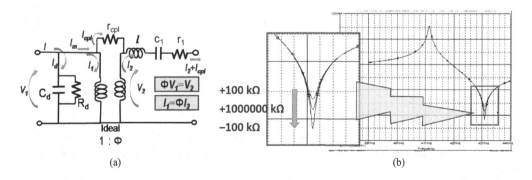

(a)   (b)

FIGURE 5.12 (a) Four-terminal (two-port) EC for a k_{31} plate, including three losses (r_1, R_d, and r_{cpl}). (b) PSpice simulation results on admittance for a k_{31}-type PZT4 $40 \times 6 \times 1$ mm^3 plate.

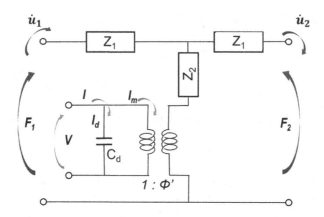

FIGURE 5.13 Six-terminal equivalent circuit for a k_{31} mode.

input voltage V and motional current I_m, damped current I_d. Supposing a general solution of the displacement as

$$u = A\cos(\omega x/v) + B\sin(\omega x/v) \tag{5.47}$$

and the boundary condition at $x = 0$,

$$A = u_1, \quad B = (\partial u/\partial x)_1(v/\omega),$$

we obtain

$$\partial u/\partial x = -(\omega/v)u_1\sin(\omega x/v) + (\partial u/\partial x)_1\cos(\omega x/v) \tag{5.48}$$

$$\partial u/\partial t = \dot{u} = j\omega\left[u_1\cos(\omega x/v) + (\partial u/\partial x)_1(v/\omega)\sin(\omega x/v)\right] \tag{5.49}$$

Now, we consider the force on the cross-section (wb). F is given by the stress (tensile is positive)

$$F = -wbX_1. \tag{5.50}$$

Since strain is given as $\partial u/\partial x = d_{31}E_z + s_{11}^E X_1$,

$$F = -\frac{wb}{s_{11}^E}\left[(\partial u/\partial x) - d_{31}E_z\right] \tag{5.51}$$

or

$$F - \Phi'V = -\frac{wb}{s_{11}^E}(\partial u/\partial x), \tag{5.52}$$

where the force constant Φ' is given by a half of Φ on the four-terminal EC, because the mechanical branch in the four-terminal model is the combination of Φ' of the two ports in the six-terminal model.

$$\Phi' = \Phi/2 = \frac{wd_{31}}{s_{11}^E} \tag{5.53}$$

When we adopt the boundary conditions at $x = 0$:

$$(\partial u / \partial x)_1 = -\frac{s_{11}^E}{wb}(F_1 - \Phi'V) \tag{5.54}$$

$$\dot{u}_1 = j\omega u_1 \tag{5.55}$$

Now Eq. (5.49) becomes

$$\partial u / \partial t = \dot{u}_1 \cos(\omega x/v) - j\frac{vs_{11}^E}{wb}(F_1 - \Phi'V)\sin(\omega x/v) \tag{5.56}$$

At $x = L$,

$$\dot{u}_2 = \dot{u}_1 \cos(\omega L/v) - j\frac{vs_{11}^E}{wb}(F_1 - \Phi'V)\sin(\omega L/v) \tag{5.57}$$

Also at $x = L$, using Eq. (5.48):

$$F_2 - \Phi'V = -\frac{wb}{s_{11}^E}(\partial u / \partial x)_2$$

$$= \left(-\frac{wb}{s_{11}^E}\right)[-(\omega/v)u_1 \sin(\omega L/v) + \dot{u}_1 \cos(\omega L/v)] \tag{5.58}$$

Now we can rewrite the relationship among F_1, F_2, $\dot{u}_1$, $\dot{u}_2$:

$$\dot{u}_2 = \dot{u}_1 \cos(\omega L/v) - j\frac{1}{Z_0}(F_1 - \Phi'V)\sin(\omega L/v) \tag{5.59}$$

$$F_2 - \Phi'V = (F_1 - \Phi'V)\cos(\omega L/v) - j\dot{u}_1 Z_0 \sin(\omega L/v) \tag{5.60}$$

$$I = j\omega C_d V + \Phi'(\dot{u}_2 - \dot{u}_1) \tag{5.61}$$

Note that the motional current is given by $\Phi'(u_2 - u_1)$. Now we can construct the six-terminal EC as shown in Figure 5.13. In order to satisfy Eqs. (5.59)–(5.61), we obtain all the components including Z_1, Z_2:

$$C_d = \frac{Lw\varepsilon_0\varepsilon_{33}^X(1 - k_{31}^2)}{b} \tag{5.62}$$

$$Z_0 = wb\rho v = wb\left(\frac{\rho}{s_{11}^E}\right)^{1/2} = \frac{wb}{v_{11}^E s_{11}^E} \tag{5.63}$$

$$Z_1 = jZ_0 \tan\left(\frac{\omega L}{2v_{11}^E}\right) \tag{5.64}$$

$$Z_2 = \frac{Z_0}{j\sin\left(\dfrac{\omega L}{v_{11}^E}\right)} \tag{5.65}$$

$$\Phi' = \frac{wd_{31}}{s_{11}^E} \tag{5.66}$$

Note that though the Mason's equivalent circuit includes the frequency-dependent Z_1 and Z_2, these impedance can be translated into a pair of L and C for each individual fundamental or higher-order harmonic mode, if required.

5.5.2.2 Application of Six-Terminal EC

Dong et al. constructed a six-terminal equivalent circuit with three (dielectric, elastic, and piezo-electric) losses, which can handle symmetric external loads for a k_{31} mode plate[8] and Langevin transducer by integrating the head and tail mass loads,[9] then estimate the optimum (i.e., minimum required input electrical energy) driving frequency at which we can drive the transducer, as demonstrated with the highest efficiency. Chapter 8 discusses this issue in detail from the drive scheme viewpoint.

In order to verify the feasibility of the EC circuit, a partial electrode configuration was designed (Figure 5.14a), which reflects intensive and extensive loss behavior on the electrode (center) and non-electrode (side) parts, respectively. The center part was electrically excited, which actuates the side non-electrode elastic load, then the vibration status can be monitored from the admittance from this center portion. The non-electrode side portions are merely the mechanical load. Figure 14b shows a combination of six-terminal ECs which models the center constant E element (i.e., intensive losses, $\tan\phi'_{11}, \tan\delta'_{33}, \tan\theta'_{31}$) and the side constant D elements (i.e., extensive losses, $\tan\phi_{11}, \tan\delta_{33}, \tan\theta_{31}$) by integrating loss factors into Eqs. (5.62)–(5.66). Note that the non-electrode part was segmented into 20 parts on each side to calculate the voltage distribution generated on the surface during the center actuation. The resonance and antiresonance frequencies and their corresponding mechanical quality factors derived from the circuits are compared with the actual sample with the load and boundary conditions.[8] The voltage distribution of non-electrode sample is simulated with the proposed equivalent circuit (Figure 5.15) and matches the experimental result on the actual sample. The voltage simulation results have the same sinusoidal distribution trend as the experimentals, and the admittance curves show a good agreement between the simulation and the measurements.

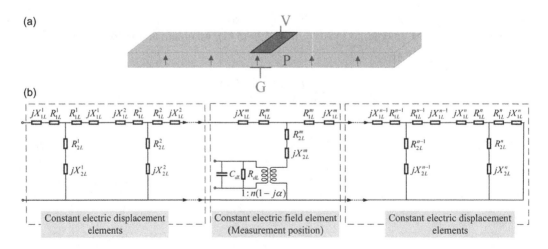

FIGURE 5.14 (a) A partial electrode configuration, and (b) its EC of a combination of six-terminal ECs which models the center constant E element (i.e., intensive losses) and the side constant D elements (i.e., extensive losses) by integrating loss factors.

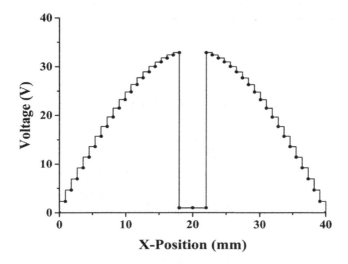

FIGURE 5.15 Voltage distribution of non-electrode sample simulated with the new six-terminal equivalent circuit.

5.6 FOUR- AND SIX-TERMINAL EQUIVALENT CIRCUITS (EC) – K_{33} CASE

The k_{33} mode requires a negative capacitance inclusion in the EC in order to reflect D-constant condition or "depolarization field". Since the sound velocity of the k_{33} mode is given by $1/\sqrt{\rho s_{33}^D}$, which is larger than $1/\sqrt{\rho s_{11}^E}$ of the k_{31} mode, the k_{33} mode is occasionally called a "stiffened mode". Equivalent circuits for the k_{33} mode are summarized for (a) two-terminal model, (b) four-terminal model, and (c) six-terminal model in Figure 5.16. Top and bottom show the difference with the negative capacitor installation; when the negative capacitance is installed in the electrical branch, $-C_d$ is directly inserted, while when it is installed in the mechanical branch, $-C_d/\Phi^2$ or $-C_d/\Phi'^2$ is inserted in series with other electrical components.

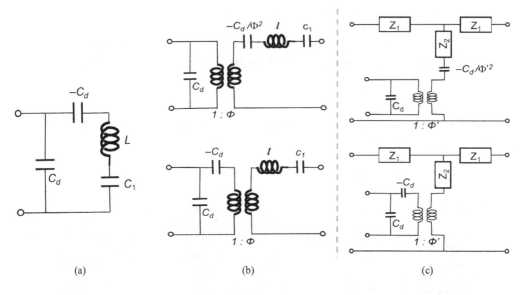

(a) (b) (c)

FIGURE 5.16 Equivalent circuits for the k_{33} mode: (a) two-terminal model, (b) four-terminal model, and (c) six-terminal model. Top and bottom show the difference with the negative capacitor installation.

Note the difference between Φ and Φ' in the four- and six-terminal models:

$$\Phi' = \Phi/2 = \frac{wbd_{33}}{Ls_{33}^D} \tag{5.67}$$

Z_1 and Z_2 in the six-terminal model are described as follows:

$$C_d = \frac{wb\varepsilon_0\varepsilon_{33}^X(1-k_{33}^2)}{L} \tag{5.68}$$

$$Z_0 = wb\rho v = wb\left(\frac{\rho}{s_{33}^D}\right)^{1/2} = \frac{wb}{v_{33}^D s_{33}^D} \tag{5.69}$$

$$Z_1 = jZ_0 \tan\left(\frac{\omega L}{2v_{33}^D}\right) \tag{5.70}$$

$$Z_2 = \frac{Z_0}{j\sin\left(\dfrac{\omega L}{v_{33}^D}\right)} \tag{5.71}$$

We can integrate three losses, $\varepsilon_{33}^{X*} = \varepsilon_{33}^X(1-j\tan\delta_{33}')$, $s_{33}^{E*} = s_{33}^E(1-j\tan\phi_{33}')$, and $d_{33}^* = d_{33}(1-j\tan\theta_{33}')$, since the k_{33} mode does not strictly have the "extensive" non-prime losses. and the electromechanical coupling factor k_{33} loss in the above six-terminal circuit components, then simulate admittance/impedance response from the circuit. Or, we may integrate three losses as R_{1L}, R_{2L} separately from X_{1L}, X_{2L}, as shown in Figure 5.14.

CHAPTER ESSENTIALS

1. Equivalency between mechanical and electrical systems:

$$M(d^2u/dt^2) + \zeta(du/dt) + cu = F(t), \text{ or } M(dv/dt) + \zeta v + c\int_0^t v\,dt = F(t)$$

$$L(d^2q/dt^2) + R(dq/dt) + (1/C)q = V(t), \text{ or } L(dI/dt) + RI + (1/C)\int_0^t I\,dt = V(t)$$

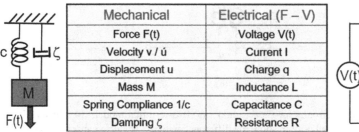

Mechanical	Electrical (F – V)
Force F(t)	Voltage V(t)
Velocity v / ú	Current I
Displacement u	Charge q
Mass M	Inductance L
Spring Compliance 1/c	Capacitance C
Damping ζ	Resistance R

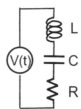

2. Equivalent circuits of a piezoelectric k_{31} plate:

$$Y = jwC_d \left[1 + \frac{k_{31}^2}{1 - k_{31}^2} \frac{\tan\left(\frac{\omega L}{2v_{11}^E}\right)}{\left(\frac{\omega L}{2v_{11}^E}\right)} \right]$$

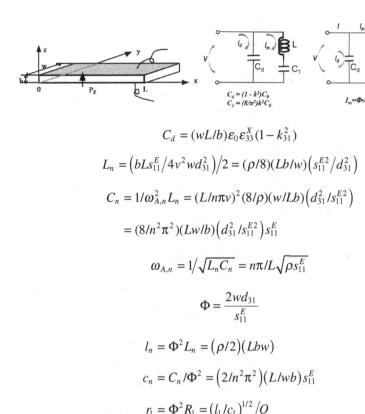

$$C_d = (wL/b)\varepsilon_0 \varepsilon_{33}^X (1 - k_{31}^2)$$

$$L_n = \left(bLs_{11}^E/4v^2 wd_{31}^2\right)/2 = (\rho/8)(Lb/w)\left(s_{11}^{E2}/d_{31}^2\right)$$

$$C_n = 1/\omega_{A,n}^2 L_n = (L/n\pi v)^2 (8/\rho)(w/Lb)\left(d_{31}^2/s_{11}^{E2}\right)$$

$$= (8/n^2\pi^2)(Lw/b)\left(d_{31}^2/s_{11}^{E2}\right)s_{11}^E$$

$$\omega_{A,n} = 1/\sqrt{L_n C_n} = n\pi/L\sqrt{\rho s_{11}^E}$$

$$\Phi = \frac{2wd_{31}}{s_{11}^E}$$

$$l_n = \Phi^2 L_n = (\rho/2)(Lbw)$$

$$c_n = C_n/\Phi^2 = \left(2/n^2\pi^2\right)(L/wb)s_{11}^E$$

$$r_1 = \Phi^2 R_1 = \left(l_1/c_1\right)^{1/2}/Q$$

3. Equivalent circuits of a piezoelectric k_{33} rod:

$$Y = \frac{j\omega C_d}{\left[1 - k_{33}^2 \frac{\tan\left(\frac{\omega L}{2v_{33}^D}\right)}{\left(\frac{\omega L}{2v_{33}^D}\right)}\right]} = j\omega C_d + \frac{j\omega C_d}{\left[-1 + 1/k_{33}^2\left\{\frac{\tan(\Omega_{33})}{(\Omega_{33})}\right\}\right]}$$

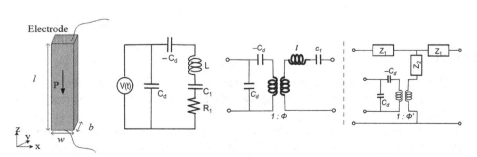

$$C_d = \frac{wb\varepsilon_0\varepsilon_{33}^X(1-k_{33}^2)}{L}$$

$$C_n = \left(8/n^2\pi^2\right)\left(wb/L\right)\left(d_{33}^2/s_{33}^{E2}\right)s_{33}^{E2}\left(1-k_{33}^2\right)$$

$$L_n = (\rho/8)\left(L^3/wb\right)\left(s_{33}^{E2}/d_{33}^2\right)$$

$$R_1 = (L_1/C_1)^{1/2}/Q$$

$$l_n = \Phi^2 L_n = (\rho/2)(Lbw)$$

$$c_n = C_n/\Phi^2 = \left(2/n^2\pi^2\right)(L/wb)s_{33}^D$$

$$r_1 = \Phi^2 R_1 = (l_1/c_1)^{1/2}/Q$$

$$\Phi' = \Phi/2 = \frac{wbd_{33}}{Ls_{33}^D}$$

$$Z_0 = wb\rho v = wb\sqrt{\frac{\rho}{s_{33}^D}} = \frac{wb}{v_{33}^D s_{33}^D}, \quad Z_1 = jZ_0\,\tan\left(\frac{\omega L}{2v_{33}^D}\right), \quad Z_2 = \frac{Z_0}{j\sin\left(\dfrac{\omega L}{v_{33}^D}\right)}$$

4. Comparison among the transversal k_{31} and longitudinal k_{33} modes:

	Transverse Effect (k_{31})	Longitudinal Effect (k_{33}, k_t)
Electric Condition ($k//x$)	$\dfrac{\partial E}{\partial x}=0$	$\dfrac{\partial D}{\partial x}=0$
Elastic constant	s_{11}^E	$c_{33}^D = 1/s_{33}^D$
Admittance	$Y = j\omega C_d\left[1+\dfrac{k_{31}^2}{1-k_{31}^2}\dfrac{\tan(\Omega_{11})}{\Omega_{11}}\right]$	$Y = \dfrac{j\omega C_d}{1-k_{33}^2\dfrac{\tan(\Omega_{33})}{\Omega_{33}}}$
Resonance	$\tan(\Omega_{11})=\infty$	$1-k_{33}^2\dfrac{\tan(\Omega_{33})}{\Omega_{33}}=0$
Half-wave frequency ($\omega_{\lambda/2}$)	ω_R	ω_A
Equivalent circuit		

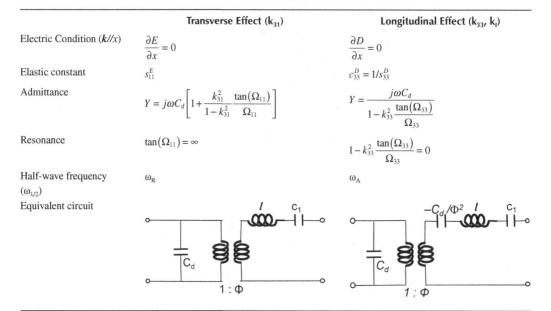

CHECK POINT

1. (T/F) When we consider an equivalent electric circuit of a mechanical system in terms of LCR series connection, inverse of the spring constant c corresponds to capacitance C. True or False?

2. (T/F) Because of the damped capacitance in the equivalent circuit of a piezoelectric oscillator, the antiresonance mode comes out, in addition to the resonance mode. True or False?

3. (T/F) The permittivity under mechanically clamped condition is larger than that under mechanically free condition. True or False?

4. (T/F) The elastic compliance under open-circuit condition is larger than that under short-circuit condition. True or False?

5. How can you describe the quality factor Q in a series connected LCR circuit?

6. The extensive losses are interrelated with the intensive losses in terms of the **K**-matrix in the 1D model.

$$\begin{bmatrix} \tan\delta \\ \tan\phi \\ \tan\theta \end{bmatrix} = [\mathbf{K}] \begin{bmatrix} \tan\delta' \\ \tan\phi' \\ \tan\theta' \end{bmatrix}, \text{ where } [\mathbf{K}] = \frac{1}{1-k^2} \begin{bmatrix} 1 & k^2 & -2k^2 \\ k^2 & 1 & -2k^2 \\ 1 & 1 & -1-k^2 \end{bmatrix}.$$

Provide the inverse [**K**]-matrix.

7. (T/F) The piezoelectric resonance is only the mechanical resonance mode, and the antiresonance is not the mechanical resonance. True or False?

8. (T/F) The fundamental antiresonance mode of the k_{33} mode has an exact half-wave length vibration on the rod specimen. True or False?

9. (T/F) The fundamental resonance mode of the k_{31} mode has an exact half-wave length vibration on the plate specimen. True or False?

10. Provide the relationship between the mechanical quality factor Q_M at the resonance frequency with the intensive elastic loss in the k_{31}-type specimen.

11. Provide the relationship between the mechanical quality factor Q_M at the resonance frequency and the extensive three losses in the k_t-type specimen.

12. (T/F) The strain distribution in a high k_{33} rod specimen is more uniform at the resonance mode than that at the antiresonance mode. True or False?

13. When $\left(\tan\delta_{33}' + \tan\phi_{11}'\right)/2 < \tan\theta_{31}'$ is satisfied, which is larger Q_A or Q_B for the k_{31}-type specimen?

CHAPTER PROBLEMS

5.1 Knowing the mechanical system (mass M, spring c, and damper ζ) and the electric circuit (inductance L, capacitance C, and resistance R) equivalency, as shown in (a) below, generate the electrical equivalent circuit corresponding to the mechanical system described in (b).

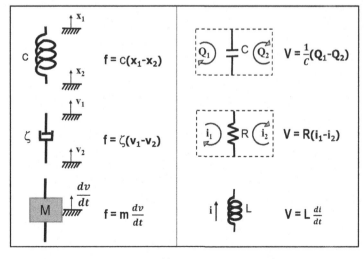

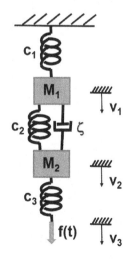

(a)   (b)

5.2 Using Mason's equivalent circuits for two length expander bars, surface and end electroded, as shown on the right, calculate the maximum step-up voltage ratio for this Rosen-type transformer under an open-circuit condition. The Rosen-type transformer is a combination of the k_{31} (thin electrode gap) and k_{33} (large electrode gap) transducers. Impedance parameters in the table can be applied.

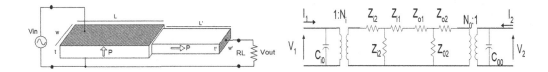

5.3 When a piezoelectric actuator is driven by a step pulse voltage with the pulse width exactly adjusted to the resonance period, the vibration displacement ΔL is generated linearly with time, and only one triangular shape displacement is realized without any vibration ringing.[6] However, when a step pulse voltage with the pulse width exactly adjusted to the resonance period is applied on an equivalent circuit (L, C, R, and C_d) of this piezo-actuator, the resulting displacement is a sinusoidal pulse (not a triangular shape). Refer to the figure on the next page. Derive these displacement shape difference, and understand the limitation of the EC model for the transient vibration analysis.

k_{31} Part	k_{33} Part
$C_{io} = \dfrac{Lw\varepsilon_0\varepsilon_{33}^X(1-k_{31}^2)}{2t}$	$C_{oo} = \dfrac{2w't'\varepsilon_0\varepsilon_{33}^X(1-k_{33}^2)}{L'}$
$Z_0 = wt\left(\dfrac{\rho}{s_{11}^E}\right)^{1/2}$	$Z_0^! = w't'\rho v_b^D = w't'\left(\dfrac{\rho}{s_{33}^D}\right)^{1/2}$
$Z_{i2} = \dfrac{Z_0}{j\sin\left(\dfrac{\omega L}{2v_a^E}\right)}$	$Z_{02} = \dfrac{Z_0^!}{j\sin\left(\dfrac{\omega L'}{2v_b^D}\right)}$
$Z_{i1} = jZ_0\tan\left(\dfrac{\omega L}{4v_a^E}\right)$	$Z_{01} = jZ_0^!\tan\left(\dfrac{\omega L'}{4v_b^D}\right)$
$N_i = \dfrac{wd_{31}}{s_{11}^E} = \dfrac{wd_{31}}{s_{11}^E}\sqrt{\dfrac{\varepsilon_0\varepsilon_{33}^X}{s_{11}^E}}k_{31}$	$N_0 = \dfrac{2w't'd_{33}}{L's_{33}^D} = \dfrac{w't'}{L'}\left(\dfrac{\varepsilon_0\varepsilon_{33}^X}{s_{33}^D}\right)^{1/2}k_{33}$
$v_a^E = \left(\dfrac{1}{\rho s_{11}^E}\right)^{1/2}$	$v_b^D = \left(\dfrac{1}{\rho s_{33}^D}\right)^{1/2}$

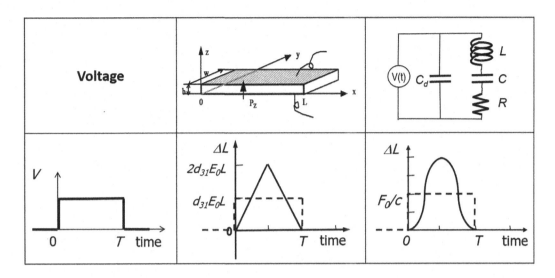

REFERENCES

1. W.P. Mason, An electromechanical representation of a piezoelectric crystal used as a transducer, *Proceedings of the Institute of Radio Engineers.* **23**, 1252 (1935).
2. D. Damjanovic, An equivalent electric-circuit of a piezoelectric bar resonator with a large piezoelectric phase-angle, *Ferroelectrics.* **110**, 129–135 (1990).
3. American National Standards Institute, IEEE Ultrasonics, Ferroelectrics, and Frequency Control Society. Standards Committee, Institute of Electrical and Electronics Engineers. *ANSI/IEEE Std 176–1987, IEEE standard on piezoelectricity*, (The Institute of Electrical and Electronics Engineers, New York, 1987) p. 56.
4. W. Shi, H.N. Shekhani, H. Zhao, J. Ma, Y. Yao, K. Uchino, Losses in piezoelectrics derived from a new equivalent circuit, *Journal of Electroceramics.* **35**, 1–10 (2015).
5. W. Shi, H. Zhao, J. Ma, Y. Yao, K. Uchino, Investigating the frequency spectrum of mechanical quality factor for piezoelectric materials based on phenomenological model, *Japanese Journal of Applied Physics.* **54**, 101501 (2015). doi: 10.7567/JJAP.54.101501.
6. K. Uchino, *Micromechatronics*, 2nd Edition, (CRC Press, Boca Raton, FL, 2019). ISBN-13: 978-0-367-20231-6.
7. W.P. Mason, *Electromechanical Transducers and Wave Filters*, (D. Van Nostrand Co. Inc., New York, NY, 1948).
8. X. Dong, M. Majzoubi, M. Choi, Y. Ma, M. Hu, L. Jin, Z. Xu, K. Uchino, A new equivalent circuit for piezoelectrics with three losses and external loads, *Sensors & Actuators: A. Physical.* **256**, 77–83 (2017). doi: 10.1016/j.sna.2016.12.026.
9. X. Dong, T. Yuan, M. Hu, H. Shekhani, Y. Maida, T. Tou, K. Uchino. Driving frequency optimization of a piezoelectric transducer and the power supply development, *Review of Scientific Instruments.* **87**, 105003 (2016). doi: 10.1063/1.4963920.

6 Heat Generation in Piezoelectrics

ABSTRACT

Heat generation in piezoelectric materials originates from three losses, namely, dielectric, elastic, and piezoelectric losses. While the observed polarization vs. electric field curve draws sharp hysteresis edge around the maximum (and minimum) electric field, the complex parameter representation with *dissipation factor tangents* in this book reveals a narrow elliptic shape (counterclockwise locus!) with rounded edges. Though this discrepancy implies the modeling inaccuracy, we adopt the complex parameter method for the loss analysis merely from the mathematical simplicity viewpoint. We discuss the heat generation mechanisms of piezoelectric actuators under (1) the off-resonance operation for actuator applications (under a large electric field, 1 kV/mm or higher) and (2) the resonance (or antiresonance) operation for ultrasonic transducer applications (under a high-vibration condition at low-electric field, 100 V/mm or lower). Heat generation at off-resonance is attributed mainly to intensive dielectric loss $\tan\delta'$, while the heat generation at resonance is mainly originated from the intensive elastic loss $\tan\phi'$.

In your kid age, the reader may remember a significant heat generated from the rubber band when you extended and shrunk it quickly. This is originated from a large stress–strain hysteresis (i.e., elastic loss) of the rubber material. I remember the failure of the device development: we developed a piezo-"crane" toy with a rubber-PZT composite sheet in the early 1980s, which could flap the wings beautifully. However, when we desired it to actually fly by increasing the flapping frequency, the wings self-heated and melted unfortunately. Another popular phenomenon is the Joule heat in a resistive wire under the current flow, which is applied for electric heaters widely.

Heat generation in piezoelectric materials originated from three losses, namely, dielectric, elastic, and piezoelectric losses. Figure 6.1a–d corresponds to the model hysteresis curves: D vs. E dielectric hysteresis curve under a stress-free condition, x vs. X elastic hysteresis under a short-circuit condition, x vs. E under a stress-free condition, and D vs. X under a short-circuit condition for

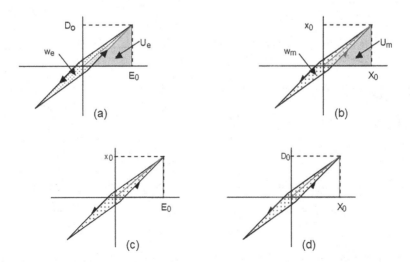

FIGURE 6.1 (a) D vs. E (stress free), (b) x vs. X (short-circuit), (c) x vs. E (stress free), and (d) D vs. X (short-circuit) curves with a slight hysteresis in each relation.

piezoelectric coupling loss, respectively. Though Figure 6.1a illustrates sharp hysteresis edge around the maximum (and minimum) electric field, similar to the actually observed hysteresis curve, the complex parameter representation with *dissipation factor tangents* in this book reveals a narrow elliptic shape (counterclockwise locus!) with rounded edges. Though this discrepancy implies the modeling inaccuracy, we adopt the complex parameter method for the loss analysis merely from the mathematical simplicity viewpoint, and evaluate the feasibility from the experimental results.

6.1   HEAT GENERATION AT OFF-RESONANCE

6.1.1   HEAT GENERATION FROM MULTILAYER ACTUATORS

Zheng et al. reported the heat generation at an off-resonance frequency from various configurations of multilayer (ML)-type piezoelectric ceramic (soft PZT) actuators.[1] Figure 6.2 shows a structure of the ML piezoelectric actuators. The temperature change with time in the actuators was monitored when driven at 3 kV/mm (high electric field) and 300 Hz (much lower frequency than the resonance frequency) (Figure 6.3a). The specimen temperature reached up to 140°C, depending on the size, showing an exponential increase with the operation time lapse. Figure 6.3b plots the saturated temperature as a function of V_e/A, where V_e is the effective volume (electrode overlapped part, *abL* in the figure) and A is the all surface area. Suppose that the temperature was uniformly generated in a

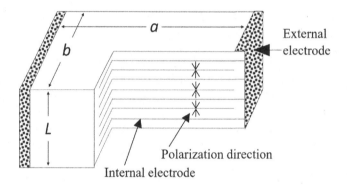

FIGURE 6.2   Structure of an ML piezoelectric actuator.

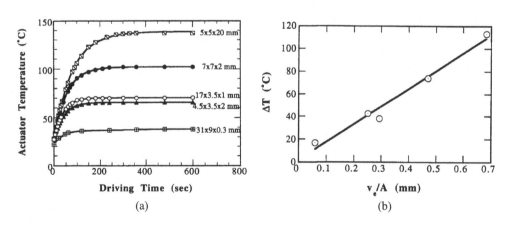

FIGURE 6.3   (a) Device temperature change with driving time for ML actuators of various sizes. (b) Temperature rise at off-resonance vs. V_e/A in various size soft PZT ML actuators, where Ve and A are the effective volume and the surface area, respectively.

bulk sample (no significant stress distribution, except for the small inactive portion of the external electrode sides), this linear relation is reasonable because the volume V_e generates the heat and this heat is dissipated through the area A. Thus, if we need to suppress the temperature rise, a small V_e/A design is preferred. Instead of one large ML, four ¼ small-MLs connected in parallel are preferred.

6.1.2 Thermal Analysis on ML Actuators

According to the law of energy conservation, the amount of heat stored in the piezoelectric, which is just the difference between the rate at which heat is generated, q_g, and that at which the heat is dissipated, q_d, can be expressed as

$$q_g - q_d = V \rho C_p (dT/dt), \tag{6.1}$$

where it is assumed that a *uniform temperature distribution* exists throughout the sample and V is the total volume, ρ is the mass density, and C_p is the *specific heat* (i.e., heat capacity per mass) of the specimen. The heat generation in the piezoelectric is attributed to losses. Thus, the rate of heat generation, q_g, is expressed as:

$$q_g = wfV_e, \tag{6.2}$$

where w is the loss per driving cycle per unit volume, f is the driving frequency, and V_e is the effective volume of active ceramic (no-electrode parts are omitted). According to the measurement conditions (no significant stress in the sample at off-resonance), this w may correspond primarily to the dielectric hysteresis loss (i.e., *P–E* hysteresis), w_e, which is expressed in terms of the intensive dielectric loss tanδ' as:

$$w = w_e = \pi \varepsilon^X \varepsilon_0 E_0^2 \tan\delta' \tag{6.3}$$

If we neglect the transfer of heat through conduction (lead wire is very thin), the rate of heat dissipation (q_d) from the sample is the sum of the rates of heat flow by radiation (q_r) and by convection (q_c):

$$q_d = q_r + q_c = eA\sigma(T^4 - T_0^4) + h_c A(T - T_0), \tag{6.4}$$

where e is the emissivity of the sample, A is the sample surface area, σ is the *Stefan–Boltzmann constant*, T_0 is the initial sample temperature, and h_c is the *average convective heat transfer coefficient*. Thus, Eq. (6.4) can be written in the form:

$$wfV_e - Ak(T)(T - T_o) = V\rho C_p (dT/dt), \tag{6.5}$$

where the quantity

$$k(T) = \sigma e\left(T^2 + T_o^2\right)(T + T_0) + h_c \tag{6.6}$$

is the overall *heat transfer coefficient*. If we assume that $k(T)$ is relatively insensitive to temperature change (if the temperature rise is not large), solving Eq. (6.5) for the rise in temperature of the piezoelectric sample yields:

$$T - T_0 = \left[wfV_e/k(T)A\right]\left(1 - e^{-t/\tau}\right), \tag{6.7}$$

where the time constant τ is expressed as:

$$\tau = \rho C_p V/k(T) A. \tag{6.8}$$

We can understand that the specimen temperature rise follows only when the heat transfer coefficient $k(T)$ is nearly constant.

As $t \to \infty$, the maximum temperature rise in the sample becomes:

$$\Delta T = wfV_e/k(T)A, \tag{6.9}$$

while, as $t \to 0$, the initial rate of temperature rise is given by

$$dT/dt = (w_T fV_e /\rho C_p V) = \Delta T/\tau, \tag{6.10}$$

where w_T can be regarded under these conditions as a measure of the total loss of the piezoelectric. The dependences of $k(T)$ calculated from Eq. (6.9) on applied electric field and frequency are shown in Figure 6.4a and b, respectively. Note that $k(t)$ is almost constant, as long as the driving voltage or frequency is not very high ($E < 1$ kV/mm, $f < 2$ kHz). The total loss, w_T, as calculated from Eq. (6.10) is given for three ML specimens in Table 6.1, three values of which are almost the same in less than 2% deviation. In parallel, we measured the $P–E$ hysteresis losses under stress-free conditions (Figure 6.1a). The w_e values obtained (Eq. (6.3)) are also listed in Table 6.1 for comparison. It is intriguing that the extrinsic $P–E$ hysteresis loss contributes more than 90% of the calculated total loss associated with the heat generated in the operating piezoelectric specimen.[1,2] We can conclude that the heat generation of the piezoelectric specimen under high-electric field off-resonance operation is primarily originated from the intensive dielectric loss factor, $\tan\delta'$.

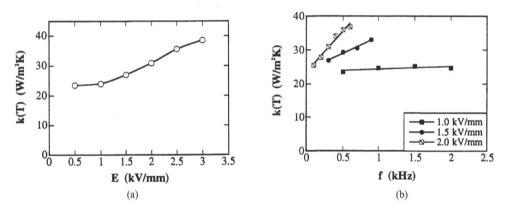

FIGURE 6.4 Overall heat transfer coefficient, $k(T)$, plotted as a function of applied electric field (a) and of frequency (b) for a PZT ML actuator with dimensions of $7 \times 7 \times 2$ mm³ driven at 400 Hz.

TABLE 6.1

Loss and Overall Heat Transfer Coefficient for PZT Multilayer Samples (under $E = 3$ kV/mm, $f = 300$ Hz)

Actuator	$4.5 \times 3.5 \times 2.0$ mm³	$7.0 \times 7.0 \times 2.0$ mm³	$17 \times 3.5 \times 1.0$ mm³
$w_T \left(\text{kJ/m}^3\right)\left[= \dfrac{\rho C_p V}{fV_e}\left(\dfrac{dT}{dt}\right)_{t\to 0}\right]$	19.2	19.9	19.7
P-E hysteresis loss (kJ/m³)	18.5	17.8	17.4
$k(T)$ (W/m² K)	38.4	39.2	34.1

6.2 HEAT GENERATION UNDER RESONANCE CONDITIONS

6.2.1 HEAT GENERATION FROM A RESONATING PIEZOELECTRIC SPECIMEN

Tashiro et al. observed the heat generation in a rectangular piezoelectric plate during a resonating drive.[3] Even though the electric field is not large, considerable heat is generated due to the large induced strain/stress at the resonance. The maximum heat generation was observed at the nodal regions for the resonance vibration, which correspond to the locations where the maximum strains/stresses are generated.

The ICAT at the Penn State University also worked on the heat generation comprehensively in rectangular piezoelectric k_{31} plates (refer to Figure 5.3) when driven at the resonance.[2] The temperature distribution profile in a PZT-based plate sample was observed with a pyroelectric infrared camera (FLIR Systems ThermaCAM S40 outfitted with 200 mm lens), as shown in Figure 6.5, where the temperature distribution profiles are shown in the sample driven at the first (28.9 kHz) and second resonance (89.7 kHz) modes, respectively. The highest temperature (bright spot) is evident at the nodal line areas of the length resonance for the specimen in Figure 6.5a and b. This observation supports that the heat generated in a resonating sample is primarily originated from the intensive elastic loss, tanϕ'. Remember that the *maximum vibration velocity* is defined as the velocity under which 20°C temperature rise occurs at the maximum temperature point (i.e., nodal line!) in the sample.

6.2.2 HEAT GENERATION AT THE ANTIRESONANCE MODE

As we discussed in Section 4.3.3, the resonance and antiresonance are both mechanical resonances with the impedance almost equal to zero or maximum, respectively. We can amplify the generating displacement significantly by the mechanical quality factor Q_A at the resonance under constant voltage drive, while by Q_B at the antiresonance under constant current drive. In the k_{31} mode, we derived the mechanical quality factors Q_A and Q_B as follows:

$$Q_{A,31} = \frac{1}{\tan\phi'_{11}} \tag{6.11}$$

$$\frac{1}{Q_{B,31}} = \frac{1}{Q_{A,31}} - \frac{2}{1 + \left(\frac{1}{k_{31}} - k_{31}\right)^2 \Omega_{B,31}^2} \left(2\tan\theta'_{31} - \tan\delta'_{33} - \tan\phi'_{11}\right) \tag{6.12}$$

Here, $\tan\delta'_{33}$, $\tan\phi'_{11}$, and $\tan\theta'_{31}$ are intensive loss factors for ε_{33}^X, s_{11}^E, and d_{31}, respectively, and $\Omega_{B,31}$ is the normalized antiresonance frequency:

$$\Omega_{A,31} = \frac{\omega_a l}{2v_{11}^E} = \frac{\pi}{2} \quad \left[v_{11}^E = 1/\sqrt{\rho s_{11}^E}\right], \tag{6.13}$$

 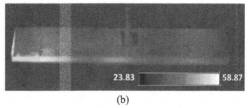

| 20.94 | 60.82 | | 23.83 | 58.87 |

(a)   (b)

FIGURE 6.5 Temperature variations in a PZT-based plate sample observed with an infrared camera. The specimens are driven at two different resonance frequencies: (a) first resonance (28.9 kHz) and (b) second resonance mode (89.7 kHz). The arrows indicate the highest temperature areas.

$$\Omega_{B,31} = \frac{\omega_b l}{2v_{11}^E}, \tag{6.14}$$

where the antiresonance (normalized) frequency $\Omega_{B,31}$ is k_{31} dependent and can be obtained from the equation, $1 - k_{31}^2 + k_{31}^2 \dfrac{\tan \Omega_B}{\Omega_B} = 0$. Because the intensive dielectric loss $\tan \theta_{31}'$ is larger than $(\tan \delta_{33}' + \tan \phi_{11}')/2$ in PZT piezo-ceramics, Q_B at antiresonance is higher than Q_A at resonance; that is, the antiresonance operation seems to be more efficient than the resonance drive.

Figure 6.6a and b shows temperature variations in a PZT-based plate specimen driven at the antiresoance (a) and resonance frequency (b) under the same vibration velocity (i.e., almost the same output mechanical energy), which clearly exhibit lower temperature rise in the antiresonace, than the resonance drive. Remember the resonance/antiresonance vibration modes for a reasonable electromechanical coupling factor $k_{31} = 30\%$, illustrated in Figure 6.6. In comparison with the resonance mode ($m = 1$), though the antiresonance mode ($m = 1$) shows the strain-zero lines (i.e., anti-node lines) a little inside from the plate edges, for this small k_{31} case, the overall vibration configurations for resonance and antiresonance modes are rather close each other. Namely, as long as the vibration velocity at the plate edge is the same, the total mechanical energy is assumed to the same. Numerical profiles

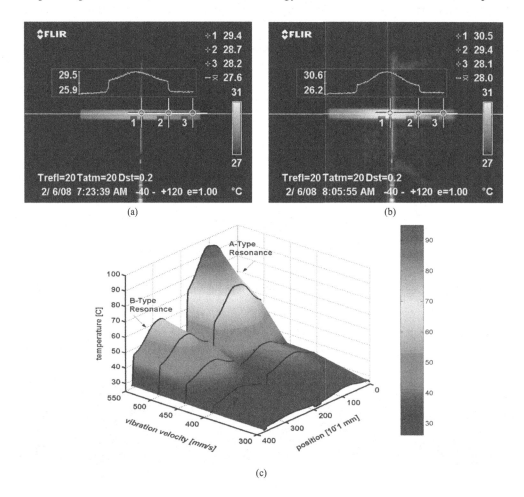

(a)

(b)

(c)

FIGURE 6.6 Temperature variations in a PZT-based plate sample observed with a pyroelectric infrared camera. The specimens are driven at (a) the antiresoance and (b) resonance frequencies. (c) Numerical temperature profile for the A- and B-type resonance modes.

of the temperature distribution for the A- and B-type resonance modes are shown in Figure 6.6c for various vibration velocity, which seems to be sinusoidal curves in terms of the length position coordinate. Under the same vibration velocity, 550 mm/s RMS, the resonance nodal line area shows the temperature up to 100°C, while the antiresonance mode shows the maximum around 60°C; dramatical heat generation reduction under the antiresonance drive. Note here that the piezoelectric loss tan (theta)' contributes to reduce the temperature rise at the antiresonance mode vibration, because of the opposite sign in comparison with dielectric or elastic losses.

6.2.3   THERMAL ANALYSIS ON THE RESONANCE MODE

6.2.3.1   Heat Transfer Modeling

We develop a 1D heat transfer model for the k_{31} mode piezoelectric rectangular plate. In comparison with the off-resonance model, where the heat is generated primarily from the dielectric loss and the uniform temperature distribution profile due to no particular stress in the speciment, the resonance case generates the heat originated from the elastic loss on a sinusoidal stress distribution of a specimen. Because the initial heat source is sinusoidally distributed in the specimen, we need to integrate the *thermal diffusivity* of the piezo-ceramic in order to analyze the temperature distribution of the sample.

We set the following assumptions for developing the heat diffusion equation:

1. 1D heat conduction in the specimen.
2. Heat generation is proportional to strain squared (i.e., elastic energy), distributed on the specimen.
3. Heat dissipation via convection (to air) and radiation (heat conduction via lead wire is neglected).

Using a temperature parameter $T(x,t)$, which is defined as temperature of a sliced volume Δx from the position coordinate x to $(x + \Delta x)$ at time t, the following equation is assumed:

$$\frac{\partial T(x,t)}{\partial t} = \frac{\kappa}{c_p \rho} \frac{\partial^2 T(x,t)}{\partial x^2} + \frac{q_g(x)}{c_p \rho} - \frac{h_d P}{c_p \rho A}\left[T(x,t) - T_{\text{air}}\right] \qquad (6.15)$$

where κ is *thermal conductivity* (unit: W/m K), c_p is specific heat (unit: J/kg K), and a new parameter $\frac{\kappa}{c_p \rho} = \alpha$ is called *thermal diffusivity* (unit: m²/s).[4] The first term of the right-hand side of Eq. (6.15) describes the temperature distribution in respect of the position x, which changes the shape with time. The second term corresponds to the temperature increment caused by heat generation per mass (devided by $c_p \rho$). Third term indicates the heat dissipation proportional to the temperature difference ΔT from the ambient temperature.

Regarding the heat generation, we further assume that q_g is expressed in proportion to the square of the strain, as visualized in Figure 6.7 in all frequency range between the resonance and antiresonance:

$$q_g(x) = g_h \cos^2\left(\frac{\pi x}{L}\right) \qquad (6.16)$$

The heat dissipation h_d can be approximated in proportion to $[T - T_{air}]$, similar to *heat transfer coefficient* introduced in Eq. (6.6).

6.2.3.2   Temperature Distribution Profile Change with Time

We prepared a piezoelectric k_{31} rectangular plate specimen with $80 \times 14 \times 2$ mm³ in size with a hard PZT composition (APC 841, American Piezo Company, USA) for both admittance and thermal imaging observation purposes. We used a constant vibration velocity method for the measurements

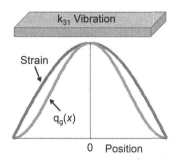

FIGURE 6.7   Heat generation modeling in proportion to strain square.

under 300 mm/s RMS. Figure 6.8a shows the temperature distribution profile change with time after driving. You can notice that the plate edge temperature increases significantly with time lapse, primarily due to the thermal diffusion in the PZT from the nodal highest point to the edge lowest temperature point. The saturated temperature distribution profile for the k_{31} specimen is shown in Figure 6.8b, which can be used for calculating the total thermal dissipation energy g_h. A small temperature dent at $x = 0$ originated from the heat dissipation by the center sample-holding rod whose heat conduction is neglected in the simulation.

If we adopt the most general definition of the mechanical quality factor as

$$Q_m = 2\pi \frac{\text{Energy Stored/Cycle}}{\text{Energy Lost/Cycle}}, \tag{6.17}$$

and assume the mechanical energy stored is constant proportional to the square of the vibration velocity, we can obtain Q_m from

$$Q_m = 2\pi f \left(\frac{\rho V_{\text{rms}}^2}{h_g} \right). \tag{6.18}$$

Refer to Eq. (7.7) for the derivation. Note that this is the unique approach to determine the mechanical quality factor with using merely the thermal data, without using any electrical energy information.

Shekhani et al. measured the admittance spectrum on the above PZT sample ($80 \times 14 \times 2$ mm^3) with the resonance frequency at 20.04 kHz at room temperature and obtained $Q_m = 507$ by a 3-dB-down method on an admittance spectrum. Then, the sample was excited under the vibration velocity of 400 mm/s for 30 second, which corresponds to the heat dissipation of 11.6 W/m^2. Figure 6.9 shows the Q_m obtained at three frequencies slightly above the resonance frequency

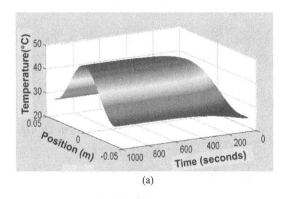

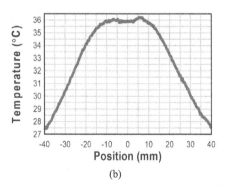

(a)   (b)

FIGURE 6.8   (a) Temperature distribution profile change with time after driving. (b) The saturated temperature distribution profile for k_{31} specimen with $80 \times 14 \times 2$ mm^3.

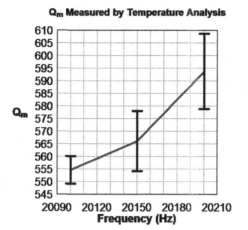

FIGURE 6.9 Change in Q_m with frequency (fr ≈ 20.006 kHz).[4]

(20.04 kHz).[4] The Q_m values around 550 agree with the extraporated values from the above 507. Thus, we can conclude that this thermal method can determine the Q_m reasonably at any frequency (not only the resonance and the antiresonance frequencies!). An increase of Q_m with increasing the frequency is obvious, which will be discussed further in Sections 7.2.4 and 8.3.2.

Example Problem 6.1

In order to discuss the high-power level in the vibration level of piezoelectric ceramics, we use the vibration velocity, rather than the vibration amplitude.

1. What is the reason of the "vibration velocity" usage?
2. What is the definition of the "maximum vibration velocity"?

Hint

1. Vibration amplitude is dependent on the specimen size.
2. How high temperature is still safe for humans?

Solution

1. Consider a k_{31}-type rectangular piezo-plate (length: L), for example. The sound velocity along the plate-length vibration and the fundamental resonance frequency are given by

$$v_{11} = 1/\sqrt{\rho s_{11}^D} \,, \tag{P6.1.1}$$

$$f_R = v_{11}/2L. \tag{P6.1.2}$$

Since the resonance amplification factor α ($\propto Q_m$) is given by

$$\alpha = Y_m/Y_0 \,, \quad [Y_m : \text{maximum admittance at } f_R; Y_0 : j\omega C_d] \tag{P6.1.3}$$

the vibration amplitude and vibration velocity at the plate edge are expressed by

$$u = \alpha |d_{31}| E_3 L, \tag{P6.1.4}$$

$$v = \frac{\partial u}{\partial t} = \omega_R \alpha d_{31} E_3 L = \pi \alpha \, v_{11} |d_{31}| E_3. \tag{P6.1.5}$$

The vibration amplitude depends directly on the size (proportional to length); while the vibration amplitude is size-independent, because the resonance frequency is inversely proportional to the length. Therefore, the vibration velocity is a better parameter to be used for the high-power performance evaluation, irrelevant to various sizes of the specimens.

2. The *maximum vibration velocity* was defined in the ICAT, above which the maximum temperature of the specimen at the node area exceeds 20°C higher than the room temperature. This is merely due to the human safety viewpoint for the home appliance applications, such as laptop computers and mobile phones. The 50°C component surface temperature may cause the finger burn for small kids. However, some of the manufacturers take 30°C–40°C higher than the room temperature, in order to show better maximum vibration velocity data on the company catalog. When you read the publication, the reader needs to take caution on their definition of "maximum vibration velocity".

6.3 THERMAL DIFFUSIVITY IN PIEZOELECTRICS

6.3.1 TEMPERATURE DISTRIBUTION PROFILE VS. THERMAL DIFFUSIVITY

Figure 6.10 demonstrates the saturated temperature distribution profile difference between two compositions, PZT-5H, and PZT-19 under the same vibration velocity operating condition. These two compositions exhibit significant difference in thermal conductivity: 0.14 vs. 2 W/m K (more than ten times difference). The profile curve of PZT-5H fits a sinusoidal line $\left(g_h \cos^2\left(\dfrac{\pi x}{L}\right) \right)$ beautifully, while PZT-19 shows considerable edge temperature rise. You can understand easily that this profile difference is originated primarily from the thermal conductivity or diffusivity difference. Taking the total thermal energy dissipated from the specimen by integrating the temperature rise with respect to the position coordinate, we can expect a similar mechanical quality factor Q_m value for these two samples. However, you can notice that the peak temperature at the nodal point is significantly lower for the larger thermal conductivity material, which means we can excite the vibration more under higher voltage for the PZT-19, since the maximum vibration velocity is defined by the highest temperature rise 20°C above room temperature.

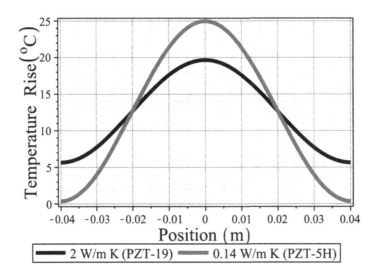

FIGURE 6.10 Saturated temperature distribution profile difference between PZT-5H and PZT-19.

6.3.2 THERMAL DIFFUSIVITY MEASUREMENTS

As demonstrated in Section 6.3.1, the thermal diffusivity or conductivity properties are essential in designing high-power piezo-devices, but we lack these data at present. Many methods exist to measure thermal diffusivity using either steady state or transient techniques. Steady-state methods yield large experimental error and inaccuracies. Transient techniques, that is, the laser flash method, are expensive and require specialized equipment and advanced data analysis. Thus, our Penn State group developed a novel experimental setup to evaluate thermal diffusivity.[5] Figure 6.11 shows our thermal diffusivity measurement setup. In this experiment, hot isothermal and insulating boundary conditions are imposed on a flat disk sample. The transient temperature profile of the insulated side of the sample is analytically similar to a classic time constant formulation. The thermal diffusivity is proportional to the inverse time constant. This method hosts a variety of advantages over other methods such as accuracy comparable to other methods, low cost, integrated modeling of interface effects, and small sample size. Several materials with low to medium thermal diffusivity (0.1→3 mm²/s) have been measured. The diameter of the sample is 32 mm and its thickness ranges from 2 to 6.5 mm. The thermal diffusivity measurements in this experiment have an accuracy of 5% or better for some standard materials such as Fused Quartz and Pyrex 7740, in comparison to the literature values.[5]

Using the newly developed equipment, we measured the thermal diffusivity on hard PZT disk specimens (APC 841, American Piezo Company, USA). Thermal diffusivity and thermal conductivity are summarized for poled disks, open-circuit and short-circuit conditions, and depoled samples in Table 6.2 (unreported data). We expect the thermal conductivity increase by the poling process, because of the better domain alignment. The polarization direction may keep the soft-phonon mode vibration direction, which may enhance the phonon–temperature coupling. This is verified from

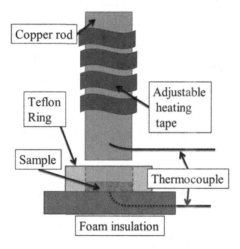

FIGURE 6.11   Thermal diffusivity measurement setup.

TABLE 6.2
Thermal Diffusivity and Thermal Conductivity for Hard PZT under Different Conditions

Hard PZT	Thermal Diffusivity $\alpha_{experiment}$ (10 mm2/s)	+/−	Thermal Conductivity $\kappa_{experiment}$ (W/m K)	+/−
Open circuit	5.02	0.23	1.4	0.06
Closed circuit	8.25	0.78	2.3	0.23
Depoled	4.32	0.34	1.2	0.10

TABLE 6.3
Thermal Properties of NKN-Cu in Comparison with Hard PZT

Thermal Properties	c_p (J/g/K)	κ (W/m/K)
Hard-PZT	0.42	1.25
NKN-Cu	0.58	3.10

the difference between the "depoled" (with lowest thermal conductivity) and the "poled" specimens (either open or closed-circuit conditions). We know the elastic compliance relation, $s_{33}^D = \left(1 - k_{33}^2\right)s_{33}^E$, that is, the PZT becomes elastically stiffened under the "depolarization field". Though this is not actually true for the open-circuit status at this low-frequency measurement, slight stiffening at least may be expected for the open-circuit. Since the elastically soft E-constant status seems to have higher coupling between the phonon and temperature, the higher thermal conductivity is expected, which is actually proved in the experiment in Table 6.2.

Pb-free piezo-ceramics such as (Na,K)NbO$_3$-based materials show much higher maximum vibration velocity than the PZTs.[6] Much larger thermal conductivity in the NKN-based material than the PZTs, as shown in Table 6.3 (unreported data), may also contribute to this good high-power performance in NKN-ceramics.

Example Problem 6.2

Two material disks with different thermal conductivities, κ_1 and κ_2, are laminated with the thickness L_1 and L_2, as shown in Figure 6.12, then the outer surfaces A and C (area S is common) are maintained at T_1 and T_2 ($T_1 > T_2$), respectively. Calculate the temperature T and the thermal energy flow Q on the contact surface B.

Hint
When the two-surface temperature of a disk specimen (thickness L) is maintained at T_1 and T_2 ($T_1 > T_2$), the thermal conductivity κ is defined as

$$Q = \kappa t S \frac{T_1 - T_2}{L} \tag{P6.2.1}$$

where Q is the quantity of heat transferred from high-temperature surface to low-temperature surface via surface S in time duration t.

Solution

Supposing that $T_1 > T_2$ and the temperature on the contact surface B is T, the thermal energy flow Q via the surface area S in time duration t should be expressed by the following equations under a steady state:

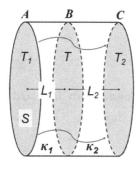

FIGURE 6.12 Thermal conductivity model.

$$Q = \kappa_1 t\, S\, \frac{T_1 - T}{L_1} = \kappa_2 t\, S\, \frac{T - T_2}{L_2} \tag{P6.2.2}$$

Thus, the temperature T is obtained from $\kappa_1 \dfrac{T_1 - T}{L_1} = \kappa_2 \dfrac{T - T_2}{L_2}$, then

$$T = \frac{\dfrac{\kappa_1}{L_1} T_1 + \dfrac{\kappa_2}{L_2} T_2}{\dfrac{\kappa_1}{L_1} + \dfrac{\kappa_2}{L_2}} \tag{P6.2.3}$$

Accordingly, the normalized energy flow q per unit area and unit time defined by $Q/S\,t$ is calculated as

$$q = \frac{\kappa_1 \kappa_2}{L_1 \kappa_2 + L_2 \kappa_1} (T_1 - T_2) \tag{P6.2.4}$$

The effective thermal conductivity of the series connection composite material is expressed by

$$\kappa_{\text{eff}} = \frac{1}{\dfrac{L_1}{\kappa_1} + \dfrac{L_2}{\kappa_2}} \tag{P6.2.5}$$

6.4   HEAT GENERATION UNDER TRANSIENT RESPONSE

ML piezoelectric actuators are occasionally used under a *pulse drive* method, such as delta (including a rectangular pulse) or a step or pseudo-step function voltage. Refer to Chapter Problem 5.4 and textbook *Micromechatronics 2nd Edition*.[7] When a piezoelectric actuator is driven by a sharp step voltage, the vibration displacement ΔL is generated linearly with time, and the displacement ringing will not decay easily. Only when the rise time of a pseudo-step function is adjusted exactly to the actuator resonance period, the displacement overshoot and ringing are completely suppressed. We discuss in this subsection on the additional heat generation, relating to the transient "ringing" vibrations.

Diesel engines are recommended rather than regular gasoline cars from the energy conservation and global warming viewpoint. The energy efficiency, measured by the energy required to realize unit drive distance for a vehicle (MJ/km), is of course better for high-octane gasoline than diesel oil. However, since the electric energy required for purification is significant, the gasoline is inferior to diesel. As well known, the conventional diesel engine, however, generates toxic exhaust gases such as SO_x and NO_x. In order to solve this problem, new diesel injection valves were developed by Siemens, Bosch, and Toyota with piezoelectric ML actuators. Figure 6.13 shows a common rail-type diesel injection valve. In order to eliminate toxic gas emission and increase the diesel engine efficiency, high-pressure fuel and quick injection control are required. Figure 6.14 shows an example of diesel fuel injection timing chart. In this one cycle (typically engine shaft rotation 60–100 Hz), the multiple injections should be realized in a very sharp shape. For this purpose, piezoelectric actuators, specifically ML types, were adopted.[8] The piezoelectric actuator is a key component to quick responsivity of the injection control valve.

However, because of very-high responsivity of the ML actuator, additional vibration ringing is observed after the trapezoidal voltage application, as shown in Figure 6.15, which was monitored under unipolar $V_{0-p} = 100\,V$ with 50% duty ratio of the pulse at 60 Hz. It is noteworthy that the vibration ringing can be significantly suppressed with adjusting the rise time. More interestingly, as summarized in Table 6.4, the temperature rise is suppressed significantly in proportion to the ringing suppression. In conclusion, when we drive a piezoelectric device under a pulse drive with some vibration overshoot and ringing, in addition to the normal dielectric loss under a high-electric field drive, the resonating ringing generates more heat via the elastic loss.

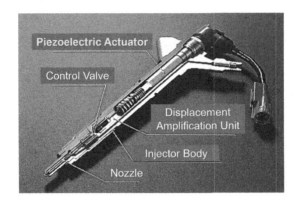

FIGURE 6.13 Common rail-type diesel injection valve with a piezoelectric ML actuator. (Courtesy by Denso Corporation.)

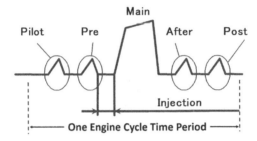

FIGURE 6.14 Diesel fuel injection timing chart in one cycle (about 60 Hz cycle).

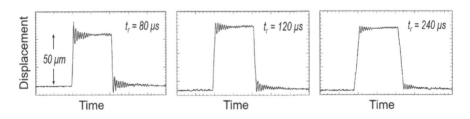

FIGURE 6.15 Vibration ringing after the trapezoidal pulse voltage application with different rise time.

TABLE 6.4
Heat Generation of the ML Actuator under Different Rise Time of Trapezoidal Voltage Drive (100 V Unipolar, 50% Duty Ratio at 60 Hz)

Rise time (µs)	72	85	97	102	146	240	800
Temp. rise (°C)	23.7	21.6	20.5	20.2	20.3	18.6	16.6
Voltage (V_{rms})	72.9	72.8	72.7	72.7	72.6	72.1	71.3

Unreported data.

CHAPTER ESSENTIALS

1. Three losses are essential to analyze the heat generation in piezoelectrics: dielectric $\tan\delta'$, elastic $\tan\phi'$, and piezoelectric $\tan\theta'$ losses.
2. Heat generation and loss factors:
   - Heat generation at off-resonance: high-voltage drive and no significant stress distribution in a piezoelectric specimen $\rightarrow$ primarily dielectric loss $\tan\delta'$
   - Heat generation at resonance: low-voltage drive but sinusoidal stress/strain distribution in a piezoelectric specimen $\rightarrow$ primarily elastic loss $\tan\phi'$
   - Heat generation under a pulse drive: high-voltage drive and vibration overshoot and ringing in a piezoelectric specimen $\rightarrow$ both dielectric and elastic losses
3. Heat generation at antiresonance in PZTs: high-voltage drive but small current with sinusoidal stress/strain distribution (similar vibration mode to the resonance) in a piezoelectric specimen $\rightarrow$ primarily elastic loss subtracted by the piezoelectric loss $\rightarrow$ lower temperature rise under the same output mechanical vibration as the resonance mode.
4. Heat flow equation ($u(x,t)$ is the temperature parameter):

$$\frac{\partial u(x,t)}{\partial t} = \frac{\kappa}{c_p\rho}\frac{\partial^2 u(x,t)}{\partial x^2} + \frac{q_g(x)}{c_p\rho} - \frac{h_d P}{c_p\rho A}\left[u(x,t)-T_{air}\right]$$

where κ is thermal conductivity (unit: W/m K), c_p is specific heat (unit: J/kg K), and aparameter $\dfrac{\kappa}{c_p\rho}=\alpha$ is called thermal diffusivity (unit: m²/s). When we consider the off-resonance operation, the thermal conductivity term can be neglected; while at the resonance (due to the temperature distibution in a specimen), thermal conductivity/diffusivity is essential to discuss the *maximum vibration velocity.*

CHECK POINT

1. (T/F) When we consider an off-resonance drive of a ML piezo-actuator, heat generation comes primarily from the elastic loss $\tan\phi'$. True or False?
2. (T/F) When we consider a resonance drive of a k_{31} piezoelectric transducer, heat generation comes primarily from the elastic loss $\tan\phi'$. True or False?
3. (T/F) The heat generation is originated from three losses (dielectric, elastic, and piezoelectric) in piezoelectrics. These three losses are always added to produce the final temperature rise. True or False?
4. When we excite the fundamental resonance mode on a k_{31}-type rectangular PZT plate under a voltage drive, which part of the specimen generates the maximum temperature rise? Answer simply.
5. (T/F) In order to suppress the temperature rise in piezoelectric actuators/transducers, a tube design is better than solid rod design. True or False?
6. (T/F) To generate the same vibration velocity of a piezoelectric transducer, the resonance drive is the most efficient, rather than antiresonance drive in PZT-based materials. True or False?
7. (T/F) From the energy efficiency viewpoint, the optimized drive frequency exists in-between the resonance and antiresonance frequencies in the PZT resonator. True or False?

CHAPTER PROBLEMS

6.1   Heat flow equation ($T(x,t)$ is the temperature parameter) is given by:

$$\frac{\partial T(x,t)}{\partial t} = \frac{\kappa}{c_p\rho}\frac{\partial^2 T(x,t)}{\partial x^2} + \frac{q_g(x)}{c_p\rho} - \frac{h_d P}{c_p\rho A}\left[T(x,t)-T_{air}\right]$$

where κ is thermal conductivity (unit: W/m K), c_p is specific heat (unit: J/kg K), and a parameter $\dfrac{\kappa}{c_p \rho} = \alpha$ is called thermal diffusivity (unit: m²/s). Based on the data in Table 6.3, for the thermal conductivity of NKN-Cu and hard PZT, simulate the saturated temperature distribution profile difference between NKN-Cu and hard PZT specimens, like Figure 6.10.

Hint

In addition to a large difference in thermal conductivity κ, a large difference even in the mass density ρ (check the mass density of NKN and PZT by yourself) provides a significant difference in the thermal diffusivity α. You may obtain larger difference in the temperature profile than Figure 6.10.

6.2   Thermal conductivity κ was measured in hard PZT ceramics. Table 6.2 summarizes the experimental results, which indicates that κ under short-circuit > κ under open-circuit > κ in "depoled" sample. Explain the physical reason of these differences by searching required references.

Hint

Remember the fact that ferroelectric-originated piezoelectric materials exhibit the phase transition associated with phonon mode "softening". First, because of the random orientation of domains in the "unpoled" sample, it is expected that there will be the most phonon scattering at the domain boundaries in this specimen. Therefore, it will have the lowest thermal conductivity. Second, with increasing the degree of domain orientation, less scattering is expected in the "poled" specimens and thermal conductivity will be larger for the poled material (κ_{33}^E and κ_{33}^D). Third, because the elastic compliance under the "short-circuit" condition (s_{33}^E) is softer than the elastic compliance under "open-circuit" conditions (s_{33}^D) via the relation $s_{33}^E\left(1 - k_{33}^2\right) = s_{33}^D$, the lattice and domain wall dynamic motion are expected to be larger in the polarization direction under the short-circuit condition, probably originated from no-"depolarization field". The internal electric field in piezo-materials stabilizes the domain wall motion in general. The larger lattice vibration and domain wall dynamics in the short-circuit condition may also introduce a larger thermal conductivity in short-circuit condition (κ_{33}^E) due to increased phonon mode transport. This speculation can also suggest the analogous relations among two coupling factors, electromechanical k_{33} and electrothermal k_{33}^κ.

REFERENCES

1. J. Zheng, S. Takahashi, S. Yoshikawa, K. Uchino, J.W.C. de Vries, Heat generation in multilayer piezoelectric actuators, *Journal of the American Ceramic Society.* **79**, 3193–3198 (1996).
2. K. Uchino, J. Zheng, A. Joshi, Y.H. Chen, S. Yoshikawa, S. Hirose, S. Takahashi, J.W.C. de Vries, High power characterization of piezoelectric materials, *Journal of Electroceramics.* **2**, 33–40 (1998).
3. S. Tashiro, M. Ikehiro, H. Igarashi, Influence of temperature rise and vibration level on electromechanical properties of high-power piezoelectric ceramics, *Japanese Journal of Applied Physics.* **36**, 3004–3009 (1997).
4. H.N. Shekhani, K. Uchino, Characterization of mechanical loss in piezoelectric materials using temperature and vibration measurements, *Journal of the American Ceramic Society.* **97** (9), 2810–2814 (2014).
5. H.N. Shekhani, K. Uchino, Thermal diffusivity measurements using insulating and isothermal boundary conditions, *Review of Scientific Instruments.* **85**, 015117 (2014).
6. E.A. Gurdal, S.O. Ural, H.Y. Park, S. Nahm, K. Uchino, High power characterization of $(Na_{0.5}K_{0.5})NbO_3$ based lead-free piezoelectric ceramics, *Sensors and Actuators A: Physical.* **200**, 44 (2013).
7. K. Uchino, *Micromechatronics*, 2nd Edition, (CRC Press, Boca Raton, FL, 2019). ISBN-13: 978-0-367-20231-6.
8. A. Fujii, *Proceedings of Smart Actuators/Sensors Study Committee*, JTTAS, Dec. 2, Tokyo (2005).

7 High-Power Piezo Characterization System

ABSTRACT

There are various methods for characterizing loss factors and high-power characteristics in the piezo-electric materials. These are pseudo-static, admittance spectrum, and transient/burst mode methods. The admittance/impedance spectrum method is further classified into (1) constant voltage, (2) constant current, (3) constant vibration velocity, and (4) constant input energy methods. Piezoelectric resonance can be excited by either electrical or mechanical driving. In the k_{31} mode, as long as the surface is elec-troded, the sound velocity is v_{11}^{E} originated from s_{11}^{E}, while no-electrode, they are v_{11}^{D} and s_{11}^{D}. With the surface electrode, a short-circuit condition realizes the resonance and an open-circuit condition provides the antiresonance mode. In order to measure the D-constant parameters (s_{11}^{D} and its extensive elastic loss tan ϕ_{11}) directly, we need to use a non-electrode (NE) sample under mechanical driving methods.

First of all, the reader needs to have a suitable power supply to drive a piezoelectric actuator, which is a sort of prerequisite prior to reading further this chapter. The required power supply specifica-tions should be better than the followings, in particular, in a multilayer (ML) actuator characterization: (a) Maximum Voltage: 200 (V), (b) Maximum Current: 10 (A), (c) Frequency Range: 0–500 (kHz), (d) Output Impedance: < 1 (Ω). You may use a conventional impedance analyzer, on which you should recognize that maximum voltage is only 30 V and the maximum current is less than 0.2 A with the output impedance 50 Ω. You should not believe the obtained admittance value for a large capacitance ML piezo-actuator measurement.

Piezoelectric resonance can be excited by either electrical or mechanical driving, as shown in Figure 7.1. "High-Power Piezo Characterization System" (HiPoCS™) introduces various character-ization methodologies of loss factors chronologically, including continuous admittance/impedance spectrum measurement, burst/transient response method, thermal monitoring method, and so on. Mechanical quality factors (Q_A at resonance and Q_B at antiresonance) are primarily measured as a function of vibration velocity.

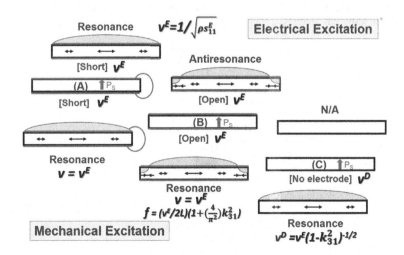

FIGURE 7.1 Resonance and antiresonance mode excitation under electrical and mechanical driving methods.

7.1   LOSS MEASURING TECHNIQUE I – PSEUDO-STATIC METHOD

We can determine "intensive" dissipation factors, tan δ', tan ϕ', and tan θ', separately from (a) D vs. E (stress free), (b) x vs. X (short-circuit), (c) x vs. E (stress free), and (d) D vs. X (short-circuit) curves (see Figure 7.2). Using a stress applying jig shown in Figure 7.3, Zheng et al. measured the x vs. X and D vs. X relationships.[1] Regarding the loss parameter calculation processes from the hysteresis loop areas in Figure 7.2, refer to Section 4.1.2.1. Figure 7.4 summarizes intensive loss factors, tan δ', tan ϕ', and tan θ', as a function of electric field or compressive stress, measured for a "soft" PZT-based ML actuator. Note first that the piezoelectric loss tan θ' is not negligibly small, as believed by the previous researchers, but rather large, comparable to or even larger than the dielectric and elastic losses; tan θ' (=0.08)> (1/2) [tan δ' (=0.06) + tan ϕ' (=0.08)] in PZTs. This relationship is discussed again in Section 7.2.

Then, using the *K-matrix*, Eqs. (4.38) and (4.39), we calculated the "extensive" losses as shown in Figure 7.5. Note again that the magnitude of the piezoelectric loss tan θ is comparable to the dielectric and elastic losses and increase gradually with the field or stress; now tan θ (=0.05) < (1/2) [tan δ (=0.05) + tan ϕ (=0.07)]. Also it is noteworthy that the extensive dielectric loss tanδ increases significantly with an increase of the intensive parameter, that is, the applied electric field, while the extensive elastic loss tan ϕ is rather insensitive to the intensive parameter, that is, the applied compressive stress. When similar measurements are conducted under constrained conditions; that is,

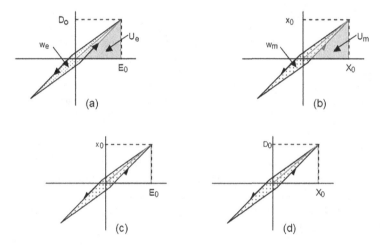

FIGURE 7.2   (a) D vs. E (stress free), (b) x vs. X (short-circuit), (c) x vs. E (stress free), and (d) D vs. X (short-circuit) curves with a slight hysteresis in each relation.

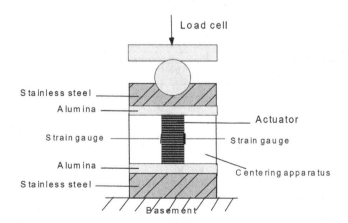

FIGURE 7.3   Stress applying jig construction for an ML piezoelectric sample.[1]

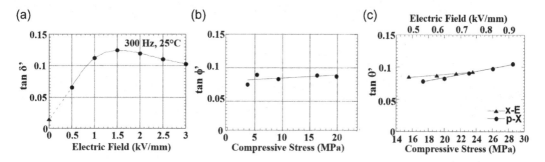

FIGURE 7.4 Intensive loss factors, tan δ' (a), tan ϕ' (b), and tan θ' (c) as a function of electric field or compressive stress, measured for a "soft" PZT ML actuator.[1]

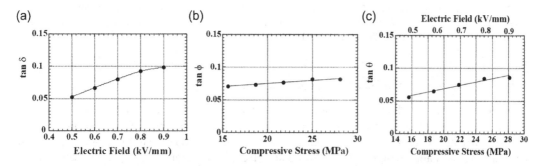

FIGURE 7.5 Extensive loss factors, tan δ (a), tan ϕ (b), and tan θ (c) as a function of electric field or compressive stress, measured for a "soft" PZT ML actuator.[1]

D vs. E under a completely clamped state, and x vs. X under an open-circuit state, respectively, we can expect smaller hystereses, that is, extensive losses, tan δ and tan ϕ, theoretically. However, they are rather difficult in practice because of the migrating charge compensation on the specimen surface. Note the experimental result difference from Figure 9.47, where small AC modulation is superposed on a large DC bias stress. Here large AC modulation results are shown.

7.2 LOSS MEASURING TECHNIQUE II – ADMITTANCE/ IMPEDANCE SPECTRUM METHOD

7.2.1 RESONANCE UNDER CONSTANT VOLTAGE DRIVE

Because commercially then-available "Impedance Analyzer" did not generate high voltage/current for measuring the high-power piezoelectric performance, Uchino (as an NF Corporation Deputy Director) commercialized "Frequency Response Analyzer" (500 V, 20 A, 1 MHz maximum) from NF Corporation, Japan. Uchino reported the existence of the critical threshold voltage and the maximum vibration velocity for a piezoelectric, above which the piezoelectric drastically increases the heat generation and becomes a ceramic heater.[2] Though this measurement technique is simple yet sophisticated, there have been problems in heat generation in the sample around the resonance range, further in significant distortion of the admittance frequency spectrum when the sample is driven by a constant voltage, due to the nonlinear behavior of elastic compliance at high-vibration amplitude.[2] Figure 7.6a exemplifies the problem, where the admittance spectrum is skewed with a jump around the maximum admittance point. Thus, we cannot determine the resonance frequency or the mechanical quality factor precisely from these skewed spectra. Thus, HiPoCS, this version, did not have a capability to measure the piezoelectric loss. This NF high-voltage/current power supply is still a key device in the present HiPoCS (III).

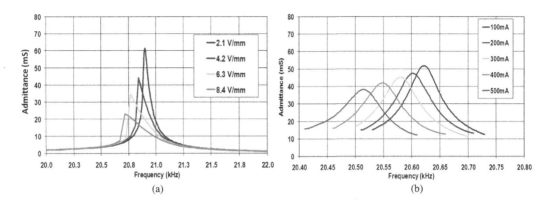

FIGURE 7.6 Experimental admittance frequency spectra under (a) constant voltage and (b) constant current condition in PZT. Note the skew-distorted spectrum with a jump under a constant voltage condition. (Data taken by Michael R. Thibeault, The Penn State University.)

7.2.2 RESONANCE UNDER CONSTANT CURRENT DRIVE

In order to escape from the problem with a constant voltage measurement, we proposed a constant current measurement technique.[3,4] Since the vibration amplitude is primarily proportional to the driving current (not the voltage) at the resonance, a constant current condition guarantees almost constant vibration amplitude through the resonance frequency region, escaping from the spectrum distortion due to the elastic nonlinearity. As demonstrated in Figure 7.6b, the spectra exhibit symmetric curves, from which we can determine the resonance frequency and the mechanical quality factor Q_A precisely.

Although the traditional constant voltage measurement was improved by using a constant current measurement method, the constant current technique (HiPoCS (II)) is still limited to the vicinity of the resonance. In order to identify a full set of high-power electromechanical coupling parameters and the loss factors of a piezoelectric, both resonance and antiresonance vibration performance (in particular, Q_A and Q_B) should be precisely measured simultaneously. Basically, Q_A can be determined by the constant current method around the resonance (A-type), while Q_B should be determined by the constant voltage method around the antiresonance (B-type). The mechanical quality factor Q_m (or the inverse value, loss factor $\tan \phi'$) is obtained from $Q_m = \omega_R/2\Delta\omega$, where $2\Delta\omega$ is a full width of the $1/\sqrt{2}$ (i.e., 3 dB down) of the maximum admittance value at ω_R. HiPoCS (II) improved the capability for measuring the mechanical quality factor Q_m at the resonance region but did not provide information on the piezoelectric loss $\tan \theta'$ easily, because we need to switch the measuring systems between constant-current and constant-voltage types.

We reported the degradation mechanism of the mechanical quality factor Q_m with increasing electric field and vibration velocity.[4] Figure 7.7 shows the vibration velocity dependence of the mechanical quality factors Q_A and Q_B, and corresponding temperature rise for A (resonance) and B (antiresonance)-type resonances of a longitudinally vibrating PZT ceramic k_{31} transducer through the transverse piezoelectric effect d_{31} (the sample size is inserted).[3] Q_m is almost constant for a small electric field/vibration velocity, but above a certain vibration level, Q_m degrades drastically, where temperature rise starts to be observed.[3] In order to evaluate the mechanical vibration level, we used the vibration velocity at the rectangular plate tip. Refer to Example Problem 6.1 to learn the vibration velocity. The *maximum vibration velocity* is defined at the velocity where a 20°C temperature rise at the nodal point from room temperature occurs. Note that even if we further increase the driving voltage/field, additional energy will convert to merely heat (i.e., PZT becomes a ceramic heater!) without increasing the vibration amplitude. Thus, the reader can understand that the maximum vibration velocity is a sort of material's constant which ranks the high-power density performance

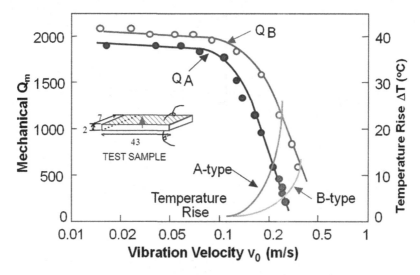

FIGURE 7.7 Vibration velocity dependence of the mechanical quality factors Q_A and Q_B, and corresponding temperature rise for *A*- (resonance) and *B* (antiresonance)-type resonances of a longitudinally vibrating PZT ceramic transducer k_{31}.[4]

of the piezo-materials. Note that most of the commercially available hard PZTs exhibit the maximum vibration velocity around 0.3 m/s, which corresponds to roughly 5 W/cm³ (i.e., 1 cm³ PZT can generate maximum 5 W mechanical energy).

Though we used a different equipment for measuring Q_A (constant current) and Q_B (constant voltage), when we compare the change trends in Q_A and Q_B, Q_B is higher than Q_A in all vibration level. (This is true in PZTs.) A similar result was already discussed in the thermal analysis in Chapter 6. Accordingly, the heat generation in the *B*-type (antiresonance) mode is superior to the *A*-type (resonance) mode under the same vibration velocity level (in other words, the maximum vibration velocity is higher for Q_B than for Q_A).

Figure 7.8b depicts an important notion on elastic loss tan ϕ' (= $1/Q_m$) in the k_{31} piezoelectric plate at its resonance frequency, where the damped and motional resistances, R_d and R_m, in the equivalent electrical circuit of a PZT sample (Figure 7.8a) are separately plotted as a function of vibration velocity.[3] We have not integrated the piezoelectric loss factor previously in the 1990s, but as long as we discuss merely on the resonance mode, the following discussion still maintains. Note that R_m, which we speculate to be mainly related to the *extensive mechanical loss* (90° domain wall motion), is insensitive to the vibration velocity, while R_d, related to the *extensive dielectric loss* (180° domain wall motion), increases significantly around a certain critical vibration velocity. Thus, the resonance loss at a small vibration velocity is mainly determined by the extensive mechanical loss which provides a high mechanical quality factor Q_m, and with increasing vibration velocity, the extensive dielectric loss contribution significantly increases. Note that the dielectric loss increases exponentially above a certain critical vibration velocity, which is related with the coercive field (or stress) of the piezo-material. After R_d exceeds R_m, we started to observe heat generation. Since we did not include the piezoelectric loss in the equivalent circuit previously, updated discussion will be required in this argument. Also, the microscopic domain dynamics model will be discussed in Chapter 9.

7.2.3 Resonance/Antiresonance under Constant Vibration Velocity

Mezheritsky derived the mechanical quality factor for Q_B from the combination of three loss factors.[5] However, Zhuang and Uchino derived an expansion-series approximation of the mechanical quality factors at both resonance and antiresonance modes in the user-friendly forms, as introduced

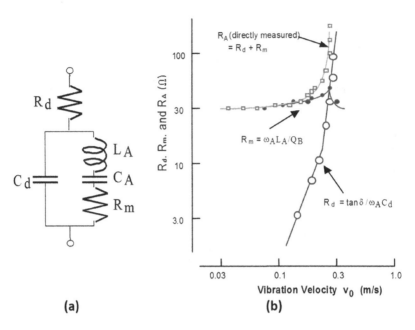

(a)   **(b)**

FIGURE 7.8 (a) Equivalent circuit of a piezoelectric sample for the resonance under high-power drive. (b) Vibration velocity dependence of the resistances R_d and R_m in the equivalent electric circuit for a longitudinally vibrating PZT ceramic k_{31} plate.[3]

in Section 4.3.3. We adopt these useful *Uchino–Zhuang formulea*:[6,7] The mechanical quality factors Q_A and Q_B for the rectangular k_{31} mode are expressed in terms of the intensive losses as

$$Q_{A,31} = \frac{1}{\tan \phi'_{11}}, \tag{7.1}$$

$$\frac{1}{Q_{B,31}} = \frac{1}{Q_{A,31}} - \frac{2}{1+\left(\dfrac{1}{k_{31}}-k_{31}\right)^2 \Omega_{B,31}^2}\left(2\tan\theta'_{31} - \tan\delta'_{33} - \tan\phi'_{11}\right) \tag{7.2}$$

where $\tan\delta'_{33}$, $\tan\phi'_{11}$, and $\tan\theta'_{31}$ are intensive loss factors for ε_{33}^X, s_{11}^E, and d_{31}, and $\Omega_{B,31}$ is the normalized antiresonance frequency given by

$$\Omega_{A,31} = \frac{\omega_a l}{2v_{11}^E} = \frac{\pi}{2}\ \left[v_{11}^E = 1/\sqrt{\rho\ s_{11}^E}\right],\quad \Omega_{B,31} = \frac{\omega_b l}{2v_{11}^E},\quad 1 - k_{31}^2 + k_{31}^2\frac{\tan\Omega_B}{\Omega_B} = 0 \tag{7.3}$$

The key is that the values Q_A and Q_B can be different, and if we precisely measure both the values, the information on the piezoelectric loss $\tan\theta'$ can be obtained. Thus, we proposed a simple, easy, and user-friendly method to determine the piezoelectric loss factor $\tan\theta'$ in k_{31} mode through admittance/impedance spectrum analysis.

In order to identify both mechanical quality factors Q_A and Q_B precisely to adopt the abovementioned methodology, both resonance and antiresonance vibration performances should be measured simultaneously with the same equipment. Basically, Q_A can be determined by the constant current method around the resonance (*A*-type), while Q_B should be determined by the constant voltage method around the antiresonance (*B*-type). Thus, we developed HiPoCS Version III shown in Figure 7.9, which is capable of measuring the impedance/admittance curves by keeping the following various conditions: (1) constant voltage, (2) constant current, (3) constant vibration velocity of a piezoelectric

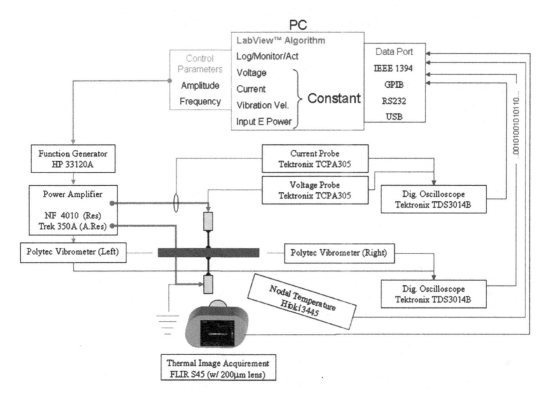

FIGURE 7.9 Setup of HiPoCS Version III.[8]

sample, and (4) constant input power.[8] Key equipment includes NF Corporation's power supply which satisfies: maximum voltage: 200 (V), maximum current: 10 (A), frequency range: 0–500 (kHz), and output impedance: < 1 (Ω). In addition, the system is equipped with an infrared image sensor to monitor the heat generation distributed in the test specimen. Figure 7.10 shows an interface display of HiPoCS Version III, demonstrating a rectangular k_{31} plate measurement under a constant vibration velocity condition. In order to keep the vibration velocity constant (i.e., stored/converted mechanical energy is constant), the current is almost constant and the voltage is minimized at the resonance; while the voltage is almost constant and the current is minimized at the antiresonance frequency. The apparent power is shown in the **top** of Figure 7.10 (for a specimen of 80 mm L), and more detailed results are shown in Figure 7.11 (for a shorter specimen), which clearly indicates that the antiresonance operation requires less power than the resonance mode for generating the same vibration velocity or stored mechanical energy. We can conclude that the PZT transducer should be operated at the antiresonance frequency, rather than the resonance mode, from the energy efficiency viewpoint.

The admittance can be calculated from the voltage and current data at each frequency from the **bottom** of Figure 7.10, which is plotted in Figure 7.12 (data is for a specimen of 20 mm L). Note that this admittance curve was obtained by keeping the same vibration velocity in all the frequency range below the resonance to above the antiresonance frequencies. In order to obtain the Q_A and Q_B, 3 dB-down and -up methods are used for the resonance and antiresonance frequencies, as shown in Figure 7.12.

7.2.4 REAL ELECTRIC POWER METHOD

Because the conventional admittance spectrum method can provide the mechanical quality factors only at two frequency points (i.e., resonance Q_A and antiresonance Q_B), we had a frustration for knowing the Q_m at any frequency. A unique methodology for characterizing the quality factor in

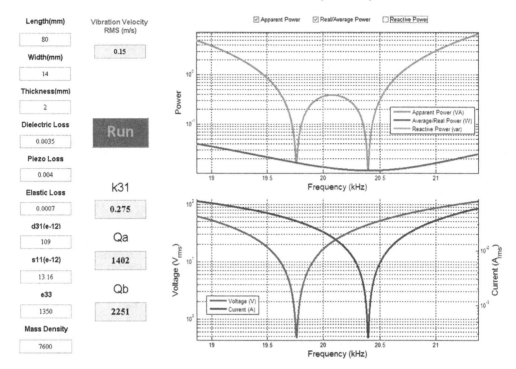

FIGURE 7.10   Voltage and current change with frequency under the constant vibration velocity condition.[8]

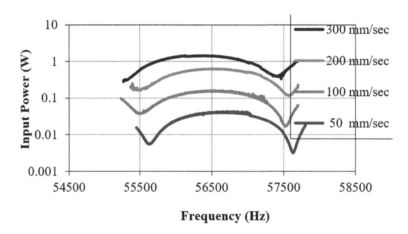

FIGURE 7.11   Frequency spectra of power under constant vibration velocity, conducted across the resonance and antiresonance frequencies.

piezoelectric materials has been developed in the ICAT by utilizing real electrical power measurements (including the phase lag), for example, $P = V \cdot I \cdot \cos \varphi$, rather than the apparent power $V \cdot I$, as shown in Figure 7.10.[9]

The mechanical quality factor, Q_m, can be defined in general as

$$Q_m = 2\pi \frac{\text{Energy Stored/Cycle}}{\text{Energy Lost/Cycle}}.$$

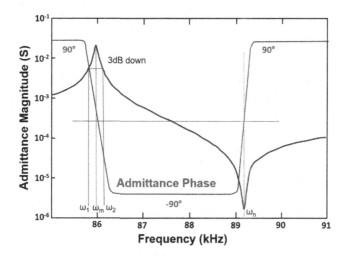

FIGURE 7.12 Admittance magnitude and admittance phase spectra for a rectangular piezo-ceramic plate for a fundamental longitudinal mode (k_{31}) through the transverse piezoelectric effect (d_{31}).

The ratio of elastic stored energy of an oscillator to the power being dissipated electrically provides the mechanical quality factor:

$$Q_m = 2\pi f_r \frac{U_e}{P_d}, \tag{7.4}$$

where U_e is the maximum stored mechanical energy and P_d is the dissipated power, measured in this experiment by P_d = electrically spent energy.[9] Because the compliance of a piezoelectric material exhibits nonlinearity, the maximum kinetic energy is used to define the stored energy term. For a longitudinally vibrating k_{31} plate (see Figure 5.3 or the inserted figure in Figure 7.7 with the plate center $x = 0$), the kinetic energy as a function of displacement, u_x, is

$$U_e = \frac{1}{2} A \int_{\frac{-L}{2}}^{\frac{L}{2}} \rho \left(\frac{\partial u_x}{\partial t} \right)^2 dx \tag{7.5}$$

Using the geometry of the rectangular plate with length L and assuming sinusoidal forcing at a frequency around the fundamental resonance and antiresonance frequencies, as long as k_{31} is not large < 40%, the spatial vibration is approximated as

$$u_x = (x,t) = V_{RMS} \sqrt{2} \sin\left(\frac{\pi x}{L} \right) \sin(2\pi f t) \tag{7.6}$$

The maximum kinetic energy can be calculated as

$$U_e = \frac{1}{2} A \int_{\frac{-L}{2}}^{\frac{L}{2}} \rho \left(V_{RMS} \sqrt{2} \sin\left(\frac{\pi x}{L} \right) \right)^2 dx = V_{RMS}^2 \rho A \frac{L}{2} \tag{7.7}$$

where V_{RMS} is the vibration velocity at the edge of the plate (*rms* value of m/s), and ρ, A, and L are the mass density, cross-section area, and length of the specimen. The relation between mechanical quality factor and real electrical power and mechanical vibration is based on two concepts: (1) at equilibrium the power input is the power lost and (2) the stored mechanical energy can be predicted using the known vibration mode shape, as we did in Eq. (7.7). We can derive the following equation

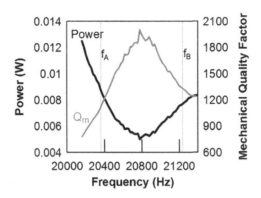

FIGURE 7.13 Mechanical quality factor measured using real electrical power (including the phase lag) for a hard PZT APC 851 k_{31} plate.

from these concepts, which allows the calculation of the mechanical quality factor at any frequency from the real electrical power (P_d) and tip RMS vibration velocity (V_{RMS}) measurements for a longitudinally vibrating piezoelectric resonator (k_t, k_{33}, k_{31}):

$$Q_{m,l} = 2\pi f \frac{\frac{1}{2}\rho V_{RMS}^2}{P_d / Lwb} \tag{7.8}$$

The change in mechanical quality factor was measured for a PZT (APC 851) ceramic plate (k_{31}) under constant vibration condition of 100 mm/s RMS tip vibration velocity (i.e., stored mechanical energy constant). The experimental key in the HiPoCS Version III usage is to determine the phase difference φ precisely to obtain the cos (φ) value. The required power and mechanical quality factor Q_m are shown in Figure 7.13. The quality factor obtained at the resonance is within 2% agreement with results from the impedance spectrum method (3dB-down bandwidth). This technique reveals the behavior of the mechanical quality factor at any frequency between the resonance and the antiresonance frequencies. Moreover, very interestingly, the mechanical quality factor reaches a maximum value between the resonance and the antiresonance frequency, the point of which may suggest the optimum condition for the transducer operation merely from an efficiency viewpoint and also for understanding the behavior of piezoelectric material properties under high-power excitation. The vibration mode at this particular frequency corresponds to the mechanical resonance of the specimen under an external impedance (capacitive) connected, not under short- or open-circuited. A practical application example for a Langevin transducer will be introduced in Chapter 8.

7.2.5 DETERMINATION METHODS OF THE MECHANICAL QUALITY FACTOR

Let us review the precise determination method of the mechanical quality factor from the admittance or impedance spectrum. The admittance spectrum on the k_{31} mode is shown in Figure 7.12. Figure 7.14a and b shows admittance circle, and its magnified vision around the antiresonance frequency. Admittance circle is a plot of *conductance G* in horizontal axis and *susceptance jB* in vertical axis by sweeping the frequency. Though the admittance circle is useful for the Q_A around the resonance peak, the impedance circle (*impedance R* vs. *reactance X*) is better for the Q_B around the antiresonance peak, as illustrated in Figure 7.14c.

7.2.5.1 Resonance/Antiresonance Frequency Definitions

The resonance frequency is defined by ω_R, the most right on the motional admittance circle ω_0 [the intersect (higher G) between the admittance circle and the susceptance $B = 0$, ω_R', is another

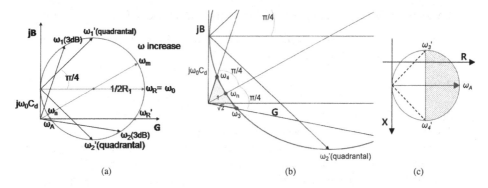

FIGURE 7.14 (a) Admittance circle. (b) Admittance circle magnified around the antiresonance. (c) Impedance circle for the antiresonance peak.

definition as the resonance], while the antiresonance frequency is defined by ω_A, the intersect (lower G) between the admittance circle and the susceptance $B = 0$. The antiresonance frequency is also defined by ωA, the most right on the motional impedance circle in Figure 7.14c. Popularly used maximum and minimum frequencies of the admittance magnitude (i.e., absolute value in the admittance spectrum), ω_m and ω_n, are not exactly the resonance and antiresonance frequencies, precisely speaking. Note the relationship: $\omega_m < \omega_R = \omega_0 < \omega_A < \omega_n$.

7.2.5.2 Mechanical Quality Factor Determination

a. Q_A *Determination*: see Figure 7.14a
- 3 dB-down method around ω_m

$$\left(Q_A^{-1}\right) = (\omega_2 - \omega_1)/\omega_m \tag{7.9}$$

- *Quadrantal frequency method* around ω_R (from the admittance circle)

$$\left(Q_A^{-1}\right)' = (\omega_2' - \omega_1')/\omega_R \tag{7.10}$$

Note ω_1(3dB) < ω_1'(quadrantal) < ω_2(3dB) < ω_2'(quadrantal). The difference between (Q_A^{-1}) and $(Q_A^{-1})'$ can be estimated as

$$\left(Q_A^{-1}\right)\Big/\left(Q_A^{-1}\right)' = 1 + 1/2M^2 \tag{7.11}$$

where $M = |Y_m|/|Y_d| = 1/R_1\omega_R C_d = Q_A K$ and $K = C_1/C_d$ (1/K: capacitance ratio). When we consider $Q_A \approx 1,000$, the deviation of Q_A values among these ways is less than 1 ppm (negligibly small).

b. Q_B *determination*: see Figure 7.14b and c
- 3 dB-*up method* around ω_n

$$\left(Q_B^{-1}\right) = (\omega_4 - \omega_3)/\omega_n \tag{7.12}$$

- *Quadrantal frequency method* around ω_A (from the impedance circle)

$$\left(Q_B^{-1}\right)' = (\omega_4' - \omega_3')/\omega_A \tag{7.13}$$

In summary in Figure 7.14, admittance circle is useful for the Q_A around the resonance peak, while impedance circle is better for the Q_B around the antiresonance peak.

7.2.6 Determination of the Three Losses from the Mechanical Quality Factors

A method for determining the piezoelectric loss is summarized for a piezoelectric k_{31} mode plate sample here[7] (refer to Section 4.3.3 and a review paper Ref. [10] for other modes):

1. Obtain $\tan \delta'$ from an impedance analyzer or a capacitance meter at a frequency away (lower range) from the resonance/antiresonance range. Typically at 10 kHz or lower.
2. Obtain the following parameters experimentally from an admittance/impedance spectrum around the resonance (A-type) and antiresonance (B-type) range (3 dB bandwidth or quadrantal frequency method): ω_a, ω_b, Q_A, Q_B, and the normalized frequency $\Omega_b = \omega_b L / 2v$.
3. Obtain $\tan \phi'$ from the inverse value of Q_A (quality factor at the resonance) in the k_{31} mode.
4. Calculate electromechanical coupling factor k_{31} from the ω_a and ω_b with the IEEE Standard equation in the k_{31} mode:

$$\frac{k_{31}^2}{1-k_{31}^2} = \frac{\pi}{2}\frac{\omega_b}{\omega_a}\tan\left[\frac{\pi(\omega_b-\omega_a)}{2\omega_a}\right];\qquad(7.14)$$

5. Finally obtain $\tan \theta'$ by the following equation in the k_{31} mode:

$$\tan\theta' = \frac{\tan\delta'+\tan\phi'}{2} + \frac{1}{4}\left(\frac{1}{Q_A}-\frac{1}{Q_B}\right)\left[1+\left(\frac{1}{k_{31}}-k_{31}\right)^2\Omega_b^2\right]\qquad(7.15)$$

As long as we obtain ω_a, ω_b, Q_A, and Q_B even from the Burst Drive method introduced in Section 7.3, the above procedure can also be used. A general problem in determining accurate piezoelectric and loss parameters is found in the k_{33} rod specimen, in which relatively large electric field leak is anticipated according to the aspect ratio (rod length/width). This problem will be discussed in Section 7.4.

Example Problem 7.1

The electromechanical coupling factor k can be obtained from the resonance and antiresonance frequencies basically. However, there are several formulae for this calculation, depending on the approximation level, which directly reflects to the accuracy of the piezoelectric loss $\tan \theta$, as indicated in Section 7.2.6. Knowing the experimental result on the impedance spectrum for a PZT 5 k_{33} rod specimen shown in Figure 7.15: $f_A = 1.3\, f_R$, calculate the k_{33} value of this PZT 5 H by using the following three formulae:

$$k_{33}^2/(1-k_{33}^2) = (\pi^2/4)(\Delta f/f_R),\qquad(\text{P7.1.1})$$

$$k_{33}^2/(1-k_{33}^2) = (\pi^2/8)(f_A^2-f_R^2)/f_R^2,\qquad(\text{P7.1.2})$$

$$k_{33}^2 = (\pi/2)(f_R/f_A)\tan\left[(\pi/2)(\Delta f/f_A)\right],\qquad(\text{P7.1.3})$$

where $\Delta f = f_A - f_R$.

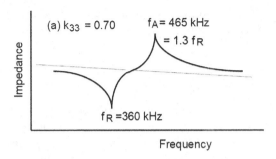

FIGURE 7.15 Impedance spectrum for a k_{33} PZT-5 rod specimen.

Solution

1. Similar to the most rough approximation for small k_{31}, $k_{31}^2/(1-k_{31}^2)=(\pi^2/4)(\Delta f/f_R)$, we use the following approximation for k_{33} mode:

$$k_{33}^2/(1-k_{33}^2)=(\pi^2/4)(\Delta f/f_R)=(\pi^2/4)\times 0.3=0.749$$

Thus, k_{33} = 0.654 (underestimation).

2. When we use a better approximation (IRE Standard, "Marutake" *Approximation*):

$$k_{33}^2/(1-k_{33}^2)=(\pi^2/8)(f_A^2-f_R^2)/f_R^2=0.851,$$

we obtain k_{33} = 0.678, higher value than the above (1) and close to the accurate (3) (slightly over).

3. When we use the accurate formula:

$$k_{33}^2=(\pi/2)(f_R/f_A)\tan\left[(\pi/2)(\Delta f/f_A)\right]=0.458,$$

we can obtain k_{33} = 0.677. Note that $k_{33}^2/(1-k_{33}^2)=-(2\pi f_A L/2v)\cot(2\pi f_A L/2v)$ is another accurate formula for the k_{33} mode. Different from the k_{31} mode formula $\left(k_{31}^2/(1-k_{31}^2)=(\pi/2)(f_A/f_R)\tan\left[(\pi/2)(\Delta f/f_R)\right]\right)$, in the k_{33} mode formula, the denominator is f_A (not f_R) because the antiresonance is the primary vibration mode [the resonance is the primary vibration mode in the k_{31} case].

The reader should understand that the k value deviates from the accurate one according to the approximation formula.

7.3 LOSS MEASURING TECHNIQUE III – TRANSIENT/BURST DRIVE METHOD

When we use a constant voltage method on the admittance spectrum measurement in a PZT k_{31} specimen, we observe a significant heat generation just around the resonance peak range, while under a constant current method, a significant heat is generated just around the antiresonance peak range. As demonstrated in Figure 6.6 under the same vibration velocity, the temperature rise at the antiresonance is lower than that at the resonance, which is reflected from the Q_m change in Figure 7.13. Even we adopt the vibration velocity constant method (i.e., the mechanical stored energy constant), though the spectrum does not show the skewed shape, heat generation occurs inevitably during the frequency sweep process around the resonance and antiresonance range; the temperature around the nodal point decreases gradually after the resonance, approaches to the minimum in-between, then starts to increase to the antiresonance frequency.

Anyhow, the temperature modulation during the measurement process is inevitable for continuous impedance/admittance spectrum measurement under high-vibration levels, so that a sort of calibration may be required in determining the physical parameters at room temperature. Transient and burst drive methods have been introduced to solve this temperature rise problem during the measurement.

7.3.1 PULSE DRIVE METHOD

The pulse drive method is a simple method for measuring high-voltage piezoelectric characteristics, developed in the ICAT/Penn State in the early 1990s. By applying a step electric field to a piezoelectric sample, the transient vibration displacement (i.e., "ringing") corresponding to the desired mode (extensional, bending etc.) is measured under a short-circuit condition. See Figure 7.16. Because the equipment cost could be minimized (in comparison with a commercialized expensive "impedance analyzer"), this method was used previously. Notice that one-time high-voltage application (though multiple time measurements are accumulated with a certain interval technically) and the following short period (~m s) ringing vibration do not generate measurable temperature rise experimentally ($< 0.2°C$). The resonance period, stabilized displacement, and damping constant are obtained experimentally, from which the elastic compliance, piezoelectric constant, mechanical quality factor, and electromechanical coupling factor can be calculated. Using a rectangular k_{31} piezoelectric ceramic plate (length: L; width: w; and thickness: b; poled along the thickness, like the inserted figure in Figure 7.7), we explain how to determine the electromechanical coupling parameters k_{31}, d_{31}, and Q_m. The density ρ, permittivity $\varepsilon_{33}{}^X$, and size (L, w, b) of the ceramic plate must be known prior to the experiments.

1. From the stabilized displacement D_s, we obtain the piezoelectric coefficient d_{31}:

$$D_s = d_{31}E\,L \tag{7.16}$$

2. From the ringing period, we obtain the elastic compliance $s_{11}{}^E$:

$$T_0 = 2L/v_{11}^E = 2L\left(\rho s_{11}^E\right)^{1/2} \tag{7.17}$$

3. From the damping constant τ, which is determined by the time interval to decrease the displacement amplitude by $1/e$, we obtain the mechanical quality factor Q_m:

$$Q_m = (1/2)\omega_0 \tau \tag{7.18}$$

   where the resonance angular frequency $\omega_0 = 2\pi/T_0$.
4. From the piezoelectric coefficient d_{31}, elastic compliance $s_{11}{}^E$, and permittivity ε_3, we obtain the electromechanical coupling factor k_{31}:

$$k_{31} = d_{31}\Big/\left(\varepsilon_0\varepsilon_3 s_{11}^E\right)^{1/2} \tag{7.19}$$

On the other hand, the antiresonance Q_m can be obtained as follows: by removing a large electric field suddenly from a piezoelectric sample, and keeping the open-circuit immediately after the first vibration overshoot, the transient vibration displacement corresponding to the antiresonance mode. The bias electric field (and the vibration velocity) dependence of piezoelectricity can be measured. One drawback is the vibration velocity level: due to just one-time high-voltage application, induced displacement or strain level is limited. Thus, we started to use an amplified initial displacement, explained in Section 7.3.2.

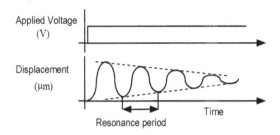

FIGURE 7.16 Pulse drive technique for measuring the electromechanical parameters.

7.3.2 BURST MODE METHOD

7.3.2.1 Background of Burst Mode Method

Pure vibration amplitude dependence of the Q_M cannot be discussed without removing the temperature influence. In order to eliminate the temperature rise effect, the burst method fits very well. The "burst drive" stands for driving electrically a piezoelectric specimen at its resonance frequency (or antiresonance frequency) for a short period. Then electrically shutting down from the driving power supply, we monitor the vibration ringing via displacement/vibration velocity or voltage/current from the specimen; that is, a sort of mechanical excitation method. In this method, the internally generated heat is close to none (less than 0.2°C, due to a short exciting period), and observations would be a direct function of the ambient temperature on piezoelectric properties. If you understand that the initial burst period is merely to excite a large mechanical vibration (vibration velocity large enough for the measurement), this method is actually a "mechanical excitation" technique, introduced in Figure 7.1, so that we can measure the piezo-material's performance change under a different constraint condition (electrically short-circuit or open-circuit). The burst drive method was reported systematically by Umeda et al. for determining the equivalent circuit parameters of a piezoelectric transducer.[11] It was thereafter adopted to measure the properties of piezoelectric ceramic samples.[12,13]

We introduce *a short-circuit or an open-circuit* system which functions immediately after the resonance burst drive (only for 1 m s) without generating heat (< 0.2°C), which generates the ring-down of vibration amplitude for the resonance and antiresonance modes, respectively. Figure 7.17 illustrates the experimental results for the short- and open-circuits.[14] In the short-circuit, the *current and vibration velocity* are proportional, the decay rate provides the mechanical quality factor Q_A (resonance), while in the open-circuit, the *voltage and end displacement* are proportional, the decay rate gives the Q_B (antiresonance). We may assume D-constant condition between the electrode gap with no charge migration for compensating the depolization field during this short (1 ms) measuring time. Note that resonance to antiresonance frequency jumps in Figure 7.17b owing to a sudden electrical open-circuit, from the initial resonance excitation. This is a good demonstration to visually show the reader that the antiresonance is a "naturally excited" mechanical resonance mode. We learned the vibration mode difference in Section 4.3.1 and Figure 4.6, where both resonance and antiresonance are mechanical resonance modes, and the resonance frequencies f_A and f_B are expressed as $f_A = \left(v_{11}^E/2L\right)$ and $f_B = \left(v_{11}^E/2L\right)\left(1+\left(\frac{4}{\pi^2}\right)k_{31}^2\right)$ (when k_{31} is not large). Though the sound velocity and the elastic compliance $s_{11}{}^E$ are E-constant parameters in both cases, because of the electric constraint difference (short- or open-circuit), the vibration modulation is introduced due to the D-constant condition only along the electrode gap 1D direction. If the specimen is not covered by electrode, we should observe the sound velocity $v_{11}^D = v_{11}^E\left(1-k_{31}^2\right)^{-1/2}$ and elastic compliance under D-constant condition, $s_{11}{}^D$. However, it is difficult by using the conventional IEEE Standard specimen configurations. We will introduce new piezo-specimen configurations to measure both

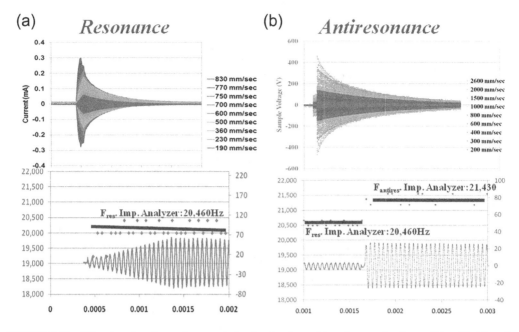

FIGURE 7.17 Vibration ring-down characteristics for the short- (a) and open-circuit condition (b). Note the sudden change of the ringing frequency for the open-circuit.

E-constant and D-constant parameters with or without electrode on the piezoelectric specimens in Section 7.3.3.

Regarding the modification of HiPoCS Ver. III, a short-circuit and open-circuit relays should be integrated. With the new blocking circuit added to our burst method characterization with the HiPoCS, we can monitor the mechanical quality factor drop also in the antiresonance.[14]

7.3.2.2 Force Factor and Voltage Factor

We derive the *force factor* (A_{31}) and *voltage factor* (B_{31}) comprehensively, first, in terms of material properties and sample geometry from the constitutive equations for the k_{31} piezoelectric sample.[14] The force factor is the relationship between *current and vibration* velocity in resonance, and it is related to the piezoelectric stress coefficient. This force factor already appeared in Chapter 5 as the transformer step up ratio Φ. The force factor analysis for the k_{31} has been presented previously by Takahashi,[12] but its explicit derivation has not. While, the *voltage factor* (B_{31}) is the relationship between open-circuit *voltage and displacement* in antiresonance, related with the converse piezoelectric coefficient (h_{31}^*), which has not been applied nor analyzed in bulk piezoceramics in the antiresonance condition so far.

The derivations of the *force factor* (A_{31}) and *voltage factor* (B_{31}) assume a k_{31} mode PZT plate specimen with thickness/electrode gap $b \ll$ width $w \ll$ plate length L ($L = 80\,\mathrm{mm}$ in this case), fully electroded, and poled along the thickness. Another assumption is that most of the vibration occurs in the length direction, traditionally corresponding to the x direction ($x = 0$ is at the plate center in this case). In general, the mode shape of a piezoelectric resonator with stress-free boundary conditions, undergoing vibration in 1D, with losses, and having finite displacement can be described as

$$u(x,t) = u_0 f(x) \sin \omega t, \tag{7.20}$$

where $f(x)$ is a function symmetric about the origin normalized to the displacement at the ends of the piezoelectric resonator, where $f(0) = 0$. Then, according to the fundamental theorem of calculus in terms of strain ($\partial u / \partial x$), we obtain:

$$\int_{-L/2}^{L/2} \frac{\partial u}{\partial x} dx = u(L/2,t) - u(-L/2,t) = 2u_0 \sin \omega t. \tag{7.21}$$

The constitutive equation describing the electric displacement of a piezoelectric k_{31} resonator is

$$D_3(t) = e_{31}^* \frac{\partial u}{\partial x} + \varepsilon_{33}^{x1} \varepsilon_0 E_3(t). \tag{7.22}$$

where e_{31}^* and ε_{33}^{x1} are the effective piezoelectric stress coeffcient defined as $e_{31}^* = d_{31}/\varepsilon_{33}^E$ and the longitudinally (1D) clamped relative permittivity, respectively. The actual piezoelectric stress coefficient e_{31} (non-star) is defined from the two- and three-directional clamp condition (free in our case).

- For the electrical boundary condition of zero electric potential case (short-circuit), the external field is equal to zero. Therefore, $D_3(t) = e_{31}^* \frac{\partial u}{\partial x}$ and $\dot{D}_3 = e_{31}^* \frac{\partial^2 u}{\partial x \partial t}$. Now, the current can be written as

$$i(t) = \int_{A_e} \dot{D}_3 \, dA_e = w \int_{-L/2}^{L/2} \dot{D}_3 \, dx, \text{ and } i_0 = 2e_{31}^* u_0 w\omega_A = 2e_{31}^* wv_0. \tag{7.23}$$

The last transformation uses vibration velocity v_0 at the plate edge ($x=\pm L/2$), given $\omega u_0 = v_0$ for sinusoidal time varying displacement. Thus, the *force factor* (A_{31}), defined as the ratio between short-circuit current and edge vibration velocity can then be written as

$$A_{31} = \frac{i_0}{v_0} = 2e_{31}^* w = 2w \frac{d_{31}}{s_{11}^E}. \tag{7.24}$$

- For open-circuit conditions, $D_3 = 0$, so the constitutive equation described in Eq. (7.22) can be written as

$$E_3(x,t) = \frac{e_{31}^*}{\varepsilon_{33}^{x1} \varepsilon_0} \frac{\partial u}{\partial x}. \tag{7.25}$$

Assuming the variation of strain in thickness is negligible, the electric field across the thickness is uniform. Therefore, the $E_3(x, t) = -V(t)/b$. Integrating across the length of the resonator

$$\int_{-L/2}^{L/2} E_3(x,t) \, dx = -\int_{-L/2}^{L/2} (V(t)/b) \, dx = \frac{e_{31}^*}{\varepsilon_{33}^{x1} \varepsilon_0} \int_{-L/2}^{L/2} \frac{\partial u}{\partial x} \, dx.$$

This equation can be rewritten using Eq. (7.21), assuming free natural vibration at the antiresonance frequency in open-circuit conditions,

$$\frac{LV_0}{b} = \frac{e_{31}^*}{\varepsilon_{33}^{x1} \varepsilon_0} 2u_0, \text{ and } V_0 = \frac{2be_{31}^*}{L\varepsilon_{33}^{x1} \varepsilon_0} u_0.$$

Thus, the *voltage factor* (B_{31}), the ratio between open circuit voltage and displacement, can be written as

$$B_{31} = \frac{V_0}{u_0} = \frac{2b}{L} \frac{e_{31}^*}{\varepsilon_{33}^{x1} \varepsilon_0} = \frac{2b}{L} \frac{g_{31}}{s_{11}^D} = \frac{2b}{L} h_{31}^*. \tag{7.26}$$

Note again that effective "*" parameters, e_{31}^*, h_{31}^*, and ε_{33}^{x1} are not exactly the same as the non-"*" parameters, because of the different constraint conditions.

7.3.2.3 Piezoelectric Parameter Determination Procedure

By applying the burst mode at resonance (short-circuit) or antiresonance (open-circuit) conditions, the force factor (A_{31}) or the voltage factor (B_{31}) can be directly obtained from the ratio between short-circuit current and edge vibration velocity or from the ratio between open-circuit voltage and displacement. The mechanical quality factors (or elastic loss factors) and the real properties of the material can also be measured. For a damped linear system oscillating at its natural frequency, the quality factor can be described using the relative rate of decay of vibration amplitude. In general,

$$Q = \frac{2\pi f}{2\ln\left(\dfrac{v_1}{v_2}\right)\Big/(t_2 - t_1)} \tag{7.27}$$

where v_1 and v_2 correspond to the velocity (or voltage) at time of t_1 and t_2, respectively. This equation is valid for both resonance and antiresonance modes. At resonance (short-circuit), the current is proportional to the vibration velocity; therefore, its decay can be used. Similarly, the voltage decay can be used at antiresonance to determine the quality factor at antiresonance, which is supported by the above formula derivation.

The first A-type resonance frequency f_A in the k_{31} resonator corresponds to the s_{11}^E according to the equation

$$s_{11}^E = \frac{1}{\left(2L\, f_A\right)^2 \rho}. \tag{7.28}$$

By utilizing the measurement of the *force factor*, the piezoelectric charge coefficient can be computed as

$$d_{31} = A_{31} s_{11}^E / 2w. \tag{7.29}$$

A more common approach to calculate this coefficient, frequently used in electrical resonance spectroscopy, is as follows: The off-resonance permittivity and resonance elastic compliance can be used to separate the piezoelectric charge coefficient from the coupling coefficient measured from the relative frequency difference between resonance and antiresonance $\left[\dfrac{k_{31}^2}{1 - k_{31}^2} = -\dfrac{\pi}{2}\dfrac{\omega_A}{\omega_R}\Big/\tan\left(\dfrac{\pi}{2}\dfrac{\omega_A}{\omega_R}\right)\right]$, which can be expressed mathematically as follows:

$$d_{31} = k_{31}\sqrt{s_{11}^E \varepsilon_{33}^X \varepsilon_0}. \tag{7.30}$$

However, this approach assumes that the ε_{33}^X does not change in the resonance condition from the off-resonance condition. The calculation of d_{31} using the force factor does not make this assumption, so it is expected to be more accurate.

By using piezoelectric stress coefficient (e_{31}^*) calculated at resonance (from the force factor) and the converse piezoelectric constant (h_{31}^*) calculated at antiresonance, the longitudinally clamped permittivity ε_{33}^{x1} can be calculated in resonance/antiresonance conditions directly by taking the ratio (*force factor A_{31}/voltage factor B_{31}*) in Eq. (7.31). Then, ε_{33}^X can be calculated using the k_{31}^2. Permittivity has never been measured directly in resonance conditions according to the author's knowledge. Though Takahashi et al. have reported permittivity in resonance conditions,[12] they assumed that only the motional capacitance changes and the clamped capacitance in resonance does not change.

$$\varepsilon_0 \varepsilon_{33}^X \left(1 - k_{31}^2\right) = \varepsilon_0 \varepsilon_{33}^{x_1} = \frac{\overset{*}{e}_{31}}{\overset{*}{h}_{31}} = \frac{A_{31}}{B_{31}} \frac{b}{Lw} \tag{7.31}$$

We improved the above methodology further to measure the "dielectric loss", in addition to the permittivity, of a piezoelectric at the resonance frequency range based on the burst excitation method.[15] Daneshpajooh et al. investigated the "force" and "voltage" factors under the short- and open-circuit conditions of the burst method, in terms of the ratio of the ring-down current and voltage with the plate end vibration velocity and the displacement, and their phase lags. The force factor (A_{31}) and voltage factor (B_{31}) can be described as follows including the loss factors:

$$A_{31} = \frac{i_0}{v_0} = 2w \frac{\overset{*}{d}_{31}}{s_{11}^{E*}} = 2w \frac{d_{31}}{s_{11}^E} \left(1 + j\left(\tan\phi_{11}' - \tan\theta_{31}'\right)\right) \tag{7.32}$$

$$B_{31} = \frac{V_0}{u_0} = \frac{2b}{L} \frac{\overset{*}{d}_{31}}{s_{11}^{E*} \varepsilon_{33}^{x_1*} \varepsilon_0} = \frac{2b}{L} \frac{d_{31}}{s_{11}^E \varepsilon_0 \varepsilon_{33}^{x_1}} \left(1 + j\left(\tan\phi_{11}' + \tan\delta_{33}''' - \tan\theta_{31}'\right)\right) \tag{7.33}$$

The permittivity and dielectric loss can be expressed:

$$\varepsilon_0 \varepsilon_{33}^X \left(1 - k^2\right) = \varepsilon_0 \varepsilon_{33}^{x_1} = \text{Re}\left\{\frac{A_{31}}{B_{31}}\right\} \frac{b}{Lw} \tag{7.34}$$

$$\tan\delta_{33}''' = \text{Im}\left\{\frac{A_{31}}{B_{31}}\right\} \times \frac{Lw}{b} \varepsilon_0 \varepsilon_{33}^{x_1} \tag{7.35}$$

7.3.2.4　Vibration Velocity Dependence of Piezoelectric Parameters by Burst Drive Method

The experimental results of the burst drive method are shown in Figures 7.18 for soft PZT (PIC 184, PI Ceramics, Germany) and hard PZT (PIC 144).[14] Using the decay of vibration at resonance and antiresonance, the quality factors Q_A and Q_B were calculated. Each data point used amplitude data from two vibration measurements; therefore, the scale was readjusted as an average of the vibration velocity. Figure 7.18a shows the results; a log–log plot is used in able to more easily distinguish and compare the trends between the two compositions. Hard PZT shows stable characteristics of the quality factor, until about 150 mm/s RMS, after which a sharp degradation in the quality factors occurs. Soft PZT, however, shows an immediate decrease in its quality factors. Q_B was larger than Q_A for both the materials. The elastic loss $\tan\phi_{11}'$ is obtain directly from Q_A, and Q_B provides the mechanical coupling loss $\left(2\tan\theta_{31}' - \tan\delta_{33}' - \tan\phi_{11}'\right)$.

Regarding the resonance characterization, the current and vibration data were used to calculate the *force factor* (A_{31}) and the piezoelectric stress constant. Using the resonance frequency, the elastic compliance was calculated, and hence the piezoelectric charge coefficient d_{31} could be calculated as well. The resonance characterization for a soft PZT k_{31} specimen is shown in Figure 7.18b, where the properties (e_{31}^*, s_{11}^E, d_{31}) increase linearly with vibration velocity. On the contrary, by utilizing the displacement and open-circuit voltage at the antiresonance mode, the *voltage factor* (B_{31}) and the converse piezoelectric coefficient (h_{31}^*) were calculated (see Figure 7.18c). The coupling factor k_{31} can also be calculated using the difference between the resonance and antiresonance frequencies ($\Delta f = f_B - f_A$). The coupling factor increases with the vibration velocity slightly;

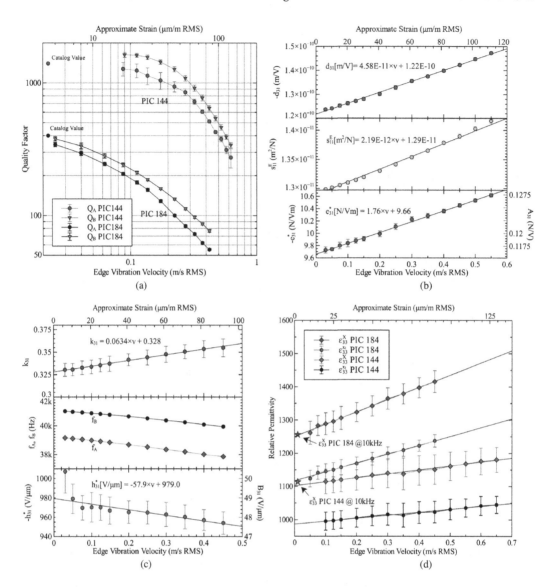

FIGURE 7.18 (a) Change in the mechanical quality factors Q_A and Q_B for hard PZT (PIC 144) and soft PZT (PIC 184). (b) Resonance characterization (force factor) of a soft PZT (PIC 184) k_{31} specimen. (c) Antiresonance analysis (voltage factor) and electromechanical coupling factor. (d) Change in dielectric permittivity with vibration velocity for hard (PIC 144) and soft PZT (PIC 184).[14]

h_{31}^* (inverse piezoelectric constant) decreases with increasing vibration velocity, contrary to the behavior of the other properties. Since f_A and f_B decrease with the vibration velocity by keeping Δf almost constant, the electromechanical coupling factor k_{31} has a much smaller dependence on vibration velocity than the other properties, that is, those determined at the resonance.

Traditional methods cannot measure the permittivity under resonance conditions. This is because the large vibration does not allow the dielectric response to exhibit a unique and distinct feature which can be characterized in order to compute the permittivity. This is also true for the dielectric loss. Therefore, researchers have used one of the two approaches to estimate the permittivity in high-power conditions: The first approach is to assume the permittivity measured at off-resonance applies to resonance conditions. This approach is problematic because the stress conditions and the frequency are different at resonance, and therefore, the property is expected to change, similar to other properties. The second approach is to assume a perturbation of the off-resonance frequency using the variation in the motional capacitance, which is proportional to d_{31}^2/s_{11}^E.

Using the force factor and the voltage factor, the permittivity under constant strain $\varepsilon_{33}^{x_1}$ can be calculated directly [Eq. (7.31)]. Using the coupling factor k_{31}, ε_{33}^X can be calculated. The permittivity vs. vibration velocity can be seen in Figure 7.18d. The off-resonance permittivity measured for the samples is in good agreement with the low vibration velocity permittivity measured through the burst technique. The off-resonance permittivity is represented as a star symbol on the figure. The clamped and free permittivities are both changing with increasing vibration velocity. The permittivity of soft PZT PIC 184 is larger than that of hard PZT PIC 144, and this is expected because a "soft" PZT exhibits larger off-resonance permittivity than a "hard" PZT. From the low vibration state to the high one, the permittivity of both compositions increases. However, the increase in soft PZT is larger, demonstrating that its properties have a larger dependence on vibration conditions. Unlike an expectation, the result shown in this study demonstrates that a majority of the change seen in the free permittivity can actually be attributed to the clamped permittivity change, because the electromechanical coupling factor k_{31} change is rather small.

Daneshpajooh et al. investigated the "force" and "voltage" factors under the short- and open-circuit conditions of the burst method, in particular, focusing on their phase lags (Eq. (7.35)). We provide the obtained material properties, including loss parameters from their phase lags, in particular, dielectric properties at the resonance frequency range. Figure 7.19 shows the vibration velocity dependence of three loss factors; (a) piezoelectric constant (d_{31}) and piezoelectric loss $(\tan\theta_{31}')$, (b) elastic compliance (s_{11}^E), and elastic loss $(\tan\phi_{11}')$, (c) and (d) free/clamped dielectric constants $(\varepsilon_{33}^X, \varepsilon_{33}^{LC})$ and intensive/extensive dielectric loss $(\tan\delta_{33}', \tan\delta_{33}''')$, respectively.[15] Though we

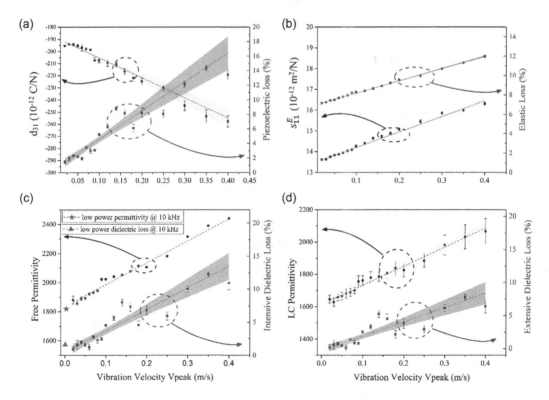

FIGURE 7.19 Vibration-level dependence of (a) piezoelectric constant (d_{31}) and piezoelectric loss $(\tan\theta_{31}')$, (b) elastic compliance (s_{11}^E) and elastic loss $(\tan\phi_{11}')$, and (c) and (d) free/clamped dielectric constants $(\varepsilon_{33}^X, \varepsilon_{33}^{LC})$ and intensive/extensive dielectric loss $(\tan\delta_{33}', \tan\delta_{33}''')$, respectively.[15]

used "extensive" dielectric loss, precisely speaking, this means "longitudinally clamped" dielelectric loss. All piezoelectric, elastic, and dielectric losses increase with vibration velocity, as well as $|d_{31}|$, s_{11}^E, ε_{33}^X, and ε_{33}^{LC} do. Note again that even the clamped permittivity and "extensive" dielectric loss change significantly with the vibration velocity!

Example Problem 7.2

After the burst drive excitation on a piezoelectric specimen with its resonance frequency ω_0, we observe the vibration displacement ring-down phenomenon. The damping time constant τ is determined by the time interval to decrease the displacement amplitude by $1/e$. Provide the relation of the mechanical quality factor Q_m with the damping factor τ and the resonance frequency ω_0.

Hint

$$Q_m = (1/2)\omega_0\tau, \tag{P7.2.1}$$

Solution

We start from the dynamic equation of the mass–spring–viscous damper model in Section 3.2.6, without external force for the transient response.

$$m\ddot{u} + \xi \cdot \dot{u} + cu = 0 \tag{P7.2.2}$$

Taking the following notations,

$$\omega_0 = \sqrt{c/m} \ \left(\text{base resonance frequency for zero damping}\right) \tag{P7.2.3}$$

$$\zeta = \xi/2m\omega_0 \ \left(\text{normalized damping factor (no dimension)}\right) \tag{P7.2.4}$$

we obtain the normalized equation,

$$\ddot{u} + 2\zeta\omega_0 \cdot \dot{u} + \omega_0^2 u = 0. \tag{P7.2.5}$$

The under-damping (small damping, $\zeta \ll 1$) solution is given by

$$u(t) = u_0 \left[\exp\left(-\zeta\omega_0 t\right)\cos\left(\sqrt{1-\zeta^2}\,\omega_0 t\right) + \frac{\zeta}{\sqrt{1-\zeta^2}} \exp\left(-\zeta\omega_0 t\right)\sin\left(\sqrt{1-\zeta^2}\,\omega_0 t\right) \right]. \tag{P7.2.6}$$

Since the cosine wave decays with time in $\exp\left(-\zeta\omega_0 t\right)$ basically, the damping time constant τ is expressed as

$$\tau = 1/\zeta\omega_0 \tag{P7.2.7}$$

On the other hand, Q_m is obtained under steady-state oscillation under a harmonic force. As we analyzed in Section 3.3, since $\dfrac{u_0\left(\omega_0\right)}{u_0(\omega=0)} = \dfrac{1}{2\zeta\dfrac{\omega}{\omega_0}}$, by putting $\dfrac{u_0}{u_0(\omega=0)} = \dfrac{1}{\sqrt{\left[1-\left(\dfrac{\omega}{\omega_0}\right)^2\right]^2 + \left(2\zeta\dfrac{\omega}{\omega_0}\right)^2}} =$

$\dfrac{1}{\sqrt{2}}\dfrac{1}{2\zeta\dfrac{\omega}{\omega_0}}$, the 3 dB-down or $1/\sqrt{2}$ frequencies can be obtained from

$$1-\left(\frac{\omega}{\omega_0}\right)^2 = \pm 2\zeta\frac{\omega}{\omega_0}, \text{depending on } \frac{\omega}{\omega_0} < 1 \text{ or } \frac{\omega}{\omega_0} > 1 \tag{P7.2.8}$$

Thus, we obtain two roots: $\frac{\omega}{\omega_0} = \sqrt{1+\zeta^2} - \zeta$ or $\frac{\omega}{\omega_0} = \sqrt{1+\zeta^2} + \zeta$ for $\frac{\omega}{\omega_0} < 1$ or $\frac{\omega}{\omega_0} > 1$, respectively. Then, $\frac{\Delta\omega}{\omega_0} = 2\zeta$ or the mechanical quality factor Q_m is expressed by

$$Q_m = \frac{\omega_0}{\Delta\omega} = 1/2\zeta \tag{P7.2.9}$$

From Eqs. (P7.2.7) and (P7.2.9), we finally obtain

$$Q_m = (1/2)\omega_0\tau. \tag{P7.2.10}$$

7.4 LOSS MEASURING TECHNIQUE – SAMPLE ELECTRODE CONFIGURATIONS

In order to obtain the material's "extensive" and "intensive" losses, we need to use various sample configurations. Using a k_{31} plate sample, we can measure s_{11}^E and $\tan\phi_{11}'$; while in a k_{33} rod sample, we can measure s_{33}^D and $\tan\phi_{33}'''$. However, we cannot measure extensive $\tan\phi_{11}$ or intensive $\tan\phi_{33}'$ experimentally, using the IEEE electrical-excitation specimen configurations, but evaluate by using k_{31} or k_{33} values. Though $\tan\phi_{11}$ or $\tan\phi_{33}'$ can be evaluated with the $[K]$ matrix-like transformation in Eq. (4.39) theoretically, the experimental errors expand dramatically. Thus, the ICAT/Penn State group proposed a new *mechanical-excitation methodology* (similar to a *composite bar structure*) rather than a conventional electrical excitation, to measure both *intensive* and *extensive elastic losses* in similar configuration specimens just with different electrode designs. We introduce below the partial electrode (PE) configuration method applications for the k_{31} and k_{33} modes.

7.4.1 EXTENSIVE LOSS MEASUREMENT IN THE K_{31} MODE

Majzoubi et al. proposed partial electrode (PE) configurations to measure extensive $\tan\phi_{11}$ directly in the k_{31} plate and examine its self-consistency with past indirect methods.[16] For this purpose, rectangular piezoelectric plates with various electrode patterns, including full electrode (FE), partial electrode with short- and open-circuit between sided electrodes (PE-short, and PE-open), and non-electrode (NE), are designed and prepared. The center portion (10%) is for the actuator to mechanically excite both the side major portions. Their schematic views are shown in Figure 7.20a. The measurements for the admittance spectrum (monitored via the center actuator portion) of these samples (hard PZT, PIC 144, $L = 40\,\text{mm}$, PI Ceramics, Germany) are summarized in Figure 7.20b.

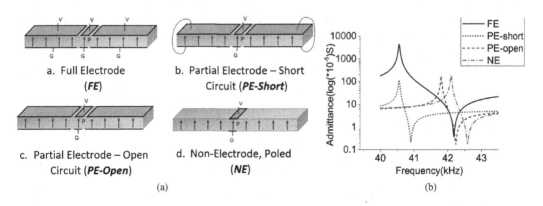

FIGURE 7.20 (a) Schematic view of different electrode patterns on a rectangular k_{31} PZT plate. (b) Admittance spectrum of FE, PE-short, PE-open, and NE samples.[16]

FE sample behaves as a typical IEEE Standard rectangular PZT k_{31} plate. For NE piezo-electric, it behaves pure D-constant performance (just has a small portion electrode for excitation), and its losses would mostly correspond to the extensive ones, which we could measure directly for the first time (refer to Figure 7.1). Higher resonance and antiresonance frequencies are observed, in comparison to FE sample, because $v_{11}{}^D > v_{11}{}^E$. PE-short, corresponds to the pure E-constant situation and the resonance mode. As shown in Figure 7.20b, the resonance frequency of this sample is almost the same as the FE one, and the antiresonance frequency occurs in much lower frequency. In contrast, the antiresonance frequency of PE-open is in the same range as FE sample, and the resonance frequency occurs in much higher frequency. This case attributed to the antiresonance mode. Since we consider the k_{31} mode and also there is voltage distribution on the bare surface, or no voltage distribution on the electrode surface, the mechanical resonance and antiresonance would happen between the pure D- or E-constant condition, which are actually the NE and PE-short cases.

The coupling factor of hard PZT PIC 144 derived from resonance and antiresonance frequencies of FE sample is $k_{31} = 0.31$, which is comparable with the one derived from the equation of $s_{11}^D = s_{11}^E \left(1 - k_{31}^2\right)$. In the aforementioned equation, s_{11}^D and s_{11}^E are calculated from antiresonance frequency of NE sample $\left(s_{11}^D = 1.06 \times 10^{-11} \text{ m}^2/\text{N}\right)$ and resonance frequency of PE-short sample $\left(s_{11}^E = 1.17 \times 10^{-11} \text{ m}^2/\text{N}\right)$. The coupling factor can be calculated 0.32 for this case. This good agreement of these k_{31} values verifies the feasibility of our *Partial-Electrode* sample configurations for measuring the piezoelectric performances under various electric boundary conditions. The mechanical quality factors for both resonance and antiresonance modes, measured from the impedance curve, are shown in Table 7.1.[16] We observe that Q_A and Q_B differ significantly in the FE sample, but the difference is much less in the partial electrodes, as expected. The PE-open and NE samples have higher mechanical quality factors (lower mechanical losses), in comparison to short-circuit one.

Regarding the loss factors, initially from the FE sample data on Q_A and Q_B, we determined the intensive losses as:

$$\tan \delta'_{33} = 2.3 \times 10^{-3}, \tan \phi'_{11} = 7.194 \times 10^{-4}, \tan \theta'_{31} = 2.2 \times 10^{-3}.$$

Using the [K]-matrix, we obtained the extensive losses:

$$\tan \delta_{33} = 2.1 \times 10^{-3}, \tan \phi_{11} = 5.76 \times 10^{-4}, \tan \theta_{31} = 6.65 \times 10^{-4}.$$

On the other hand, the extensive losses were directly measured from the Q_B on the NE sample as $\tan \phi_{11} = 5.92 \times 10^{-4}$. The slight difference in obtained values between the FE and NE methods was most likely to be originated both from neglecting the 10% electrode and also error propagation in indirect calculations.

TABLE 7.1

Resonance and Antiresonance Frequencies and Mechanical Quality Factors[16]

		FE	PE-short	PE-open	NE
f_r	[kHz]	40.69	40.64	[41.93]	[42.43]
f_{ar}	[kHz]	42.31	[40.97]	42.29	42.85
Q_A	-	$1,390 \pm 28$	$1,350 \pm 27$	[1,720]	[1,690]
Q_B	-	$1,650 \pm 33$	[1,360]	$1,700 \pm 34$	$1,770 \pm 35$

7.4.2   Reliable Loss Measurement in the k_{33} Mode

You may recognize the experimental difficulty in the IEEE Standard for k_{33} mode physical parameter characterization.[17] Two major problems with the standard specimen configuration are: (1) standard k_{33} rod specimen exhibits small capacitance, which causes enormous experimental error due to small current measurement, and (2) due to the leakage of electric field in the k_{33}-type specimen with a high aspect ratio (length/width), extensive elastic compliance ($s_{33}{}^D$) and intensive dielectric permittivity ($\varepsilon_{33}{}^X$) are likely to be overestimated, and electromechanical coupling factor (k_{33}) is underestimated, when we use the observed frequencies and admittance values directly. Because our PE configuration can provide 100–150 times higher capacitance (in the center actuator portion), the experimental data around the antiresonance frequency, in particular, are much more reliable [see big noise current (up–down fluctuation in Figure 7.21b) signal around 92 kHz on the IEEE Standard specimen].[18] Furthermore, our methodology can directly characterize the intensive elastic compliance ($s_{33}{}^E$) and intensive elastic loss tan ϕ_{33}' (with side electrode or SE), in addition to the usual extensive loss tan ϕ_{33} (precisely speaking, tan ϕ_{33}''' without SE), just by coating the SE, with which the IEEE Standard electric excitation method cannot conduct. Figure 7.21a with various electrode patterns for k_{33} mode measurements, including no electrode (IEEE Standard), partial electrode with short-circuit (SC), which corresponds to resonance behavior of standard k_{33} specimen, and open-circuit (OC), which corresponds to antiresonance behavior of standard k_{33} specimen, and SE for intensive elastic compliance $\left(s_{33}^E\right)$ and loss tanϕ_{33}' are designed and prepared. The center portion (about 10% of the whole length) is for the actuator to mechanically excite both the side major portions. Figure 7.21b shows admittance spectra for all the configurations in Figure 7.21a. You can notice that the admittance spectrum of the IEEE standard sample has huge electrical noise near antiresonance frequency, due to small capacitance, but those of partial electrode do not have any electrical noise, due to much higher capacitance, compared to the standard specimen.

Different from parameter determination method for k_{31} PE[16], the parameter determination method for k_{33} PE requires curve fitting of experimental data to analytical solution,[19] since resonance and antiresonance frequencies and related parameters are more sensitive upon the portion of the center actuator part. First, by measuring center part near 1 kHz, intensive relative permittivity (ε_{33}^X) and loss (tanδ_{33}) can be measured. This can be done with any sample with the PE geometry. Second, extensive elastic compliance s_{33}^D and elastic triple prime loss (tanϕ_{33}''')[7] can be determined from measuring open-circuit specimen. Third, by using the same sample and attaching wires, piezoelectric constant (d_{33}) and intensive piezoelectric loss (tanθ_{33}') can be determined from measuring short-circuit specimen. Finally, by measuring SE specimen, intensive elastic compliance (s_{33}^E) and loss (tanϕ_{33}') can be determined. After these six parameters are determined, electromechanical coupling factor (k_{33}) can be calculated by using either $\left(1-k_{33}^2\right)=s_{33}^D/s_{33}^E$ or

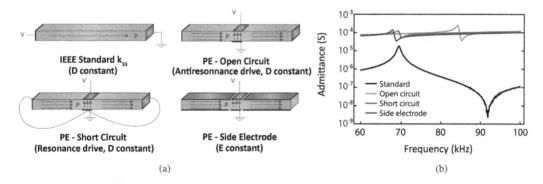

       (a)   (b)

FIGURE 7.21   (a) Schematic view of different electrode patterns on a rectangular k_{33} PZT plate. (b) Admittance spectrum of IEEE Standard, PE-short, PE-open, and SE samples for the k_{33} mode characterization.

$k_{33}^2 = d_{33}^2 / (s_{33}^E \varepsilon_0 \varepsilon_{33}^X)$. From the PE sample, we obtained $s_{33}^E = 17.7 \ p \cdot m^2 / N$ and $s_{33}^D = 9.21 \ p \cdot m^2 / N$, leading to $k_{33} = 69\%$; while the standard sample provided $k_{33} = 69\%$ from the resonance and anti-resonance frequencies. Significant difference can be found in the loss parameter accuracy: the PE gave $\tan \phi_{33}' = 0.012(\pm 0.3\%)$, $\tan \phi_{33}''' = 0.0056(\pm 0.3\%)$, $\tan \theta_{33}' = 0.018(\pm 0.3\%)$, while the standard IEEE geometry gave $\tan \phi_{33}' = 0.012(\pm 50\%)$, $\tan \phi_{33}''' = 0.0049(\pm 29\%)$, $\tan \theta_{33}' = 0.018(\pm 67\%)$.

Example Problem 7.3

When we measure the admittance spectrum for a low Q_m piezo-ceramic k_{31} mode plate specimen, we occasionally obtain a low admittance peak at the resonance frequency, as shown in Figure 7.22. In order to determine the mechanical quality factor Q_m, we try to obtain the two frequencies, ω_1 and ω_2, from the 3 dB-down or $1/\sqrt{2}$ of the peak admittance method. However, as seen in Figure 7.22, because of the low Q_m, the lower frequency ω_1 of $1/\sqrt{2}$ of the peak admittance $(0.003 \ S \times 1/\sqrt{2})$ is out of the peak range. Consider how to determine the mechanical quality factor Q_m from this sort of small peak admittance spectrum.

Hint
Instead of 3 dB-down method, can we use 0.3 dB-down method?

Solution

Q_m is obtained under steady-state oscillation under a harmonic force. As we analyzed in Section 3.3, since $\dfrac{u_0(\omega_0)}{u_0(\omega = 0)} = \dfrac{1}{2\zeta \dfrac{\omega}{\omega_0}}$, by putting $\dfrac{u_0}{u_0(\omega = 0)} = \dfrac{1}{\sqrt{\left[\left[1 - \left(\dfrac{\omega}{\omega_0}\right)^2\right]^2\right] + \left(2\zeta \dfrac{\omega}{\omega_0}\right)^2}} = \dfrac{1}{\sqrt{2}} \dfrac{1}{2\zeta \dfrac{\omega}{\omega_0}}$,

the 3 dB-down or $1/\sqrt{2}$ frequencies can be obtained from

$$1 - \left(\frac{\omega}{\omega_0}\right)^2 = \pm 2\zeta \frac{\omega}{\omega_0}, \text{ depending on } \frac{\omega}{\omega_0} < 1 \text{ or } \frac{\omega}{\omega_0} > 1 \tag{P7.3.1}$$

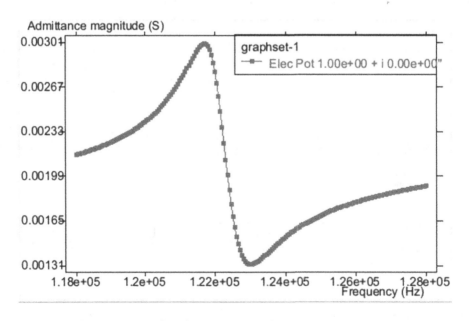

FIGURE 7.22 Admittance spectrum for a k_{31} PZT plate specimen.

Thus, we obtain two roots: $\frac{\omega}{\omega_0} = \sqrt{1+\zeta^2} - \zeta$ or $\frac{\omega}{\omega_0} = \sqrt{1+\zeta^2} + \zeta$ for $\frac{\omega}{\omega_0} < 1$ or $\frac{\omega}{\omega_0} > 1$, respectively. Then, $\frac{\Delta\omega}{\omega_0} = 2\zeta$ or the mechanical quality factor Q_m is expressed by

$$Q_m = \frac{\omega_0}{\Delta\omega} = 1/2\zeta \tag{P7.3.2}$$

When we cannot observe the above ω_1 and ω_2, we may use the −0.3 dB points. Instead of $1/\sqrt{2}$ (= 0.708), we can use −0.3 dB value 0.966:

$$\frac{u_0}{u_0(\omega=0)} = \frac{1}{\sqrt{\left[1-\left(\frac{\omega}{\omega_0}\right)^2\right]^2 + \left(2\zeta\frac{\omega}{\omega_0}\right)^2}} = 0.966 \times \frac{1}{2\zeta\frac{\omega}{\omega_0}} \tag{P7.3.3}$$

Thus, from the relation

$$1-\left(\frac{\omega}{\omega_0}\right)^2 = \pm 0.535\zeta\frac{\omega}{\omega_0}, \text{ depending on } \frac{\omega}{\omega_0} < 1 \text{ or } \frac{\omega}{\omega_0} > 1$$

we obtain two roots: $\frac{\omega}{\omega_0} = \sqrt{1+0.071\zeta^2} - 0.268\zeta$ or $\frac{\omega}{\omega_0} = \sqrt{1+0.071\zeta^2} + 0.268\zeta$ for $\frac{\omega}{\omega_0} < 1$ or $\frac{\omega}{\omega_0} > 1$, respectively. Then, $\frac{\omega_2 - \omega_1}{\omega_0} = 0.535\zeta$. Thus, when we use −0.3 dB method, obtained $\frac{\omega_0}{(\omega_2 - \omega_1)}$ value must be calibrated by the factor of 0.535/2. The value $\frac{\omega_0}{(\omega_2 - \omega_1)}$ gives roughly four times larger than Q_m.

CHAPTER ESSENTIALS

1. High-power piezoelectric characterization system – categorization
   - Pseudo-static method – hysteresis loop measurements on polarization P–electric field E, strain x–stress X, x–E, and P–X.
   - Admittance/impedance spectrum method: electrical excitation
     a. Constant voltage drive – spectrum peak skew at the resonance
     b. Constant current drive – spectrum peak skew at the antiresonance
     c. Constant vibration velocity (output power) method – symmetric spectrum curves at both resonance and antiresonance
     d. Constant input power method
   - Transient/burst drive method: mechanical excitation
     a. Pulse drive method
     b. Burst mode method – quick measurement time, no heat generation
   - Specimen electrode configuration method – partial electrode configurations for measuring both intensive and extensive loss factors.
2. Mechanical quality factor Q_m determination methods:
   - Admittance spectrum – 3 dB-down or -up methods for the resonance and antiresonance. Better accuracy is found in admittance or impedance circle usage – quadrantal frequency method – $Q_m = \frac{\omega_0}{\Delta\omega}$
   - Pulse/burst drive methods – the damping time constant τ of the decaying vibration ring-down amplitude or current/voltage provides the Q_m value. $-Q_m = (1/2)\omega_0\tau$

3. Three losses and mechanical quality factors Q_A and Q_B: k_{31} example

- Resonance: $Q_{A,31} = \dfrac{1}{\tan \phi'_{11}}$

- Antiresonance: $\dfrac{1}{Q_{B,31}} = \dfrac{1}{Q_{A,31}} - \dfrac{2}{1 + \left(\dfrac{1}{k_{31}} - k_{31}\right)^2 \Omega^2_{B,31}}\left(2\tan\theta'_{31} - \tan\delta'_{33} - \tan\phi'_{11}\right)$

where $\tan \delta'_{33}$, $\tan \phi'_{11}$, and $\tan \theta'_{31}$ are intensive loss factors for $\varepsilon_{33}{}^X$, $s_{11}{}^E$, and d_{31}, and $\Omega_{B,31}$ is the normalized antiresonance frequency given by

$$\Omega_{A,31} = \frac{\omega_a l}{2v^E_{11}} = \frac{\pi}{2} \quad \left[v^E_{11} = 1\big/\sqrt{\rho\, s^E_{11}} \right], \quad \Omega_{B,31} = \frac{\omega_b l}{2v^E_{11}}, \quad 1 - k^2_{31} + k^2_{31}\frac{\tan\Omega_B}{\Omega_B} = 0$$

4. Burst mode method can determine the permittivity and dielectric loss around the resonance frequency.
   - Force factor: ratio between (short-circuit current) and (edge vibration velocity)

$$A_{31} = \frac{i_0}{v_0} = 2e^*_{31}w = 2w\frac{d_{31}}{s^E_{11}}.$$

   - Voltage factor: ratio between (open circuit voltage) and (edge displacement)

$$B_{31} = \frac{V_0}{u_0} = \frac{2b}{L}\frac{e^*_{31}}{\varepsilon^{x_1}_{33}\varepsilon_0} = \frac{2b}{L}\frac{g_{31}}{s^D_{11}} = \frac{2b}{L}h^*_{31}.$$

   - Permittivity at the resonance region:

$$\varepsilon_0\varepsilon^X_{33}\left(1 - k^2_{31}\right) = \varepsilon_0\varepsilon^{X_1}_{33} = \frac{e^*_{31}}{h^*_{31}} = \frac{A_{31}}{B_{31}}\frac{b}{Lw}$$

5. Partial-electrode configuration:
   - k_{31} type – $s_{11}{}^D$ can be measured.
   - k_{33} type – $s_{33}{}^E$ can be measured.

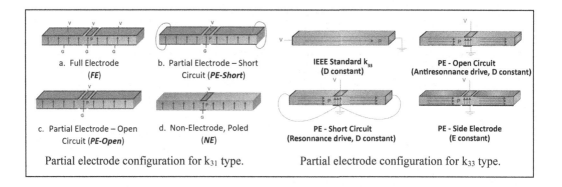

a. Full Electrode (FE)
b. Partial Electrode – Short Circuit (PE-Short)
c. Partial Electrode – Open Circuit (PE-Open)
d. Non-Electrode, Poled (NE)
Partial electrode configuration for k_{31} type.

IEEE Standard k_{33} (D constant)
PE - Open Circuit (Antiresonnance drive, D constant)
PE - Short Circuit (Resonnance drive, D constant)
PE - Side Electrode (E constant)
Partial electrode configuration for k_{33} type.

CHECK POINT

1. (T/F) When we drive the piezoelectric k_{31} plate mechanically under a sinusoidal force, the resonance frequency should be the same for both short-circuit and open-circuit conditions. True or False?

2. (T/F) When we drive the piezoelectric k_{31} plate electrically under constant high voltage condition, the admittance spectrum shows a significant skew distortion at the antiresonance frequency. True or False?

3. (T/F) The mechanical quality factors Q_A (at resonance) and Q_B (at antiresonance) should be the same in a piezoelectric k_{33} rod specimen. True or False?

4. (T/F) The mechanical quality factor Q_A at resonance is larger than Q_B at antiresonance in a PZT piezoelectric k_{31} plate specimen. True or False?

5. (T/F) To generate the same vibration velocity of a piezoelectric transducer, the resonance drive is the most efficient, rather than antiresonance drive in PZT-based materials. True or False?

6. Can you find more energy-efficient frequency for driving a PZT piezoelectric k_{31} plate specimen than its resonance or antiresonance frequencies? If yes, where do you find this most efficient frequency?

7. (T/F) The off-resonance dielectric permittivity and loss values are usually used for evaluating the piezoelectric constant and its loss factor in the admittance spectrum method, because it is theoretically impossible to measure the dielectric performance at the resonance mode. True or False?

CHAPTER PROBLEMS

7.1 The electromechanical coupling factor k can be obtained from the resonance f_A and antiresonance f_B frequencies basically. The electromechanical coupling factor k_{33} for a piezo-ceramic rod can be calculated accurately from the following equations:

$$k_{33}^2/\left(1-k_{33}^2\right) = -\left(2\pi f_A L/2v\right)\cot\left(2\pi f_A L/2v\right), \text{or} \tag{P1}$$

$$k_{33}^2 = (\pi/2)\left(f_R/f_A\right)\tan\left[(\pi/2)\left(\Delta f/f_A\right)\right], \tag{P2}$$

where $\Delta f = f_A - f_R$.

a. Verify these two equations (1) and (2) are equivalent.

b. Verify the following Marutake equation is viable as the first approximation.

$$k_{33}^2/\left(1-k_{33}^2\right) = \left(\pi^2/8\right)\left(f_A^2 - f_R^2\right)/f_R^2 \tag{P3}$$

c. Verify the following equation is viable as a rough approximation when k_{33} is small.

$$k_{33}^2/\left(1-k_{33}^2\right) = \left(\pi^2/4\right)\left(\Delta f/f_R\right) \tag{P4}$$

REFERENCES

1. J. Zheng, S. Takahashi, S. Yoshikawa, K. Uchino, J.W.C. de Vries, Heat generation in multilayer piezoelectric actuators, *Journal of the American Ceramic Society*. **79**, 3193–3198 (1996).

2. K. Uchino, H. Negishi, T. Hirose, Drive voltage dependence of electromechanical resonance in PLZT piezoelectric ceramics, *Japanese Journal of Applied Physics*. **28**(Suppl. 28-2), 47–49 (Proceeding of the FMA-7, Kyoto) (1989).

3. S. Hirose, M. Aoyagi, Y. Tomikawa, S. Takahashi and K. Uchino, High-Power Characteristics at Antiresonance Frequency of Piezoelectric Transducers, *Proceedings of Ultrasonics International '95*, Edinburgh, pp. 184–187 (1995).

4. K. Uchino, J. Zheng, A. Joshi, Y.H. Chen, S. Yoshikawa, S. Hirose, S. Takahashi, J.W.C. de Vries, High power characterization of piezoelectric materials, *Journal of Electroceramics*. **2**, 33–40 (1998).

5. A.V. Mezheritsky, *IEEE Transactions on Ultrasonics, Ferroelectrics, and Frequency Control*. **49** (2002) 484.

6. Y. Zhuang, S.O. Ural, A. Rajapurkar, S. Tuncdemir, A. Amin, K. Uchino, Derivation of piezoelectric losses from admittance spectra, *Japanese Journal of Applied Physics*. **48**, 041401 (2009).

7. Y. Zhuang, S.O. Ural, S. Tuncdemir, A. Amin, K. Uchino, Analysis on loss anisotropy of piezoelectrics with ∞mm crystal symmetry, *Japanese Journal of Applied Physics*. **49**, 021503 (2010).

8. S.O. Ural, S. Tuncdemir, Y. Zhuang, K. Uchino, Development of a high power piezoelectric characterization system (HiPoCS) and its application for resonance/antiresonance mode characterization, *Japanese Journal of Applied Physics*. **48**, 056509 (2009).

9. H.N. Shekhani, K. Uchino, Evaluation of the mechanical quality factor under high power conditions in piezoelectric ceramics from electrical power, *Journal of the European Ceramic Society*. **35**(2), 541–544 (2014).

10. K. Uchino, Y. Zhuang, S.O. Ural, Loss determination methodology for a piezoelectric ceramic: New phenomenological theory and experimental proposals, *Journal of Advanced Dielectrics*. **1**(1), 17–31 (2011).

11. M. Umeda, K. Nakamura, S. Ueha, *Japanese Journal of Applied Physics*. **38**, 3327–3330 (1999).

12. S. Takahashi, Y. Sasaki, M. Umeda, K. Nakamura, S. Ueha, Characteristics of piezoelectric ceramics at high vibration levels, *MRS Proceedings*. **604**, 15 (1999).

13. M. Umeda, K. Nakamura, S. Ueha, Effects of vibration stress and temperature on the characteristics of piezoelectric ceramics under high vibration amplitude levels measured by electrical transient responses, *Japanese Journal of Applied Physics*. **38**, 5581 (1999).

14. H. Shekhani, T. Scholehwar, E. Hennig, K. Uchino, Characterization of piezoelectric ceramics using the burst/transient method with resonance and antiresonance analysis, *Journal of the American Ceramic Society*. 1–10 (2016). DOI: 10.1111/jace.14580.

15. H. Daneshpajooh, H. Shekhani, M. Choi, K. Uchino, *Journal of the American Ceramic Society*. **101**, 1940–1948 (2018). DOI: 10.1111/jace.15338.

16. M. Majzoubi, H.N. Shekhani, A. Bansal, E. Hennig, T. Scholehwar, K. Uchino, Advanced methodology for measuring the extensive elastic compliance and mechanical loss directly in k_{31} mode piezoelectric ceramic plates, *Journal of Applied Physics*. **120**(Issue 22) (2016). DOI 10.1063/1.4971340.

17. ANSI/IEEE Std 176-1987, *IEEE Standard on Piezoelectricity*, (The Institute of Electrical and Electronics Engineers, New York, 1987), p. 56.

18. Y. Park, Y. Zhang, M. Majzoubi, T. Scholehwar, E. Henning, K. Uchino, Improvement of the standard characterization method on k_{33} mode piezoelectric specimens, *Sensors and Actuators A: Physical*, **312**, 112124 (2020). DOI: 10.1016/j.sna.2020.112124.

19. Y. Park, M. Majzoubi, Y. Zhang, T. Scholehwar, E. Hennig, K. Uchino, Analytical modeling of k_{33} mode partial electrode configuration for loss characterization, *Journal of Applied Physics*, **127**, 204102 (2020). DOI: 10.1063/1.5143728.

8 Drive Schemes of Piezoelectric Transducers

ABSTRACT

"Drive Schemes of Piezoelectric Transducers" treats the piezo-device driving methods in order to minimize the losses and maximize the transducer efficiency, in particular, at off-resonance capacitive region, at resonance or antiresonance frequency, and at inductive region between the resonance and antiresonance frequencies. In the off-resonance (pseudo-DC) drive, the inductor or "negative capacitance" usage is the key, while in the resonance drive, there are two drive ways: (1) resonance/antiresonance (resistive) drive with self-oscillating circuit and (2) power minimization (reactive) drive with impedance converter capacitor usage.

In Chapter 7 "High-Power Piezo Characterization Systems" (HiPoCS™), the reader was recommended to have a suitable power supply to drive a piezoelectric actuator. The power supply better than the followings, in particular, in a multilayer (ML) actuator characterization, is required:

- Maximum voltage: 200 (V)
- Maximum current: 10 (A)
- Frequency range: 0–500 (kHz)
- Output impedance: < 1 (Ω).

However, the above numbers are merely for the precise "characterization" purpose and over-specs for the actual drivers of the commercialized piezoelectric products. This chapter will describe the piezo-actuator drive systems with compact and energy-efficient power supplies, aiming at practical applications.

8.1 PIEZO-ACTUATOR CLASSIFICATION AND DRIVE METHOD

8.1.1 CLASSIFICATION OF PIEZOELECTRIC ACTUATORS

Piezoelectric and electrostrictive actuators are classified into two major categories based on the type of drive voltage applied to the device and the nature of the strain induced by the voltage as depicted in Figure 8.1. They are: (1) *rigid displacement devices,* for which the strain is induced unidirectionally, aligned with the applied pseudo-DC field, and (2) *resonant displacement devices,* for which an alternating strain is excited by an AC field, in particular, at the mechanical resonance frequency (*ultrasonic motors*). The first category can be further divided into two general types: *servo displacement transducers (positioners),* which are controlled by a feedback system through a position detection signal, and *pulse drive motors,* which are operated in a simple on/off switching mode. The drive/control techniques presented in this chapter will be discussed in terms of this classification scheme.

The response of the resonant displacement device is not only directly proportional to the applied voltage but also dependent strongly on the drive frequency. Although the positioning accuracy of this class of devices is not as high as that of the rigid displacement devices, ultrasonic motors are able to produce very rapid motion due to their high-frequency operation and amplified displacement. Servo displacement transducers, which are controlled by a feedback voltage superimposed

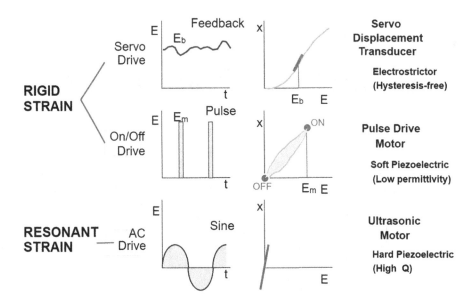

FIGURE 8.1   Classification of piezoelectric/electrostrictive actuators according to the type of drive voltage and the nature of the induced strain.

on a DC bias, are used as positioners for optical and precision machinery systems. In contrast, a pulse drive motor generates only on/off strains, suitable for the impact elements of inkjet printers or automobile fuel injection valves.

The material requirements for each class of devices are different, and certain compositions will be better suited for particular applications. The servo displacement transducer suffers most from strain hysteresis, and therefore, $Pb(Mg_{1/3}Nb_{2/3})O_3$-$PbTiO_3$ [PMN] electrostrictive materials are preferred for this application. It should be noted that even when a feedback system is employed, the presence of a pronounced strain hysteresis (i.e., time delay) generally results in much slower response speed. The pulse drive motor, for which a quick response rather than a small hysteresis is desired, requires a low-permittivity material under a current (A) – limitation of a power supply. Soft $Pb(Zr,Ti)O_3$ [PZT] piezoelectrics are preferred over the high-permittivity PMN for this application. The ultrasonic motor, on the other hand, requires a very "hard" piezoelectric with a high-mechanical quality factor, Q_m, in order to maximize the AC strain and to minimize heat generation, as you learned in Chapter 6. Note that the figure of merit for the resonant strain is characterized by $\left(\dfrac{8}{\pi^2}\right) dEL{\cdot}Q_m$ (d: piezoelectric strain coefficient, E: applied electric field, L: sample length, and Q_m: mechanical quality factor) at its fundamental resonance mode. Although hard PZT materials have smaller d coefficients, they have significantly larger Q_m values, thus providing the higher resonant strains needed for these devices.

8.1.2   Drive Frequency and Phase

Figure 8.2 shows an example admittance spectrum of a 20 mm long k_{31}-type piezoelectric (PZT) plate specimen with the resonance frequency around 86 kHz. When the operating frequency is lower than 10 kHz, this is considered as "off-resonance" (or pseudo-DC) drive, and its characteristic is purely "capacitive" with the admittance ($j\omega C$) phase lag of 90°. When the operating frequency is 86 kHz or 89 kHz, the characteristic becomes "resistive" with the phase lag of 0°, which corresponds to the resonance or antiresonance frequency. In order to induce the same level of vibration velocity, "low voltage and high current" or "high voltage and low current" is required

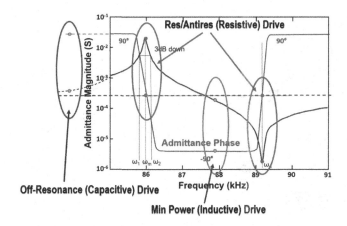

FIGURE 8.2 Application frequency ranges for piezoelectric positioners, pulse drive motors, and resonance ultrasonic motors.

at the resonance or antiresonance drive, respectively. We also introduce an operating frequency at 88 kHz in the inductive ($1/j\omega L$) region in order to minimize the required input drive power for obtaining the same mechanical vibration level. To the contrary, the pulse drive includes wide range of frequencies (pseudo-DC to multiple higher-order resonance frequencies), which exhibit linear- or parabolic- time-dependence of total displacement (rather than sinusoidal dependence), in addition to the overshoot and/or vibration ringing, depending on the pulse voltage shape applied.

8.2 OFF-RESONANCE (PSEUDO-DC) DRIVE

8.2.1 POWER SUPPLY PROBLEM FOR C-TYPE DRIVE

It is known that when we drive a piezoelectric actuator at an off-resonance (or pseudo-DC) frequency by power supplies, it is 2%–6% efficient, because of the electrical impedance mismatch. This is the major reason why the reader needs to prepare "over-spec" power supply for the "High-Power Characterization" purpose. Historically, when NEC Corporation developed a dot-matrix printer (pulse drive) with piezoelectric ML actuators (capacitance C) in the 1980s, they integrated an inductive coil (inductance L) at the interface between the piezo-component and resistive power supply, so that C and L resonate at the printer operating frequency 1 kHz (much lower than the ML actuator mechanical resonance frequency 100 kHz). This energy catch ball between C and L helped significantly with trapping the energy locally without dissipation. In a word, when dealing with capacitive piezoelectric actuators, the important consideration is not "apparent power" but actual "energy flow"!

8.2.2 NEGATIVE CAPACITANCE USAGE

Most of the conventional *linear* and *switching power systems* have been developed for driving primarily *resistive* loads such as electromagnetic motors. Figure 8.3a illustrates energy flow charts for a switching power system in driving a resistive load (top) and a capacitive load (bottom) with practical numbers. Ninety percent of the input 160 W can be spent in the resistive load (90% efficiency), while only 2% can be used in a capacitive load (2% efficiency). Ninety percent of the input power is spent out in the power supply (mostly as heat) in the capacitance drive. We had better consider much better driving scheme for solid-state capacitive components such as ML piezo-actuators. When dealing with capacitive components, it is important to consider "energy" flow ("real" power $V\cdot I\cdot cos\,\varphi$) rather than "apparent" power flow ($V\cdot I$). The key to escape from this electric impedance mismatching

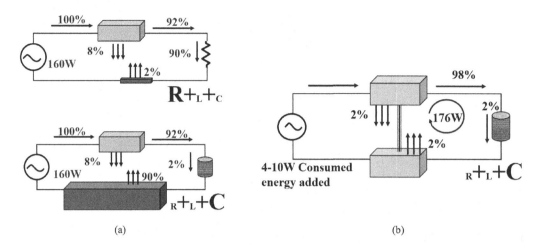

FIGURE 8.3 (a) Efficiency of a switching power system in driving a resistive load 90% or a capacitive load 2%. (b) Energy flow of a switching power system with a negative capacitance in driving a capacitive load.[1]

is to insert a reactive (or inductive) component in the driving system in series to the capacitive device. Knowing that the conventional coil inductor kills the size and weight (and Joule loss) in the power supply significantly, Knowles et al. at Qortek, PA, introduced a "negative capacitance" in the switching power system, as schematically illustrated in Figure 8.3b.[1] The negative capacitance $-C_d//T$, where C_d is the *damped capacitance* of the piezoelectric device, is inserted in series with the R-C circuit. Though the usage of power is still 2% (in case of small electromechanical coupling factor k such as 20%, or converted mechanical energy 4%), the remaining 98% will be recovered without losing much as heat in the power system because of the negative capacitance, leading to a high-efficiency 98%. Note that since the actually consumed electric energy level is just 4–10 W, depending on the mechanically consumed energy in the piezo-actuator (in addition to the energy loss in the power circuit), the total size/weight of the power system is significantly smaller than the conventional ones. Because we need to fix the negative capacitance to adjust to the operating frequency, the actual compact power supply needs to be designed after finalizing the piezo-component's specifications.

One product example can be found in "noise barrier" developed by Kobayashi Riken [private communication] (Figure 8.4a). Large-area PVDF films were used to cover the walls separating residential areas from a busy highway in Tokyo suburban area. The low-frequency traffic noise less than 1 kHz is the primary annoyance for the human. In order to eliminate acoustic traffic noise around the highway, the acoustic noise is detected by a piezoelectric microphone first, then the signal is feedbacked to piezoelectric PVDF film speakers to generate a "conjugate" acoustic wave (i.e., "antinoise") at around the targeted 200~300 Hz, which effectively cancels the noise signal. Using a negative capacitance circuit, the acoustic transmission loss was remarkably improved by 20 dB in a low-frequency (off-resonance) range, in comparison with the data obtained without this circuit (difference between red and green lines in Figure 8.4b). Because of small electromechanical coupling factor of PVDF (i.e., the damped capacitance is very large), the negative capacitance integration dramatically improved the total system efficiency.

As introduced in the beginning, the pulse drive-type dot-matrix printer firstly adopted this inductive/reactive component. Therefore, the negative capacitance component can also be applied to the pulse drive motors, since they are cyclically operated at a certain frequency such as 1 kHz. The coupling inductance L should be chosen to adjust to $1/2\pi\sqrt{LC}$ (i.e., electrical resonance frequency) exactly to 1 kHz, according to the piezoelectric ML actuator's capacitance C.

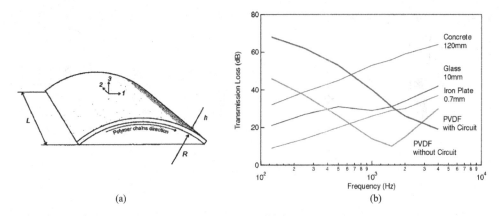

(a)   (b)

FIGURE 8.4 (a) Curved PVDF acoustic noise barrier. (b) Acoustic transmission loss spectrum with the negative capacitance circuit or without the circuit.

8.3 RESONANCE DRIVE

8.3.1 RESONANCE/ANTIRESONANCE (RESISTIVE) DRIVE

When the operating frequency is the exact resonance or antiresonance, the electric characteristic becomes "resistive" with the phase lag of 0°. Thus, the conventional (resistive output impedance) power supplies are possibly used, as long as the voltage and current specifications are satisfied. However, in order to induce the same level of vibration velocity (i.e., mechanically converted energy), low voltage and high current or high voltage and low current is required at the resonance or antiresonance drive, respectively. Note that both resonance and antiresonance are indeed the mechanical resonance.

Remember also that the mechanical quality factor at the resonance Q_A is smaller than that at the antiresonance Q_B in PZTs [see Figure 8.2]. The frequency dependence of the electromechanical conversion efficiency for this device is shown in Figure 8.5 for various applied loads, simulated

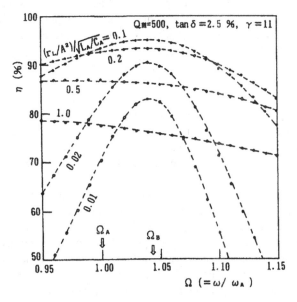

FIGURE 8.5 Frequency dependence of the electromechanical conversion efficiency of a longitudinally vibrating PZT ceramic bar transducer under various loads. [$\Omega = \omega/\omega_A$; A: resonance; and B: antiresonance].

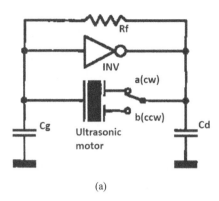
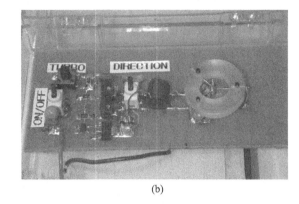

(a)　　　　　　　　　　　　　　　　　　　(b)

FIGURE 8.6　Ultrasonic motor driven by its own oscillation with electric elements. (a) Schematic diagram. (b) Actual circuit developed for driving a metal-tube motor.

from the equivalent circuit in Figure 7.8a.[2] The efficiency exhibits the maximum at the antiresonance frequency when we include two loss factors (dielectric and elastic). The analysis including the piezoelectric loss is discussed in Section 8.3.2. The difference between the resonance (A-type) and antiresonance (B-type) frequencies is also highlighted on this graph. When the load is not large, a significant variation in the efficiency with frequency is observed. As the load increases, the efficiency curve becomes flatter. When we consider the driving conditions that apply for each state, that is, a constant electric field, E, for the resonance mode and a constant electric displacement, D, for the antiresonance mode, the lower loss in the antiresonance mode makes sense. Recall that the strain hysteresis is significantly less when the strain is considered as a function of the electric displacement (or polarization) as compared with its electric field dependence. Moreover, antiresonance operation requires a low-driving current and a high-driving voltage, in contrast to the high current and low voltage required for resonance mode operation, thus allowing for the use of a conventional, inexpensive low-current power supply.

Recently, compact ultrasonic motors much less than 5 mm ϕ have widely been installed in electronic devices such as mobile cameras, phones, and aerial drones, where the power supplies should also be miniaturized. Thus, self-oscillating circuits are popularly utilized. One example is shown in Figure 8.6, where the circuit oscillation frequency is adjusted exactly to the resonance of the ultrasonic motor.

8.3.2　Power Minimization (Reactive) Drive

As we already discussed in Section 7.2.4 Real Electric Power Method and Section 5.3.2 Equivalent Circuit with Three Losses, the antiresonance drive is more efficient than the resonance drive in the PZT-based piezoelectric ceramic devices. Further a certain frequency between the resonance and antiresonance exhibits the best real power efficiency for generating the same output mechanical vibration velocity.

Yuan et al. proposed an innovative driving scheme of a Langevin piezoelectric transducer[3] under its reactive frequency range, which takes advantage of the maximum efficiency frequency between the resonance and antiresonance. In this approach, first, a constant vibration velocity measurement system is used to find the optimum driving frequency, which is defined as the point where the real input electric power ($V \cdot I \cdot \cos \varphi$) is the lowest for a given output mechanical vibration level. Figure 8.7 shows the admittance spectrum observed on this Langevin transducer (HEC#45402, Honda Electronics, Toyohashi, Japan) under a constant vibration velocity condition, that is, the mechanical vibration level of the head mass is constant for all the measurement. Using the admittance magnitude and phase, we can calculate the actual input power (W) as in Figure 8.8a, which

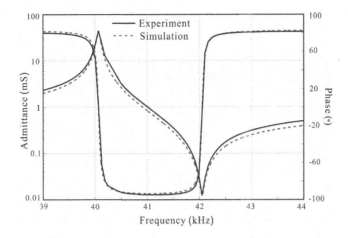

FIGURE 8.7 Admittance and corresponding phase comparisons between simulation and experimental result.

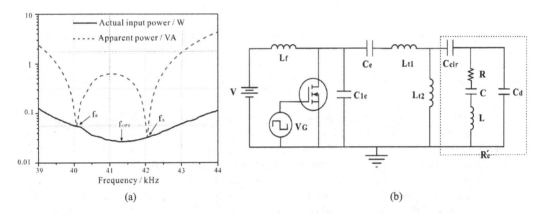

FIGURE 8.8 (a) Actual input power and apparent power of the Langevin transducer under constant vibration velocity (30mm/s) drive. (b) Class E inverter with impedance converter.

shows the resonance frequency 40.07 kHz, antiresonance frequency 42.05 kHz, while the lowest input power (i.e., maximum mechanical quality factor) at 41.27 kHz. The transducer has a *reactive behavior* at the optimum frequency (the phase lag is rather close to −90°).[4] Second, in this approach, an equivalent circuit of the transducer based on Butterworth-Van Dyke (BVD) model (C_d and L, C, R motional branch) is established (Figure 8.8b right-hand side), where the head and tail mass loads are already integrated. Third, these BVD equivalent circuit parameters are used to design *Class E inverter* driving circuits. Using MATLAB®, a Class E inverter is precisely designed (Figure 8.8b left-hand side) to drive the transducer at the resonance frequency (resistive). Finally, an impedance converter (basically capacitance C_{cir}) is added (slightly modified on the Class E inverter) when driving at the optimum frequency (reactive). The optimum driving frequency of this Langevin piezoelectric transducer was $f_{opt} = 41.27$ kHz, that is, between resonance frequency and antiresonance frequency. The impedance at this frequency is $Z_{opt} = 144.2 + j2065$ (Ω), which is mostly an inductive load. According to the analysis of the Class E inverter, R_e should be resistive load, therefore, a capacitor C_{cir} is connected in series with this Langevin piezoelectric transducer (BVD equivalent circuit is on the right-hand side of Figure 8.8b) to accomplish a resistive load R_e' and the value of C_{cir} can be calculated by the impedance of optimum frequency:

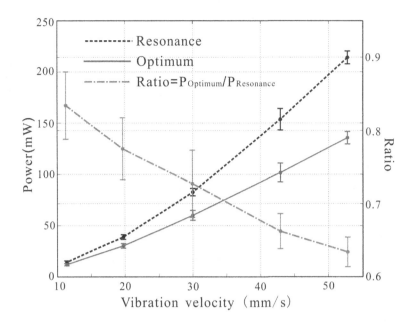

FIGURE 8.9 Powers comparison of two driving methods and ratio of powers in different vibration velocities.[3]

$$C_{cir} = \frac{1}{\omega_{opt} \cdot img(Z_{opt})}, \tag{8.1}$$

where ω_{opt} is the optimum driving frequency and $\omega_{opt} = 2\pi f_{opt}$, and img($Z_{opt}$) is the imaginary part of Z_{opt}. The total impedance R_e' of the Langevin transducer and capacitor C_{cir} is 143 Ω. An impedance converter circuit is applied to change the impedance of R_e' to be the same as R_e,[3] while the driving frequency set to the experimentally determined optimum frequency of 41.27 kHz. The topology of the impedance converter circuit and Class E inverter is shown in Figure 8.8b. Figure 8.9 summarizes the required electric power (*real power V·I·cos φ*) for generating the same vibration velocity (measured at the edge of the Langevin head mass) at both driving frequencies, resonance (40.07 kHz) and the optimum condition (41.27 kHz), and the ratio of ($P_{optimum}/P_{resonance}$). The required power for the optimum frequency driving method is reduced by 39% compared with the resonance frequency driving method, and smaller heat generation is also revealed according to the experiments. Note that our driving scheme becomes more attractive with increasing the driving power level (i.e., the above ratio decreases (more attractive) with an increase in vibration velocity).

Example Problem 8.1

Determine the power supply specifications for an ML ceramic actuator drive step by step according to the following process:

a. The ML actuator has 100 ceramic layers, each 100 μm thick with an area of (5 × 5) mm². The relative permittivity is 10,000. Calculate the capacitance of the actuator.

b. Assuming it has a density ρ = 7.9 g/cm³ and elastic compliance s_{33}^D = 13 × 10⁻¹² m²/N, calculate the resonance frequency of the actuator. The ML length should be 10 mm in total. The electrode weight load may be ignored for this calculation.

c. Determine the current required, if 60 V is to be applied to the actuator as quickly as possible.

d. The cut-off frequency (1/RC) must be higher than the ML mechanical resonance frequency. Determine the required output impedance of the power supply.

Solution

   a.

$$C = n\varepsilon_0\varepsilon(A/t) \qquad\qquad (P8.1.1)$$

$$= 100(8.854\times10^{-12}\text{ F/m})(10{,}000)\left[(5\times5\times10^{-6}\text{ m}^2)/(100\times10^{-6}\text{ m})\right]$$

$$C = 2.21\times10^{-6}\text{ (F)}$$

Note: ML actuators have a capacitance of higher than 1 μF.
   b. The resonance frequency for the thickness vibration with $L = 10$ mm (neglecting the coupling with width vibrations) is given by:

$$f_R = \frac{1}{2L\sqrt{\rho s_{33}^D}} \qquad\qquad (P8.1.2)$$

$$= \frac{1}{2[100]\left[100\times10^{-6}\text{ (m)}\right]\sqrt{\left[7.9\times10^3\left(\text{kg/m}^3\right)\right]\left[13\times10^{-12}\left(\text{m}^2/\text{N}\right)\right]}}$$

$$f_R = 156\text{ (kHz)}$$

Note: The response speed of the power supply must be greater than the actuator's resonance frequency.
   c. The relationship between the actuator voltage and the charging current is given by:

$$I = Q/\tau_R = CV f_R \qquad\qquad (P8.1.3)$$

$$= \left[2.21\times10^{-6}\text{ (F)}\right]\left[60\text{ (V)}\right]\left[156\times10^3\text{ (Hz)}\right]$$

$$I = 21\text{ (A)}$$

Note: Ideally a significant current is required from the power supply, even if just for a relatively short period (6 μs). The *apparent power* is estimated to be [60 (V) × 21 (A)]. So, we see that more than 1 (kW) is needed for the resonance drive. A power of 12 (W) is needed for the 2 (kHz) drive in a dot-matrix printer. The *real power* consideration was made in the foregoing Section 8.3.2.
   d. Assuming: $\omega_R = 2\pi f_R = 1/RC$:

$$R = 1/\left[2\pi f_R C\right] \qquad\qquad (P8.1.4)$$

$$1/\left[2\pi\left[156\times10^3\text{ (Hz)}\right]\left[2.21\times10^{-6}\text{ (F)}\right]\right]$$

$$R = 0.46\text{ }[\Omega]$$

The output impedance of the power supply should be less than 1 Ω.
   The power supply specifications for the ML resonance measurement should be:

Maximum voltage: 200 (V), *maximum current*: 10 (A), *frequency range*: 0–500 (kHz), *output impedance*:<1 (Ω).

You may use a conventional impedance analyzer, on which you should recognize that maximum voltage is only 30 V and maximum current less than 0.2 A with the output impedance 50 Ω. Therefore, your measured admittance value on the ML device at its resonance is one order of magnitude smaller than the expected value, because the peak current cannot be supplied from the analyzer power system. Thus, through NF Corporation, Japan, Uchino's team initially developed and commercialized new power supplies for driving ML ultrasonic motors with the specs: 300 V/10 A and output impedance less than 0.2 Ω, which have practically accelerated the progress in the piezoelectric actuators, in particular, in high-power applications (refer to Chapter 1).

8.4 FUNDAMENTAL CIRCUIT COMPONENTS

Though several circuit components have been already introduced in the foregoing chapters, fundamental electric components are described in the section for the reader's sake, if you do not have a background of electric circuit designing.

8.4.1 POWER MOSFET

Let us start reviewing *Power MOSFET* (Metal-Oxide-Semiconductor Field Effect Transistor), which is a transistor that uses an electric field to control the electrical behavior of the device. The symbol of the MOSFET is shown in Figure 8.10a, and its drain current–drain-source voltage characteristics are plotted in Figure 8.10b, which corresponds to 500 V, 50 A Class *Power MOSFET*. Under keeping the drain-source voltage (V_{DS}) at 5 V, the drain current can be significantly changed from 0 to 60 A with changing the gate-source voltage (V_{GS}) from 0 to 5.5 V. Thus, the current on–off switch is obtained by controlling the gate-source voltage with on–off signal.

8.4.2 SWITCHING REGULATOR

On the basis of the current switching effect of the MOSFET, we can design *switching regulators*, which is also called *step-down buck choppers*. The switching regulator is occasionally used as a step-down voltage (step-up current) converter (popularly *DC–DC converter*) without using an electromagnetic transformer, which kills the size and weight in a power system. This is the key

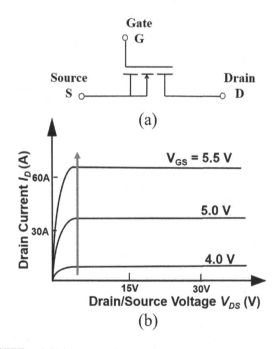

FIGURE 8.10 (a) MOSFET and (b) its output characteristics.

component for the piezoelectric energy harvesting circuit design to charge up the electric energy into a rechargeable battery. Figure 8.11 shows a simplest switching regulator composed of a MOSFET, inductor L, diode, and the resistive load R (capacitor C is occasionally added to stabilize the ripple wave form) under a DC constant voltage (battery). Figure 8.11a and b illustrates the operations for the MOSFET on and off stages, respectively. When the gate-source voltage with a rectangular wave form (reasonably high-carrier frequency with the *duty ratio* $d = \left(\dfrac{T_{on}}{T_{on} + T_{off}} \right)$) is applied, the MOSFET behaves as an ON-and-OFF switch. During ON stage, E–FET–L–R is the current flow route, so that $v_0 = E$, but v_R does not reach to E so quickly because of the inductor (which needs to accumulate the electrical energy first). While, during OFF state, $v_0 = 0$, but v_R does not reach to 0 so quickly; moreover, because the inductor will release the electrical energy, or generate the *reverse electromotive force*, the current still continues to flow in the route L – R – D now. Note first that the average voltage of a rectangular wave form (0 – E) with the duty ratio (d) is estimated as $d \cdot E$. Voltage wave forms of v_0, v_L, and v_R of the step-down buck chopper are illustrated as a function of time in Figure 8.11c, where you can find first v_0 as exactly the "similar" rectangular wave form to the MOSFET gate voltage with 0 to E voltage height (unipolar). On the contrary, v_L shows the same wave form, but a negative bias voltage equal to v_R. Since v_R is obtained as the subtraction $v_0 - v_L$, we can conclude that the v_R is almost constant around the average voltage $d \cdot E$. More details on the v_R behavior is described using Figure 8.11c. Without using a capacitance, the v_R exhibits a ripple mode of an exponential curve with time constant $\tau = (L/R)$ in an L–R circuit. In order to minimize the ripple level, the carrier time period (inverse of the carrier frequency) should be chosen much less than the circuit time constant (L/R) first. An additional smoothing capacitance C helps more with realizing almost constant output voltage $d \cdot E$.

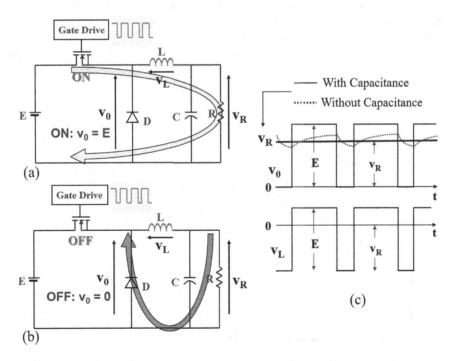

FIGURE 8.11 Switching regulator: (a) on and (b) off operations. (c) Voltage wave forms of v_0, v_L and v_R of the step-down buck chopper.

8.4.3 On–Off Signal Generator

We occasionally use rectangular pulse voltage in the pulse motor application. You learn in this section how to synthesize the desired rectangular signal with a certain duty ratio. Figure 8.12 illustrates the principle of *pulse width modulation* (PWM) based on a triangular carrier wave. When we use a triangular carrier wave v_C, a certain input signal level v_S is easily converted into an ON–OFF signal with a certain duty ratio by subtracting these two voltage values (Figure 8.12c). The subtraction operational amplifier is called "Comparator" (Figure 8.12a), and its practical device example, a low-power complementary metal–oxide–semiconductor (CMOS) clocked comparator, is shown in Figure 8.12b.

So far, we demonstrated only a mono-polar drive (0 to E V) switching regulator. However, we occasionally use a bipolar drive ($-E$ to $+E$ V) power supply, in particular, in DC to AC converters. The key is to use a bridge circuit illustrated in Figure 8.13a. In order to obtain "positive" E, we control Tr_1^+ (on and off) by keeping Tr_2^- (on), while to obtain "negative" E, we control Tr_1^- (on and off) by keeping Tr_2^+ (on). Figure 8.13b shows a DC–DC voltage converter from a small signal voltage v_S to amplified voltage v_{ave} utilizing $\pm$ triangular carrier. Note here that when we denote the duty ratio as d, the average output voltage $v_{ave} = k \cdot E$, where $k = 2d - 1$. Figure 8.13c shows the principle of AC voltage pulse width modulation. We will start from a triangular carrier signal v_C with $\pm v_C$. We now consider two sine input signals, v_{Sa} and v_{Sb} (Figure 8.13c (**1**)), each of which generates a pulse-width modulated wave shown in v_a or v_b at the terminal a or b (Figure 8.13c (**2**) and (**3**)), respectively. Since the final output voltage v_0 is provided by the subtraction of $v_a - v_b$, we obtain the pulse-width modulated $\pm$ signals (Figure 8.13c (**4**)).

8.4.4 Piezoelectric Transformer

Because conventional inductive *coil transformers* kill the size/weight of analog power systems significantly, *piezoelectric transformers* are alternative components nowadays. Since the piezo-transformer is one of the best high-power piezoelectric components commercially available, as discussed in Chapter 10, basic operating principle is introduced.

One of the bulkiest and most expensive components in solid-state actuator systems is the power supply with an electromagnetic transformer. Electromagnetic transformer loss occurring through

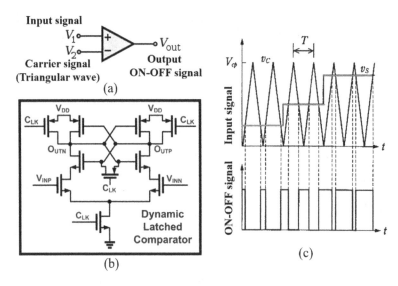

FIGURE 8.12 (a) Comparator, (b) low-power CMOS clocked comparator, and (c) duty ratio realization with a triangular carrier signal.

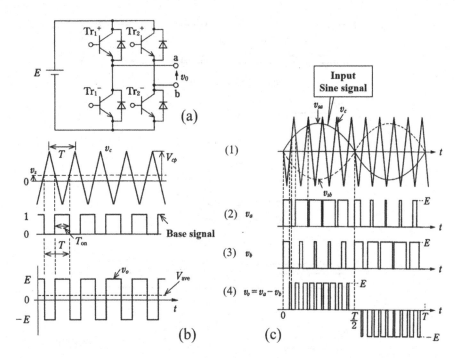

FIGURE 8.13 (a) Bridge circuit, (b) DC voltage converter, and (c) principle of AC voltage pulse width modulation.

the skin effect, thin wire loss, and core loss all increase dramatically as the size of the transformer is reduced. Therefore, it is difficult to realize miniature low-profile electromagnetic transformers with high efficiency. The piezoelectric transformer (PT) is an attractive alternative for such systems due to its high efficiency, small size, and lack of electromagnetic noise. They are highly suitable as miniaturized power inverter components, which might find application in lighting up the cold cathode fluorescent lamp behind a color liquid crystal display or in generating the high voltage needed for air cleaners.

The original design to step up or step down an input AC voltage using the converse and direct piezoelectric effects of ceramic materials was proposed by Rosen.[5] This type of transformer operates by exciting a piezoelectric element like the one pictured in Figure 8.14a at its mechanical resonance frequency. An electrical input is applied to one part of the piezoelectric element (at the top left electrode), which produces the fundamental mechanical resonance. This mechanical vibration is then converted back into an electrical voltage at the other end (right edge electrode) of the piezoelectric plate. Since the electric field level is similar among the input and output parts (because almost symmetrical stress distribution is expected in the input half and output half parts), the voltage step-up ratio (r) without load (i.e., open-circuit condition) is primarily given by the electrode gap ratio:

$$r \propto k_{31}k_{33}\,Q_m\,(l/t),\qquad(8.2)$$

where l and t are the electrode gap distances for the input and output portions of the transformer, respectively (Figure 8.14a). Note from this relationship how the length-to-thickness ratio, the electromechanical coupling factors, and/or the mechanical quality factor Q_m are the primary means of increasing the step-up ratio. Try Chapter Problem 5.3 for learning the detailed derivation process of the step-up voltage ratio. This transformer was utilized on a trial basis in some color televisions during the 1970s.

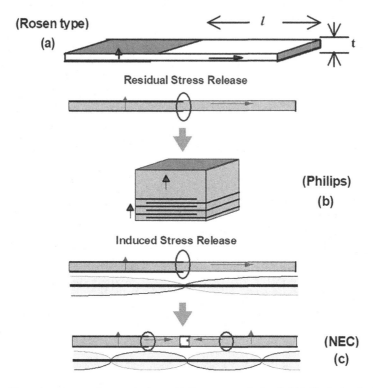

FIGURE 8.14   Piezoelectric transformer designs: (a) Rosen-type design,[5] (b) ML design by Philips, NEC,[6] and (c) third resonant mode type developed by NEC.[7]

In spite of its many attractive features, the original Rosen transformer design had a serious reliability problem. Mechanical failure tends to occur at the center of the device where the residual stress from the poling process is most highly concentrated. This happens to also be coincident with the nodal point of the vibration mode where the highest induced stress occurs. Two recently developed transformers pictured in Figures 8.14b and c are designed to avoid this problem and are commercially produced for use as backlight inverters in liquid crystal displays. Both of the newer designs make use of more mechanically tough ceramic materials. The NEC and Philips Components' transformer shown in Figure 8.14b further alleviates the problem by using an ML structure to avoid the development of residual poling stress in the device.[6] Another NEC design pictured in Figure 8.14c makes use of an alternative electrode configuration to excite a third resonance excitation in the rectangular plate to further redistribute the stress concentrations in a more favorable manner.[7] The reader is requested to understand that high-power piezoelectric devices need to be developed comprehensively from the three aspects: high-power materials development, device design, and drive scheme.

CHAPTER ESSENTIALS

   1. Classification of ceramic actuators:

Actuator Type	Drive	Device	Material Type
Rigid displacement	Servo	Servo-displacement transducer	Electrostrictor
	On/Off	Pulse drive motor	Soft piezoelectric
Resonant displacement	AC	Ultrasonic motor	Hard piezoelectric

2. Application frequency ranges for piezoelectric positioners, pulse drive motors, and resonance ultrasonic motors:
   a. Off-resonance (capacitive) drive
   b. Resonance/antiresonance drive
   c. Minimum energy (reactive) drive

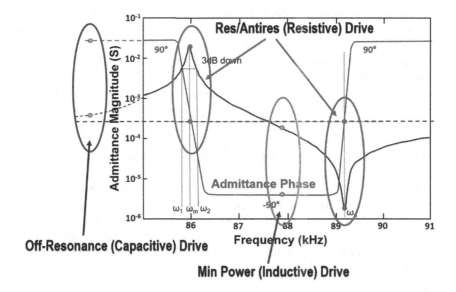

3. Off-resonance (pseudo-DC) drive:
   a. Inductor usage
   b. Negative capacitance usage
4. Resonance drive:
   a. Resonance/antiresonance (resistive) drive – self-oscillating circuit
   b. Power minimization (reactive) drive – impedance converter capacitor usage
5. Fundamental circuit components:
   a. Power MOSFET
   b. Switching regulator
   c. On–off signal generator – pulse width modulation
   d. Piezoelectric transformer.

CHECK POINT

1. The electric field vs. strain relation exhibits a hysteresis curve in a PZT specimen during a cyclic measurement process. Is it clockwise or counterclockwise trace?
2. (T/F) When we measure the admittance spectrum on a k_{31}-type piezoelectric plate specimen, the admittance minimum point corresponds to the resonance point. True or False?
3. (T/F) The voltage supply has a small output impedance, while the current supply has a large output impedance. True or False?
4. When we measure the admittance spectrum on a k_{31}-type piezoelectric plate, the phase changes from +90° to −90° around the resonance point with an increase in drive frequency. What is the phase lag at the exact resonance point?
5. When we measure the admittance spectrum on a k_{31}-type piezoelectric plate, the phase changes from +90° to −90° around the resonance point with an increase in drive frequency. How do you call the frequencies that provide the phase ±45°?

6. Describe the resonance frequency f for the following electrical circuit:

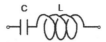

7. (T/F) When a piezoelectric actuator is driven by a rectangular pulse voltage, the mechanical ringing is completely suppressed when the pulse width is adjusted exactly to a half of the resonance period of the sample. True or False?

8. When a piezoelectric actuator is driven under a pseudo-DC condition by a switching amplifier, the efficiency is very low. What electric component (impedance converter) is recommended to be coupled in the driving circuit in order to increase the efficiency?

9. (T/F) Negative capacitance is utilized to drive a "capacitive" actuator device efficiently. True or False?

10. What is the full name for abbreviation PWM?

11. (T/F) Driving a PZT transducer at the antiresonance frequency shows much better efficiency than the resonance frequency drive. True or False?

12. Can you find a better frequency to drive a PZT transducer from the efficiency viewpoint, other than at the resonance or antiresonance frequencies? Answer simply.

13. When a piezoelectric actuator is driven around the resonance/antiresonance frequency range under a "reactive" condition by an E-Class inverter for the maximum efficiency, what electric component (impedance converter) is recommended to be coupled in the driving circuit?

CHAPTER PROBLEMS

8.1 Using a rectangular piezo-ceramic plate (length: l, width: w, and thickness: b, poled along the thickness), we can find the following parameters: k_{31}, d_{31}, and Q_m. Explain the fundamental principles for both resonance and pulse drive methods. The density, ρ, and dielectric constant, $\varepsilon_{33}{}^X$, of the ceramic can be known prior to the following experiments.

a. Using an impedance analyzer, the admittance for the mechanically free sample (i.e., one that is supported at the nodal point at the center of the plate) is measured as a function of the drive frequency, f, and the admittance curve is obtained. Explain how to determine the k_{31}, d_{31}, and Q_m values from these data. Also verify that the following approximate equation can be used for a low-coupling piezoelectric material:

$$k_{31}^2 / \left(1 - k_{31}^2\right) = \left(\pi^2/4\right)\left(\Delta f / f_R\right), \text{ where } \left(\Delta f = f_A - f_R\right).$$

b. Using a pulse drive technique, the transient displacement change is measured as a function of time, including the displacement ring-down performance. Explain how to determine the k_{31}, d_{31}, and Q_m values from these data. Use the relationship: $Q_m = (1/2) \omega_0 \tau$.

REFERENCES

1. http://www.qortek.com/en/products/piezo-drivers/polydrive-low-cost-lab-driver/.
2. S. Hirose, S. Takahashi, K. Uchino, M. Aoyagi, Y. Tomikawa, Measuring Methods for High-Power Characteristics of Piezoelectric Materials, *Proceedings of MRS '94 Fall Meeting*, 360, 15 (1995).
3. T. Yuan, X. Dong, H. Shekhani, C. Li, Y. Maida, T. Tou, K. Uchino, Driving an Inductive Piezoelectric Transducer with Class E Inverter, *Sensors & Actuators A: Physical, A.* **261**, 219 (2017).

4. X. Dong, T. Yuan, M. Hu, H. Shekhani, Y. Maida, T. Tou, K. Uchino, Driving frequency optimization of a piezoelectric transducer and the power supply development, *Review of Scientific Instruments*. **87**, 105003 (2016). DOI: 10.1063/1.4963920.

5. C.A. Rosen, Ceramic Transformers and Filters, *Proceedings of Electronic Component Symposium*, pp. 205–211 (1956).

6. NEC, Thickness Mode Piezoelectric Transformer, US Patent No. 5,118,982 (1992).

7. S. Kawashima, O. Ohnishi, H. Hakamata, S. Tagami, A. Fukuoka, T. Inoue, S. Hirose, *Proceedings of IEEE International Ultrasonic Symposium*, Nov. (1994).

9 Loss Mechanisms in Piezoelectrics

ABSTRACT

High-power piezoelectric material's development, based on the loss mechanisms, are introduced in this chapter "Loss Mechanisms in Piezoelectrics". Practical high-power "hard" $Pb(Zr, Ti)O_3$ (PZT)-based materials are initially described, which exhibit vibration velocities close to 1 m/s (rms), leading to the power density capability ten times that of the commercially available "hard" PZTs. Based on the macroscopic phenomenological study results, we consider microscopic crystallographic and semi-microscopic domain wall dynamics models to understand the experimental results. Impurity dipole alignment categories include: (1) random alignment – higher coercive field – "domain wall pinning" model, (2) unidirectionally fixed alignment – internal bias field, and (3) unidirectionally reversible alignment – pinched/double hysteresis. We detail an internal bias field model to explain the low-loss and high-power origin of these materials. The discussion is expanded to Pb-free piezoelectrics, which have been developed recently. The latter part of this chapter is devoted to the phenomenological handling of the domain wall dynamics, based on Ginzburg–Landau functional.

In the foregoing chapters, we stick on the phenomenological and piezoelectric characterization discussions of piezoelectric losses. In this chapter, we consider microscopic or crystallographic origins of loss mechanisms in ferroelectrics/piezoelectrics from the materials science viewpoint. Losses are considered to consist of four portions: (1) domain wall motion, (2) fundamental lattice portion, which should also occur in domain-free monocrystals, (3) semi-microstructure grain portion, which occurs typically in polycrystalline samples, and (4) conductivity portion in less-resistive samples. However, in typical piezoelectric ceramics cases such as lead zirconate titanates (PZTs), the losses originated from the domain wall motion exceed the other three contributions significantly.

9.1 HISTORY OF HIGH-POWER PIEZOELECTRICS

9.1.1 RENAISSANCE OF PIEZOELECTRIC TRANSFORMERS

After the failure of the first piezoelectric transformer production to vacuum-tube color television applications in the early 1970s, due to the mechanical collapse of the piezo-transformer component at the nodal point (Refer to Section 8.4.4), this component development stopped roughly for a quarter century. However, due to the urgent requirement for high step-up voltage transformers for the compact laptop backlight inverter applications, piezoelectric transformers have been focused again in the late 1990s. During the quarter century, the PZT fracture toughness had been significantly improved by the fine raw powder preparation technology, and by new transformer structures and electrode configurations, now, much higher energy density devices are being sought from further miniaturization viewpoint in portable electronic instruments.

The conventional piezo-ceramics had the limitation in the *maximum vibration velocity* (v_{max}) under the resonance drive, since the additional input electrical energy is converted into heat, rather than into mechanical energy. The typical rms value of v_{max} for commercially available "Hard" PZT materials, defined by the temperature rise of 20°C from room temperature, was around 0.3 m/s (i.e., ~5 W/cm^3) for rectangular samples operating in the k_{31} mode (like a Rosen-type transformer) in the 1970s, when the author started to work on the piezo-transformers.[1] However, because of the necessity of much higher power density ~40 W/cm^3 in laptop computers, etc., materials developments have been accelerated. $Pb(Mn, Sb)O_3$ (PMS) – PZT ceramics with the v_{max} of 0.62 m/s were developed for NEC transformers in the 1990s, which satisfied then-requirement for the inverter applications. The details will be discussed in Section 9.4.[2]

9.1.2 MULTILAYER PULSE DRIVE MOTORS

In parallel, pulse drive motor applications, such as dot-matrix and inkjet printers and automobile fuel injection valves, have been booming up in the 1980s, using multilayer (ML) piezoelectric actuators, which was invented by the author in 1978. In the off-resonance drive case, in general, soft PZTs are preferably used due to high piezoelectric constant k_{33}. However, because of pulse/step voltage and cyclical (~1 kHz) drive, associated overshoot/ringing vibration generated additional heat, as described in Section 6.4, we needed to develop new semi-soft, semi-hard PZTs with lower losses (or higher Q_m) for these applications. This will also be discussed in Section 9.5.

9.2 MICROSCOPIC ORIGINS OF EXTENSIVE LOSSES

We discuss first origins of the losses of piezoelectrics, followed by practical high-power piezoelectrics developments in terms of ion doping and grain size effects. After the bias field and stress dependences of losses, the final section is devoted to the domain wall dynamics theory, aiming at the future research directions. Note again that the author focuses on the loss mechanism analysis primarily from the domain wall dynamics viewpoint, which is actually a biased idea. Researchers in the following generation are requested to expand the discussion more comprehensively.

9.2.1 ORIGIN OF THE MULTIDOMAIN STRUCTURE

We refresh the reader with the phenomenology of ferroelectricity from the Landau free energy F (Devonshire Theory)[3,4] in $1 - D$ represented in terms of the order-parameter, polarization P (electromechanical coupling is initially neglected) in the simplest formula, that is, second-order transition. We start from the following two-term expression:

$$F(P,T) = (1/2)\alpha P^2 + (1/4)\beta P^4 \tag{9.1}$$

The coefficients α and β depend, in general, on the temperature, but only α is assumed to be temperature dependent in the following discussion (Refer to Section 2.5.2 for the details). The phenomenological formulation should be applied for the whole temperature range over which the material is in the paraelectric (high temperature) and ferroelectric (low temperature) states. Because the spontaneous polarization should be zero in the paraelectric state, the free energy should be zero in the paraelectric phase at any temperatures above its phase transition temperature (i.e., *Curie temperature*). To stabilize the ferroelectric state with a spontaneous polarization, the free energy for a certain polarization P should be lower than "zero". Otherwise, the paraelectric state is maintained. Thus, at least, the coefficient α of the P^2 term must be negative for the spontaneously polarized state to be stable; while in the paraelectric state, it must be positive passing through zero at a certain temperature T_0, called *Curie–Weiss temperature*. From this concept, we assume α as a linear relation:

$$\alpha = (T - T_0)/\varepsilon_0 C \tag{9.2}$$

where C is taken as a positive constant called the *Curie–Weiss constant*, and T_0 is equal to the actual transition temperature T_C (*Curie temperature*), in the second-order transition. On the contrary, the first-order phase transition model gives $T_C > T_0$, and the spontaneous polarization P_S shows a discontinuous jump at the Curie temperature.

The equilibrium polarization under an electric field E should satisfy the condition:

$$(\partial F / \partial P) = E = \alpha P + \beta P^3 \tag{9.3}$$

With no electric field applied, Eq. (9.3) $[P(\alpha + \beta P^2) = 0]$ provides two cases: (1) $P = 0$ and (2) $P^2 = -\alpha/\beta$. The trivial solution in (1) case corresponds to a paraelectric state (high-temperature phase). The (2) finite polarization solutions $\left(P_S = \pm\sqrt{-\alpha/\beta} \right)$ correspond to a ferroelectric state, when $\alpha < 0$ or low-temperature phase.

Figure 9.1 shows the temperature dependence of the Landau free energy curve. One potential minimum at a high temperature will split into two minimum branches at a low-temperature phase. This "Y"-shape phase splitting is called "bifurcation". The critical point corresponds to the Curie temperature T_C (in this case also T_0). Because the free energy curve is symmetric with respect to the polarization P when the external field $E = 0$, the probability of the state $+P_S \left(= +\sqrt{-\alpha/\beta} \right)$ or $-P_S \left(= -\sqrt{-\alpha/\beta} \right)$ should be equal. Thus, with decreasing the temperature, passing through T_C, $+P_S$ domains and $-P_S$ domains may arise locally in a specimen with equal volumetric ratio, leading to the multidomain states. The total polarization may be zero because of the compensation between $+P_S$ and $-P_S$ domains. This state is called "depolarized/depoled" state.

In order to generate the ferroelectricity and piezoelectricity, we need the "poling" process; that is, by applying a reasonable external electric field, the polarization direction is aligned to one direction ($+P_S$ or $-P_S$). From Eq. (9.3), the electric field dependence of the free energy curve can be calculated, which is shown in the inserted figure of Figure 9.1. Under the external electric field, the energy curve becomes asymmetric, which promotes the polarization switching to one direction. When we apply $+E$, we expect the $+P_S$ polarized structure, which is called "poling" process. Though we cannot expect perfectly polarized state thermodynamically, large portion of the ferroelectric specimen will be polarized. Schematic polarization reorientation process or poling process in a polycrystalline specimen is visualized in Figure 9.2. First, the polycrystal is composed of many small single crystals (each is called "grain") with random crystal orientations. Thus, the complete alignment of the polarization is impossible. Further, due to this crystallographic misorientation, some residual stress exists in the specimen, which promotes multidomain status even under a high electric field. We start from the initially negatively poled status "1". You can notice some domains in each grain. With increasing the electric field up to the "coercive" field E_C "2" (where the free energy at $-P_S$ reaches zero), the largest number of domains come up and the total polarization becomes almost zero. When we further increase the field "3", the domain rapidly disappears to become close to monodomain state in each grain. The slope of the strain vs. electric field around "3" corresponds

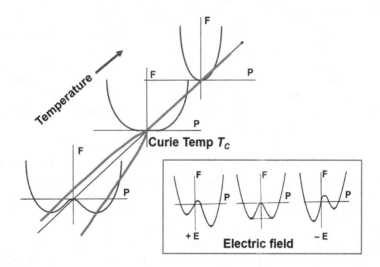

FIGURE 9.1 Temperature and electric field dependence of the Landau free energy curve.

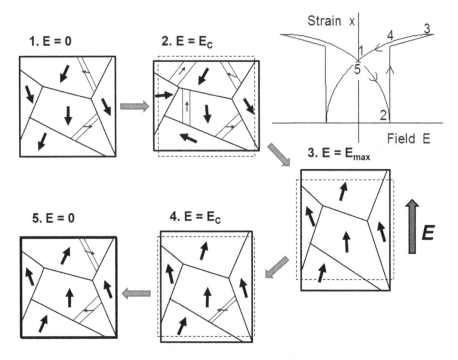

FIGURE 9.2 Domain structure change with the external electric field in polycrystalline ferroelectrics.

to the piezoelectric constant. Now if we decrease the field down to the coercive field "4", we may start to observe some domain generation in grains, then finally at zero field "5", we observe similar domains in each grain as in the state "1", though the polarization directions are opposite to the state "1".

Once the polarization model is extended to 3D and the first-order phase transition, the situation becomes more complicated. Let us take barium titanate, $BaTiO_3$ (BT), as a classic example of a ferroelectric. BT has the perovskite crystal structure. In its high temperature, paraelectric (nonpolar) phase, there is no spontaneous polarization in a cubic symmetry ($O_h - m3m$). Below the transition temperature, which is designated by T_C and called the *Curie temperature* (~130°C for $BaTiO_3$), spontaneous polarization develops. The crystal structure becomes slightly elongated, assuming a tetragonal symmetry ($C_{4v} - 4mm$). There are also two lower temperature phase transitions for $BaTiO_3$, one from a tetragonal to an orthorhombic phase (~0°C) and the other from an orthorhombic to a rhombohedral phase (~–90°C). Figure 9.3 illustrates these successive phase transitions from cubic, tetragonal, orthorhombic, and rhombohedral phases with a decrease in temperature. In the tetragonal phase, the spontaneous polarization direction is in the perovskite z axis <001>, which cants to the face diagonal <011> axis in the orthorhombic phase, then finally in the lowest rhombohedral phase, to the body diagonal <111> axis.

We focus for the moment on the tetragonal room temperature phase. There exist six equivalent spontaneous polarization directions: $[100], [\bar{1}00], [010], [0\bar{1}0], [001]$, and $[00\bar{1}]$. Thus, we expect the 180° and 90° domain walls to distinguish the different domain areas. Figure 9.4 demonstrates the electric field-induced 90° domain reorientation in single crystal barium titanate (BT) at room temperature (cut in parallel to the perovskite axis directions). From the negatively poled specimen ($-P_S$) (mostly downward polarization domain pattern), we applied the positive electric field. Figure 9.4a shows the domain pattern around the coercive field range, where the most complicated domain pattern comes out. The reader can recognize first the 45°-cant stripes, which indicate the domain wall directions between two domain areas. The dark stripes correspond to the rightward domain areas,

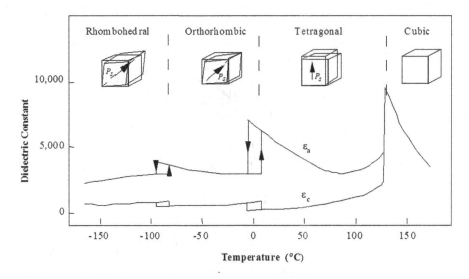

FIGURE 9.3   Various phase transitions in barium titanate (BT).

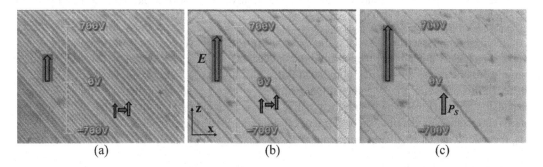

FIGURE 9.4   Electric field-induced 90° domain reorientation in barium titanate (BT) at room temperature. The arrow length corresponds to the voltage magnitude (a. 300, b. 600, c. 700 V).

while the bright stripes to the upward domain areas. Because of the volumetric compensation of these domains, the strain/displacement curve exhibits the minimum point around the coercive field. With increasing the applying field, the dark stripes narrow and diminish finally at the maximum field to approach to almost monodomain state (notice the change from Figure 9.4b and c).

The strains induced parallel ($\Delta l / l = x_3$) and perpendicular ($\Delta l / l = x_1$) to the applied electric field in a 7/62/38 (Pb, La) (Zr, Ti)O_3 (PLZT) ceramic are shown in Figure 9.5. For a cycle with a small maximum electric field (curve a in the figure), the field-induced strain curve appears nearly linear and reflects primarily the "converse piezoelectric effect". As the amplitude of the applied electric field is increased, however, the strain curve becomes more hysteretic (curve c), and finally transforms into the characteristic symmetric "butterfly" shape (curve e) when the electric field exceeds a certain critical value known as the *coercive field*. This is caused by the switching of ferroelectric domains under the applied electric field, resulting in a different state of polarization. Strictly speaking, this composition of PLZT undergoes this change in polarization state in two stages:

1. Individual domain reorientation in each grain, and
2. Overall multidomain reorientation and domain wall movement within the entire polycrystalline structure (which may be regarded as just an assembly of randomly oriented tiny crystallites).

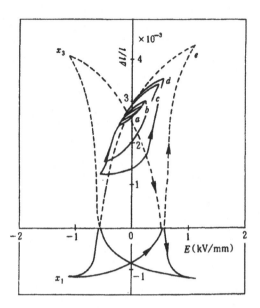

FIGURE 9.5 Electric field-induced longitudinal and transverse strain in a piezoelectric PLZT 7/62/38 ceramic.

Another important issue to point out is the *Poisson's ratio* σ, which is defined by the ratio of the transverse shrinkage x_1 (or x_2) over the longitudinal extension x_3 ($\sigma = -x_1/x_2$). The piezostriction Poisson's ratio is rather close to the elastically induced Poisson's ratio in practice, which are about 1/3 (0.26 ~ 0.35). In this particular PLZT composition exhibits $\sigma = 0.31$. You should understand that volumetric expansion is always associated with the positive electric field application on the PZTs. Note that the pure reorientation model of tetragonal micro crystals should result in the Poisson's ratio 1/2, indicating that additional lattice distortion should be taken into account to explain the volumetric expansion.

Example Problem 9.1

Barium titanate (BaTiO$_3$) has a rhombohedral crystal symmetry at liquid nitrogen temperature (–196°C) and the distortion angle from the cubic structure is not very large (≈1°). Determine all possible angles between the two non-180° domain walls.

Solution

The polarization of barium titanate at this temperature is oriented along the <111> directions of the perovskite cell (eight directions in total). Let us consider the representative three shown in Figure 9.6a: [111], [11$\bar{1}$], and [1$\bar{1}\bar{1}$]. Assuming a *head-to-tail alignment* of the spontaneous polarization across the domain wall [i.e., the normal component of the P_S should be continuous to satisfy

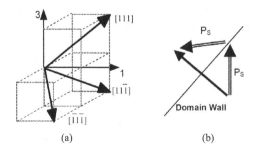

FIGURE 9.6 (a) Polarization directions and (b) spontaneous polarization vectors across a domain wall.

the Gauss Law, div $\boldsymbol{D} = 0$ (equivalent to div $\boldsymbol{P} = 0$ in a high permittivity dielectric) in highly resistive piezoelectrics], as depicted in Figure 9.6b, we expect the plane of the domain wall to be normal to one of the following directions or their equivalent directions:

$$[111] + [1\,1\bar{1}] = [220]$$

$$\text{and } [111] + [1\bar{1}\bar{1}] = [200]$$

The angle between two of the non-180° domain walls is thus calculated as follows:

1. $(002)/(200); \ (022)/(0\bar{2}2); \ (002)/(220)$

   $(002) \cdot (200) = 2^2 [\cos(90°)] = 0 \qquad\qquad \rightarrow \theta = 90° \qquad\qquad$ (P9.1.1)

2. $(022)/(220); \ (022)/(2\bar{2}0)$

   $(022) \cdot (220) = \left(2\sqrt{2}\right)^2 [\cos(\theta)] = 4 \text{ or} -4 \qquad \theta = 60° \text{ or } 120° \qquad$ (P9.1.2)

3. $(002)/(022); \ (002)/(0\bar{2}2)$

   $(022) \cdot (220) = 2\left(2\sqrt{2}\right)[\cos(\theta)] = 4 \text{ or} -4 \qquad \theta = 45° \text{ or } 135° \qquad$ (P9.1.3)

We can observe three possibilities, 45°, 60°, and 90°, among the domain wall angles.

From the domain observation under a microscope, we can speculate the polarization direction of the domain (except for the + or − information) and the crystal symmetry (tetragonal, rhombohedral, etc.). Using Figure 9.4, consider how the spontaneous polarization directions are assigned as shown in that figure by arrows.

9.2.2   The Uchida-Ikeda Model

In this section, we discuss the *Uchida-Ikeda theory*, by which the polarization and strain curves for piezoelectric polycrystalline specimens are described and predicted in terms of domain reorientation, the crystal structure, and the coercive field. This is one of the classic phenomenological approaches without considering the details of domain wall dynamics.

Let us start from barium titanate ($BaTiO_3$) single crystal, as an example, which has a tetragonal symmetry at room temperature. X-ray diffraction of the crystal reveals a slight elongation along the [001] direction of the perovskite unit cell with $c/a = 1.01$. Therefore, if an electric field is applied on an "a" plane single crystal (electrode on the top and bottom of the BT plane), a 90° domain reorientation from an "a" to a "c" domain is induced, resulting in a strain of "1%" in the field direction (refer to Figure 9.4). However, the situation is much more complicated in the case of a polycrystalline specimen (refer to Figure 9.5). Uchida and Ikeda treated this problem statistically, assuming the grains (or small crystallites) are randomly oriented.[5,6]

There will be no remnant polarization in a homogeneous "unpoled" polycrystalline sample. Let this state be the basis for zero strain. If an electric field, E_3 is applied to this sample, a net polarization P_3 will be induced. The strains x_1, x_2, and x_3 will also be generated, where $x_1 = x_2 = -\sigma x_3$ and σ is the *Poisson's ratio* (in terms of electric field induced strains). Let the spontaneous polarization and the *principal strain* of the individual crystallites (i.e., single crystal values) be designated by P_S and S_S, respectively. For uniaxial crystal symmetries, such as the tetragonal and rhombohedral, the principal strain, S_S, is in the direction of P_3 and is defined for each symmetry as follows:

$$S_S = \left[(c/a) - 1\right]\left(\text{tetragonal crystal}\right) \tag{9.4}$$

$$S_S = (3/2)[\pi/2 - \alpha] = (3/2)\delta\left(\text{rhombohedral crystal}\right) \tag{9.5}$$

The distinction between the principal strain and the spontaneous strain should be noted here. They are *not* interchangeable terms, but in fact define two very different strains. Using $BaTiO_3$ as an example, we see that the principal strain is $S_S = 0.01$, but the spontaneous strains in the tetragonal phase are defined as:

$$x_{3(S)} = \left[(c/a_0) - 1 \right]$$ (9.6)

$$x_{1(S)} = x_{2(S)} = \left[(a/a_0) - 1 \right]$$ (9.7)

where a_0 is the lattice parameter of the paraelectric phase. When the appropriate lattice parameters are used in Eq. (9.6), the spontaneous strains $x_{3(S)}$ and $x_{1(S)}$ are 0.0075 and -0.0025, respectively. The principal strain $S_S \approx (x_{3(S)} - x_{1(S)})$, and the Poisson's ratio $\sigma = -(x_{1(S)}/x_{3(S)}) = 1/3$. To the contrary, when we consider just the realignment of tetragonal crystals, the Poisson's ratio should be 1/2 theoretically.

First, assuming an angle, θ, between the direction of the spontaneous polarization, P_S, of a microscopic volume, dv, in a ceramic and the direction of the electric field E_3 (refer to Figure 9.7), the polarization contribution to the electric field direction, P_3, is given by $P_S \cos \theta$. By integrating on all volumes:

$$P_3 = \frac{\int P_S \cos \theta \, dv}{\int dv} = P_S \overline{\cos \theta}$$ (9.8)

where $\overline{\cos \theta}$ is the average value of $\cos \theta$ in all the volume elements of the ceramic. The average strain is determined from the orientation of the strain ellipsoid:

$$x_3 = S_S \left[\frac{\int \cos^2 \theta \, dv}{\int dv} - \frac{1}{3} \right] = S_S [\overline{\cos^2 \theta} - (1/3)]$$ (9.9)

Since the strain is the second-rank tensor, we need to multiply the direction cosine ($\cos \theta$) twice to calculate the contribution along the z axis. It is assumed in the Uchida-Ikeda model that the spontaneous strains associated with the microscopic regions change only in their orientation with no change in volume, hence the Poisson's ratio is apparently $\sigma = 0.5$. This also implies that:

$$x_1 = x_2 = -x_3/2$$ (9.10)

However, this assumption does not agree well with the experimental data shown in Figure 9.5 $(|x_1/x_3| \approx 1/3)$.

In order to arrive at an expression for the induced strain as a function of the applied electric field, the relationship between θ and E_3 must be determined. This is accomplished in the context of this

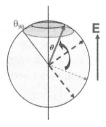

FIGURE 9.7 Polarization reorientation model for non-180° domain wall's case.

model by introducing a *characteristic angle*, θ_{90}, for non-180° domain reorientations. In tetragonal crystals, this corresponds to a 90° reorientation and in rhombohedral crystals, 71° and 109° reorientations will occur, but in order to simplify the analysis, all reorientations are represented by the former. Suppose a 90° domain rotation occurs in a small volume element, dv, in a ceramic, and as a result the domain orientation within dv becomes θ (refer to Figure 9.7). It is assumed that there exists a characteristic angle θ_{90}, such that if $\theta < \theta_{90}$, a 90° rotation will occur, whereas if $\theta > \theta_{90}$, no rotation will occur, and the region will remain in its initial state (i.e., simplest on-off model). Given a specific angle, θ_{90}, which corresponds to a certain applied field strength, E_3, Eq. (9.9) can be integrated over the range of volume for which $\theta < \theta_{90}$ is satisfied, to obtain the induced strains x_3, x_2, and x_1 as a function of θ_{90}. The quantity $(\overline{\cos^2\theta - 1/3})$ is plotted as a function of the characteristic angle, θ_{90}, in Figure 9.8, where the saturation values of the polarization and the strain of a tetragonal and rhombohedral ceramic under high electric field are summarized as follows:

$$\text{Tetragonal: } P_3 \rightarrow (0.831)P_S \qquad x_3 \rightarrow (0.368)S_S$$

$$\text{Rhombohedral: } P_3 \rightarrow (0.861)P_S \qquad x_3 \rightarrow (0.424)S_S$$

Figure 9.9a shows the measured values of induced strain for a rhombohedral PZT ceramic. When the above equations defining these trends are combined, the curve representing the relationship between θ_{90} and E_3 is generated as depicted in Figure 9.9b. It is important to note that the saturated polarization reaches higher than 83% of the single crystal value, but the strain will reach only 1/3 of the single crystal value. Further, the critical θ_{90} is around only 25° (Figure 9.9b) in PZTs practically, which corresponds to $(\overline{\cos^2\theta - 1/3}) \sim$ only 0.1–0.15 of the single crystal spontaneous strain. In conclusion, we can evaluate the single crystal spontaneous polarization roughly from the polycrystalline experimental value just by multiplying 1.15 or so (the order is almost the same), while the spontaneous strain evaluation is very difficult, because the ceramic strain value is only 1/10–1/7 (one order of magnitude smaller) of the single crystal value. Furthermore, by finding the

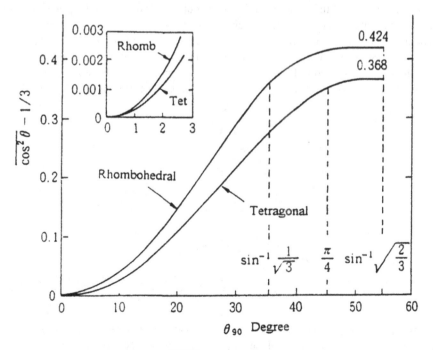

FIGURE 9.8 Strain-proportional quantity $(\overline{\cos^2\theta - 1/3})$ vs. characteristic angle, θ_{90}.[5,6]

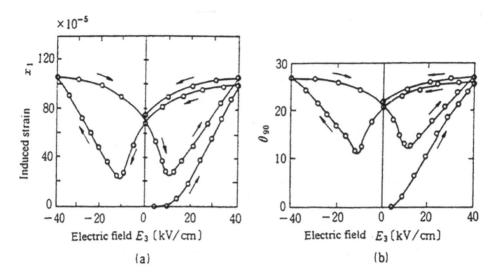

FIGURE 9.9   Field-induced transverse strain, x_1, in a Pb(Zr$_{0.57}$Ti$_{0.43}$)O$_3$ ceramic sample: (a) x_1 vs. applied electric field, E_3, and (b) calculated characteristic angle, θ_{90}, vs. E_3. Measurement at 30°C.[5,6]

polarization P_3 and field-induced strain x_3 (or x_1) as a function of the electric field, E_3, it is possible to estimate the volume in which a 180° reversal or a 90° reorientation occurs. This is because only the 90° reorientation contributes to the induced strain, whereas 180° domain reversal contributes mainly to the polarization. Curves representing the volume fraction of 180° domains that have undergone reversal and 90° domains that have rotated by 90° as a function of applied electric field are shown in Figure 9.10. We see from these curves that the 180° reversal occurs quite rapidly above a certain field (H) as compared to the slower process of 90° rotation started from a low field (G).[7] It is notable that at G on the curve, there remains some polarization and the induced strain is zero, while at H, the strain is not at a minimum, but contributions to the polarization from the 180° and 90° reorientations cancel each other so that the net polarization becomes zero. Due to a sudden change in the 180° reversal above a certain field (H), we can expect a sudden increase in the polarization hysteresis and in the dielectric loss. This may reflect rapid extensive dielectric loss increase with field as observed in Figure 7.5a. On the contrary, since the slope of 90° reorientation is gradual with almost constant slope, we can expect a constant extensive elastic loss with changing

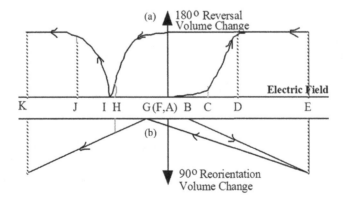

FIGURE 9.10   Electric field dependence of the volume fraction of domains: (a) 180° reversal and (b) 90° reorientation. Notice the non-coincidence of the zero fraction points I and G (coercive field), which correspond to the 180° and 90° cases, respectively.[7]

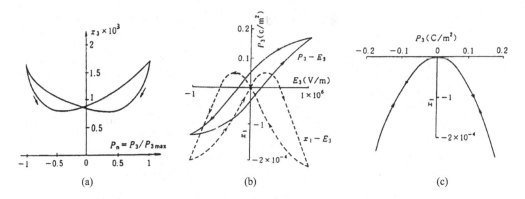

FIGURE 9.11   (a) $x_3 - P_3$ in tetragonal PLZT (6.25/50/50) ceramic at room temperature.[8] (b) $P_3 - E_3$, $x_1 - E_3$, and (c) $x_1 - P_3$ in rhombohedral $Pb(Mg_{1/3}Nb_{2/3})O_3$ ceramic at $-110°C$.[9]

the external parameter, E or X, which is observed in the extensive elastic loss in Figure 7.5b. This dramatic increase in 180° domain wall motion (i.e., the extensive dielectric loss) seems to be the origin of the apparent Q_m degradation and heat generation above the maximum vibration velocity, which was already discussed in Figure 7.8.

A plot of the induced strain, x_3, as a function of polarization for a tetragonal PLZT (6.25/50/50) ceramic, in which 180° reversal is dominant, is shown in Figure 9.11a. We see that it is characterized by a rather large hysteresis in the tetragonal ceramics probably due to the coercive field difference between 180° and 90° reorientation.[8] In contrast, materials whose polarization is dominated by non-180° domain rotations, such as the low-temperature rhombohedral phase of $Pb(Mg_{1/3}Nb_{2/3})O_3$, exhibit no significant hysteresis in their $x - P$ curves (see Figure 9.11c), though $P - E$ and $x - E$ curves show significant hysteresis (Figure 9.11b).[9] This suggests almost the same coercive fields with respect to 180° and non-180° domain reorientations, and that the $x - P$ curve fits to the electrostrictive relation $x_1 = Q_{13}P_3^2$ without hysteresis even in the ferroelectric state, as shown in Figure 9.11c.

9.2.3   Domain Wall Dynamic Models for Losses

We relate here the microscopic loss origins only with the domain wall motion-related losses, though it seems to be too simplified. Taking into account the fact that the polarization change is primarily attributed to 180° domain wall motion, while the strain is attributed to 90° (or non-180°) domain wall motion, we suppose that the extensive dielectric and mechanical losses are primarily originated from 180° and 90° domain wall motions, respectively, as illustrated in Figure 9.12. The dielectric loss comes from the hysteresis during the 180° polarization P reversal under E, while the elastic loss comes from the strain x hysteresis during the 90° polarization reorientation under X. Regarding the piezoelectric loss, we presume that it is originated from "Gauss Law", div $(D) = \sigma$ (charge). As illustrated in Figure 9.12, when we apply a tensile stress on the piezoelectric crystal, the vertically elongated cells will transform into horizontally elongated cells, so that the 90° domain wall will move rightward. However, this "ferroelastic" domain wall will not generate charges because the rightward and leftward polarizations may compensate each other. Without having migrating charges σ in this crystal, div $(P) = 0$, leading to the polarization alignment "head-to-tail", rather than "head-to-head" or "tail-to-tail". After the "ferroelastic" transformation, this polarization realignment will need an additional time lag (via electromechanical coupling effect), which we may define the piezoelectric loss. Superposing the ferroelastic domain alignment and polarization alignment (via Gauss Law) can generate actual charge under stress application. In this model, the intensive (observable) piezoelectric loss is explained by the 90° polarization reorientation under E, which can be realized by superimposing the 90° polarization reorientation under X and the 180° polarization

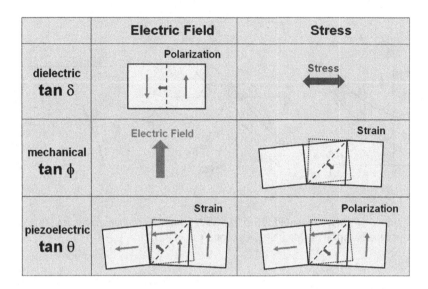

FIGURE 9.12 Polarization reversal/reorientation model for explaining "extensive" dielectric, elastic, and piezoelectric losses.

reversal under E. This is the primary reason why we cannot measure the piezoelectric loss independently from the elastic or dielectric losses experimentally. More quantitative approach is discussed in Section 9.8.

9.3 CRYSTAL ORIENTATION DEPENDENCE OF LOSSES

9.3.1 PMN-PT SINGLE CRYSTAL

Extensive studies have been made on the different characteristics between ferroelectric single crystals and polycrystalline PZT. $Pb(Mg_{1/3}Nb_{2/3})O_3 \cdot (PMN)$-$PbTiO_3 \cdot (PT)$ single crystals, for example, have significant loss anisotropy and doping dependence, as shown in Figures 9.13 and 9.14.[10] Figure 9.13a shows the maximum vibration velocities (defined by 20°C temperature rise) of single crystals and ceramics (k_{31} mode). The performance of single crystals is not as good as high-power

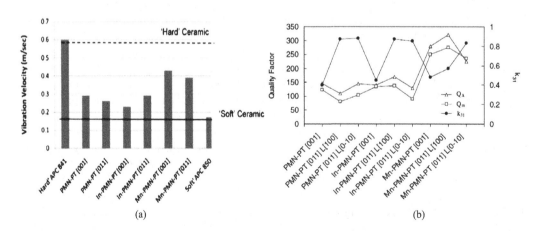

FIGURE 9.13 (a) Maximum vibration velocities of single crystals and ceramics (k_{31} mode). (b) Mechanical quality factors and electromechanical coupling factors for the k_{31} vibration mode.

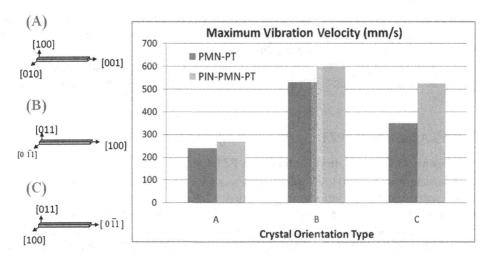

FIGURE 9.14 High-power characteristics for k_{31} mode 0.67PMN-0.33PT and 0.23PIN-0.5PMN-0.27PT samples with different crystal orientations.

hard PZT but better than soft PZTs. Mechanical quality factors and electromechanical coupling factors for the k_{31} vibration mode are plotted for various orientations of single crystals in Figure 9.13b. You may find significant crystallographic orientation dependence of mechanical losses, as well as the real parameter k_{31} change.

In addition, PMN-PT single crystals have been utilized in high-power applications and showed significant advantages compared with hard PZTs. The maximum power densities (defined by 20°C temperature rise) of different high-power piezoelectric transformers are shown in Table 9.1.[11,12] In recent years, the Generation III $Pb(In_{1/3}Nb_{2/3})O_3$ (PIN)-PMN-PT single crystals have been developed, which have higher coercivity and Curie temperature than PMN-PT.[13] Moreover, the high-power characteristics were measured in the ICAT, The Penn State University, as shown in Figure 9.14. High-power characteristics for k_{31} mode 0.67PMN-0.33PT and 0.23PIN-0.5PMN-0.27PT samples with different crystal orientations were tested in terms of the maximum vibration (defined by 20°C temperature rise). According to the results, irrelevant to the composition Type B orientation ([011] plate with [100] length direction) for k_{31} vibration mode has the best high-power performance among the three. Type C orientation ([011] plate with [0$\bar{1}$1] length direction) exhibits the second, and Type A orientation ([100] plate with [001] length direction) seems to show high losses. The ternary composition PIN-PMN-PT is better than the binary PMN-PT in the sense of quality factor, maximum vibration velocity, and heat generation.

TABLE 9.1

Maximum Power Densities and Efficiencies, Mechanical Quality Factors, and Electromechanical Coupling Coefficients for Each Material

Sample	Material	Max Power Density (W/cm³)	Max Efficiency	Q_m	k_{31}
1	APC841 hard PZT ceramic	7.7	0.66	1,100	0.30
2	PMN-PT L[100] w[010] t[001]	1.1	0.33	150	0.40
3	PMN-PT L[0$\bar{1}$1] w[100] t[011]	11.8	0.86	150	0.88
4	Mn-PMN-PT L[100] w[010] t[001]	5.2	0.88	250	0.50
5	Mn-PMN-PT L[100] w[0$\bar{1}$1] t[011]	30.1	0.96	320	0.61
6	Mn-PMN-PT L[0$\bar{1}$1] w[100] t[011]	38.1	0.93	220	0.83

In contrast, crystal orientation dependence of mechanical quality factor and dielectric loss of 0.7PMN-0.30PT crystals were reported by Zhang et al. on the k_{33} mode.[14] The lowest loss factor was found to be along their respective polar direction (poled along perovskite [001] direction), with mechanical Q_m values being >1,000 (low vibration velocity). Of particular significance is that both high electromechanical coupling (~0.9) and large mechanical Q_m ~ 600) were achieved in the [011] poled crystals.

9.3.2   LOSS ANISOTROPY IN PZT CERAMICS

9.3.2.1   Twenty Loss Dissipation Factors for a PZT Ceramic

Zhuang et al. determined all 20 loss dissipation factors for a PZT ceramic using the ICAT HiPoCS admittance spectrum method introduced in Section 7.2.3 on characterization.[15] Five specimens shown in Figure 4.7 were prepared and measured on the admittance/impedance spectra for determining the piezoelectric properties. Table 9.2 summarizes all elastic, dielectric, and piezoelectric losses determined on a soft PZT, APC 850 (APC International, State College, PA). Compared with the error range 2%–3% of the intensive loss factors, you can notice significantly large (ten times) error range of the extensive loss factors in the table, which is originated from the error propagation due to the $[K]$ matrix calculation. Because of this reason, we proposed partial-electrode specimen configurations to measure both E-constant and E-constant elastic compliances independently. Refer to Section 7.4.

Note that the following general conclusions:

1. The antiresonance Q_B is always larger than the resonance Q_A in PZTs: This is a significant contradiction with the IEEE Standard assumption, leading to the necessity of the "Standard" revision.

TABLE 9.2

The 20 Loss Factors of Soft PZT (APC 850) with Experimental Uncertainties

	$\tan\phi'_{11}$	$\tan\phi'_{12}$	$\tan\phi'_{13}$	$\tan\phi'_{33}$	$\tan\phi'_{55}$
Mean	0.01096	0.0095	0.01507	0.01325	0.0233
Uncertainty	0.00007	0.0003	0.00034	0.00033	0.0022
Relative (%)	0.6	3.2	2.2	2.5	9.6

	$\tan\phi_{11}$	$\tan\phi_{12}$	$\tan\phi_{13}$	$\tan\phi_{33}$	$\tan\phi_{55}$
Mean	0.0105	0.0104	0.0076	0.00433	0.0149
Uncertainty	0.0018	0.0028	0.0013	0.00008	0.0003
Relative (%)	17	28	17	1.7	2.1

	$\tan\delta'_{33}$	$\tan\delta'_{11}$	$\tan\delta_{33}$	$\tan\delta_{11}$
Mean	0.0143	0.0176	0.0058	0.0092
Uncertainty	0.0002	0.0004	0.0011	0.0023
Relative (%)	1.4	2.3	20	25

	$\tan\theta'_{31}$	$\tan\delta'_{33}$	$\tan\theta'_{15}$	$\tan\theta_{31}$	$\tan\theta_{33}$	$\tan\theta_{15}$
Mean	0.0184	0.0178	0.0296	0.0133	0.0004	0.0024
Uncertainty	0.0006	0.0004	0.0026	0.0081	0.0004	0.0013
Relative (%)	3.2	2.1	8.8	61	100	57

2. The intensive (prime) losses are larger than the corresponding extensive (non-prime) losses: This is understood by the boundary condition difference between "intensive" and "extensive"; that is, *Free* or *Clamped/Constrained* conditions.
3. The intensive piezoelectric losses are significantly larger than the intensive dielectric or elastic losses in PZTs. That is, $\tan\theta' > (\tan\delta' + \tan\phi')/2$: This is NOT true for Pb-free piezoelectrics, as discussed later.
4. There is apparent loss anisotropy in dielectric, elastic, and piezoelectric losses, indicating the anisotropy in domain wall mobility in the crystal.

Further specific conclusions include:

5. $\tan\delta_{33} < \tan\delta_{11}$: Polarization seems to be more stable along the spontaneous polarization, similar to the general permittivity trend ($\varepsilon_{33}^X < \varepsilon_{11}^X$).
6. $\tan\phi_{33} < \tan\phi_{11}$: Elastic compliance also seems to be more stable along the spontaneous polarization.
7. $\tan\theta_{33} < \tan\theta_{31}$: Piezoelectric constant also seems to be more stable along the spontaneous polarization.

9.3.2.2 Spontaneous Polarization Direction Dependence of Losses

Due to large "intensive" piezoelectric loss, the mechanical quality factor of Pb(Zr, Ti)O$_3$ (PZT) ceramics at antiresonance frequency is much higher than the one at resonance frequency under constant vibration velocity measurement. Thus, driving the piezoelectric resonator at antiresonance frequency is recommended to reduce the required electric power for generating the same level of mechanical vibration. Furthermore, the maximum mechanical quality factor Q_m is realized at a driving frequency between the resonance and antiresonance frequencies basically due to a suitable compensation for the dielectric and elastic losses with the piezoelectric loss. However, since the study on piezoelectric loss has not been made intensively, its physical origin is yet unclear. Assuming the origin of loss as from domain dynamics, it is essential to understand the piezoelectric loss behavior by polarization orientation.

Choi et al. explored the loss anisotropy in piezoelectric PZT ceramics.[16] To observe polarization angle dependence of losses, two different models for k_{31} and k_{33} mode vibration were prepared as shown in Figure 9.15.

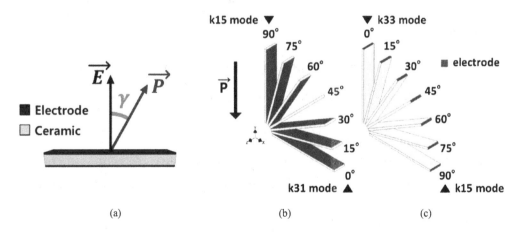

FIGURE 9.15 Sample configurations with various polarization direction. (a) Angle, (b) k_{31} type, and (c) k_{33} type.

The samples were prepared by PI Ceramics GmbH, Germany, with cutting and re-electroding, following the conventional ceramic processing. The polarization angle is defined by the angle of the polarization measured from the electric field direction (i.e., $0°$: $P_S \mathbin{/\!/} E$; $90°$: $P_S \perp E$). With similar elastic compliance and same length, the fundamental longitudinal resonance of both models happens around $100\,\text{kHz}$, while the shear vibration occurs around $1\,\text{MHz}$.

The change of piezoelectric loss factors by polarization angle was investigated using conventional characterization method with effective k_{31} and k_{33} mode structures. One percent Nb-doped PZT ceramics (PNZT) were prepared in tetragonal (Tet), rhombohedral (Rhomb), and morphotropic phase boundary (MPB) structures with $0°$, $15°$, $30°$, $45°$, $60°$, $75°$, and $90°$ polarization angles. Soft PZT was intentionally chosen to identify the difference the loss factor among different specimens (ten times larger than hard PZTs). As a result, we could find out the intensive piezoelectric loss increase with the polarization angle larger than the change in intensive dielectric and elastic losses. However, serious problems were found in the k_{33} structure with small motional capacitance, which are:

1. Error in $3\,\text{dB}$ method to define mechanical quality factor due to the small damped capacitance.
2. Large relative error from indirect calculation (via $[K]$ matrix) and large impedance of k_{33} rod.

Thus, we proposed a new analysis procedure to obtain the elastic parameters of k_{33} mode, using the effective k_{31} vibration mode. The sample capacitance of k_{31} mode structure, 0.24 nF, is L^2/t^2 times higher than 1.0 pF of the k_{33} mode sample, thus structural impedance problem can be minimized. Using three different effective k_{31} mode geometries with different polarization angle, we verified the IEEE Standard underestimates the elastic compliance s_{33}^E and overestimates the elastic loss $\tan\phi_{33}'$.[17]

9.3.2.2.1 Characterization Methodology

We have completed the crystallographic methodology to obtain all independent piezoelectric loss parameters for the polycrystalline ceramics. The proposed methodology contains:

1. Off-resonance dielectric measurements

$$\varepsilon_{33,\text{eff}}^X(\gamma) = \cos^2(\gamma)\varepsilon_{33}^X + \sin^2(\gamma)\varepsilon_{11}^X$$

$$\varepsilon_{33,\text{eff}}^X(\gamma)\tan\delta_{33,\text{eff}}'(\gamma) = \cos^2(\gamma)\varepsilon_{33}^X\tan\delta_{33}' + \sin^2(\gamma)\varepsilon_{11}^X\tan\delta_{11}'$$

2. Effective k_{31} mode analysis

$$s_{11,\text{eff}}^E(\gamma) = \cos^4(\gamma)s_{11}^E + \sin^4(\gamma)s_{33}^E + \cos^2(\gamma)\sin^2(\gamma)\left[2s_{13}^E + s_{55}^E\right]$$

$$s_{11,\text{eff}}^E(\gamma)\tan\phi_{11,\text{eff}}'(\gamma) = \cos^4(\gamma)s_{11}^E\tan\phi_{11}' + \sin^4(\gamma)s_{33}^E\tan\phi_{33}'$$
$$+ \cos^2(\gamma)\sin^2(\gamma)\left[2s_{13}^E\tan\phi_{13}' + s_{55}^E\tan\phi_{55}'\right]$$

$$d_{31,\text{eff}}(\gamma) = \cos^3(\gamma)d_{31} + \cos(\gamma)\sin^2(\gamma)[d_{33} - d_{15}]$$

$$d_{31,\text{eff}}(\gamma)\tan\theta_{31,\text{eff}}'(\gamma) = \cos^3(\gamma)d_{31}\tan\theta_{31}' + \cos(\gamma)\sin^2(\gamma)[d_{33}\tan\theta_{33}' - d_{15}\tan\theta_{15}']$$

3. Off-resonance d_{33} measurement

$$d_{33,\text{eff}}(\gamma) = \cos^3(\gamma)d_{33} + \cos(\gamma)\sin^2(\gamma)[d_{15} + d_{31}]$$

4. Effective k_{15} mode analysis
   a. With undistorted quadrant of $|Z|$ in γ_B

$$s_{55,\text{eff}}^E (\gamma_B) = \frac{s_{55,\text{eff}}^D (\gamma_B)}{\left[1 - k_{15,\text{eff}}^2 (\gamma_B)\right]} = \sin^2 (2\gamma_B)\left(s_{11}^E + s_{33}^E - 2s_{13}^E\right) + \cos^2 (2\gamma_B) s_{55}^E$$

   b. With undistorted quadrant of $|Y|$ in γ_A

$$\tan \phi_{13^\circ}', \tan \phi_{55^\circ}', \tan \theta_{33^\circ}', \tan \theta_{15^\circ}'$$

can be calculated since

$$Q_{B,\text{eff}} (\gamma_B) = f\left\{\gamma_B, \tan \phi_{13}', \tan \phi_{55}', \tan \theta_{33}', \tan \theta_{15}'\right\}$$

$$Q_{A,\text{eff}} (\gamma_A) = f\left\{\gamma_A, \tan \phi_{13}', \tan \phi_{55}', \tan \theta_{33}', \tan \theta_{15}''\right\}$$

$$\frac{1 - k_{15,\text{eff}}^2 (\gamma_B)}{k_{15,\text{eff}}^2 (\gamma_B)} \cdot \frac{1}{Q_{B,\text{eff}} (\gamma_B)} = \tan \delta_{11,\text{eff}}' (\gamma_B) - 2 \tan \theta_{15,\text{eff}}' (\gamma_B) + \frac{\tan \phi_{55,\text{eff}}' (\gamma_B)}{k_{15,\text{eff}}^2 (\gamma_B)}$$

Note the shear mode parameters could only be selectively obtained due to the spurious modes. The obtained independent parameters are shown in Table 9.3.

Figures 9.16 and 9.17 show the P–E angle dependence of effective dielectric permittivities, elastic compliances, and piezoelectric constants and their corresponding intensive loss factors for Tet, MPB, and Rhomb specimens. In our previous paper,[18] we reported effective d_{33} constant enhancement in a rhombohedral soft PZT composition around 45° cut angle. However, interestingly we did not observe the peak in d_{33} value in this composition, but in d_{31} in Rhomb composition.[16] In order to

TABLE 9.3
The Independent Parameters Obtained with Proposed Method

Tetragonal	ε_{33}^X	1,213	ε_{11}^X	1,079	s_{13}^E (pm²/N)	−3.72
	$\tan \delta_{33}'$ (%)	1.11	$\tan \delta_{11}'$ (%)	1.46	$\tan \phi_{13}'$ (%)	4.52
	s_{11}^E (pm²/N)	12.77	s_{33}^E (pm²/N)	12.84	s_{55}^E (pm²/N)	32.83
	$\tan \phi_{11}'$ (%)	0.91	$\tan \phi_{33}'$ (%)	0.90	$\tan \phi_{55}'$ (%)	1.88
	d_{31} (pC/N)	−90	d_{33} (pC/N)	233	d_{15} (pC/N)	322
	$\tan \theta_{31}'$ (%)	1.51	$\tan \theta_{33}'$ (%)	3.03	$\tan \theta_{15}'$ (%)	3.10
MPB	ε_{33}^X	1,455	ε_{11}^X	1,516	s_{13}^E (pm²/N)	−8.01
	$\tan \delta_{33}'$ (%)	1.48	$\tan \delta_{11}'$ (%)	1.62	$\tan \phi_{13}'$ (%)	4.40
	s_{11}^E (pm²/N)	16.09	s_{33}^E (pm²/N)	18.06	s_{55}^E (pm²/N)	51.41
	$\tan \phi_{11}'$ (%)	0.96	$\tan \phi_{33}'$ (%)	1.04	$\tan \phi_{55}'$ (%)	2.19
	d_{31} (pC/N)	−148	d_{33} (pC/N)	391	d_{15} (pC/N)	593
	$\tan \theta_{31}'$ (%)	1.73	$\tan \theta_{33}'$ (%)	2.63	$\tan \theta_{15}'$ (%)	2.50
Rhombohedral	ε_{33}^X	604	ε_{11}^X	950	s_{13}^E (pm²/N)	−5.13
	$\tan \delta_{33}'$ (%)	2.11	$\tan \delta_{11}'$ (%)	2.67	$\tan \phi_{13}'$ (%)	1.64
	s_{11}^E (pm²/N)	12.15	s_{33}^E (pm²/N)	14.75	s_{55}^E (pm²/N)	37.48
	$\tan \phi_{11}'$ (%)	0.74	$\tan \phi_{33}'$ (%)	0.89	$\tan \phi_{55}'$ (%)	1.30
	d_{31} (pC/N)	−77	d_{33} (pC/N)	217	d_{15} (pC/N)	383
	$\tan \theta_{31}'$ (%)	2.04	$\tan \theta_{33}'$ (%)	1.28	$\tan \theta_{15}'$ (%)	2.24

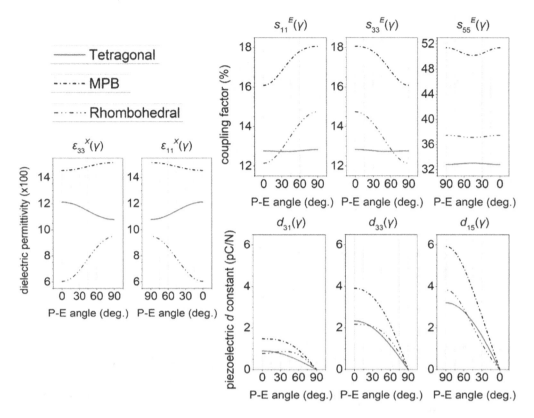

FIGURE 9.16 *P–E* angle dependence of effective dielectric permittivities, elastic compliances, and piezoelectric constants for Tet, MPB, and Rhomb specimens.

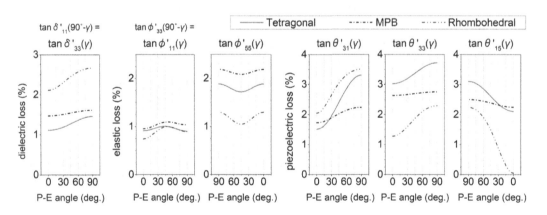

FIGURE 9.17   Intensive loss factors vs. polarization direction.

obtain the peak in the effective piezoelectric constant d at a cant angle, significantly large d_{15} seems to be essential. Compared to the permittivity perpendicular to the polarization, the permittivity along polarization is larger in Tet and smaller in Rhomb and MPB. The MPB structure has highest dielectric constant as expected. On the contrary, the dielectric loss is smaller along the polarization direction compared to the perpendicular direction, regardless of the structure. The elastic compliance along polarization direction is higher than the one perpendicular to the polarization especially in MPB and Rhomb structures. The intensive elastic loss could be considered as constant for the two directions regardless of the structure. It is interesting to note that the piezoelectric loss appeared to be smallest in Rhomb and largest in Tet structure in k_{33} and k_{15} vibration modes. Opposed to the modes, the piezoelectric loss appeared to be largest in Rhomb and smallest in Tet structure in k_{31} vibration. We believe the phenomena are strongly related with the anisotropic real parameters since both relative factors $(-d_{15}/d_{31})$ and $(-d_{33}/d_{31})$ are largest in Rhomb and smallest in Tet. It is also notable that when the relative factors are large, the corresponding piezoelectric loss is small.

More descriptions are provided on Figure 9.17: The dielectric loss is smaller when the applied field is more aligned to polarization. Though the dielectric constant is high in MPB structure, the dielectric loss appeard to be smaller than Rhomb PNZT, presumably due to the coexisting Tet phase. The elastic loss for transversal and longitudinal modes shows a maximum while the loss for the shear mode shows a minimum when the polarization angle is canted near 45°. The "intensive'" piezoelectric loss factors exhibit a similar increasing tendency with angle, regardless of the crystal structure. Higher piezoelectric loss was observed when the polarization is canted. The change of piezoelectric loss was smallest in MPB PNZT, which is the most isotropic structure. Tet PNZT showed the largest polarization orientation dependence of the piezoelectric loss under k_{31} vibration mode, while Rhomb PNZT showed the largest piezoelectric loss for the k_{15} vibration mode. The "electromechanical coupling loss" for effective k_{31} vibration mode was additionally obtained using Eq. (9.11).

$$\frac{k_{31}''}{k_{31}'} = \frac{\left(2 \tan \theta_{31}' - \tan \delta_{33}' - \tan \phi_{11}'\right)}{2}. \qquad (9.11)$$

Here, k' and k'' are the real and imaginary parameters of the electromechanical coupling factor k_{31}; $k_{31} = k_{31}' - jk_{31}''$. Note that Eq. (9.11) is a half of Eq. (4.108), where we discussed on $k_{31}{}^2$. The electromechanical coupling loss anisotropy in terms of crystal structures is shown in Figure 9.18. The value (k_{31}''/k_{31}') shows rather isotropic in MPB, intermediate in Rhomb, and the largest anisotropy in Tet. Under low-vibration excitation, the loss factor along the spontaneous polarization direction is the smallest in Tet, followed by MPB and Rhomb. This is why most of the conventional hard PZT composition was based on the tetragonal phase. However, under high-vibration velocity level, due to large fluctuation of the P_S direction, we need to consider the loss factor effectivity at a cant

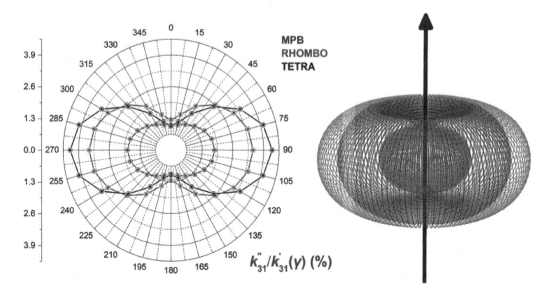

FIGURE 9.18 Electromechanical coupling loss in effective k_{31} vibration mode.

angle. The reader can easily understand that the Q_M decay with vibration velocity increase will be minimized for MPB due to the isotropic loss contour. This is why the high-power density PZT composition is now based on MPB composition, not Tet or Rhomb phases.

9.3.2.2.2 Extensive Loss Determination

Considering three-dimensionally clamped condition, extensive loss parameters were additionally obtained for effective k_{31} and k_{15} vibration mode using "K matrix". Figure 9.19 shows the calculated *extensive loss factors*. The maximum, minimum, and average measured data for each geometry

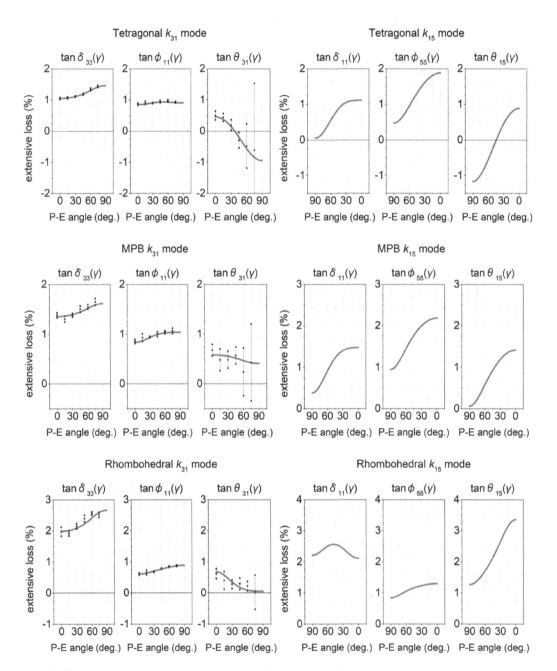

FIGURE 9.19 Extensive loss parameters for effective k_{31} and k_{15} vibrations.

for effective k_{31} mode is plotted for comparison. Considering an open-circuit condition, due to the different charge development in different modes, the effective dielectric loss varies ($\tan \delta_{11}(\gamma) \neq \tan \delta_{33}(\pi/2-\gamma)$). Note that in the short-circuited condition where the charge could be well distributed, the effective dielectric loss is mode-independent ($\tan \delta'_{11}(\gamma) = \tan \delta'_{33}(\pi/2 - \gamma)$). A similar phenomenon is observed for the elastic loss factor. The dielectric and elastic loss is mostly higher when the polarization is angled from the standard structure of each resonator, except for the shear mode in Rhomb PNZT. The piezoelectric loss is more related with the angle between the polarization and the applied electric field than dielectric or elastic loss. The piezoelectric loss is smaller when the angle is larger, meaning the compensation of dielectric and elastic loss is smaller when the polarization is canted from the applied field, from the electromechanical coupling loss perspective. This phenomenon could be a very important point for further study on domain dynamics. The changes are least in MPB structure which is most isotropic. In the effective k_{31} mode, the change of "extensive" piezoelectric loss is largest in Tet PNZT, which is least isotropic. In the effective k_{15} mode, the change of extensive piezoelectric loss is largest in Rhomb PNZT which has the strongest relative shear property. It is interesting to note that the "extensive" piezoelectric loss becomes "negative" in Tet PNZT when the polarization is strongly canted with respect to the applied electric field. The negative extensive piezoelectric loss in Tet structure could be obtained from both k_{31} and k_{15} vibration modes.

9.3.2.2.3 Negative Piezoelectric Loss

Unlike dielectric or elastic hysteresis (electric field vs. polarization or strain vs. stress), which always rotates counterclockwise for a positive compliance, the piezoelectric hysteresis (electric field vs. strain or stress vs. polarization) could rotate counterclockwise or clockwise since the hysteresis does not have energy-density units. The counterclockwise or clockwise hysteresis corresponds to the positive or negative piezoelectric loss. A schematic illustration of the hysteresis is shown in Figure 9.20.

This is the first experimental report of a "negative piezoelectric loss" at high frequencies near resonance (~100 kHz for k_{31} mode and ~1 MHz for k_{15} mode). The issue is, the piezoelectric loss hysteresis cannot be measured separately from dielectric or elastic loss. The electromechanical coupling factor is responsible for the coupled loss factors and the imaginary part becomes as Eq. (9.12) which is equivalent to Eq. (9.11):

$$\frac{k''}{k'} = \frac{-2\tan\theta + \tan\delta + \tan\phi}{2}. \tag{9.12}$$

From the equation, we can see that the piezoelectric loss compensates the elastic and dielectric loss when all parameters are positive. However, when the piezoelectric loss is negative, the phase lag is added up, resulting in an increase of the total loss. This phenomenon could be obtained by segmenting the intensive elastic loss with extensive loss parameters as shown in Figure 9.21. It is shown in most structures that the piezoelectric loss tends to compensate other losses, resulting in to lower the overall intensive elastic loss. However, in Tet PNZT with a largely canted polarization,

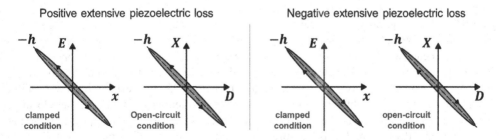

FIGURE 9.20 Hysteresis loop for positive and negative extensive piezoelectric loss.

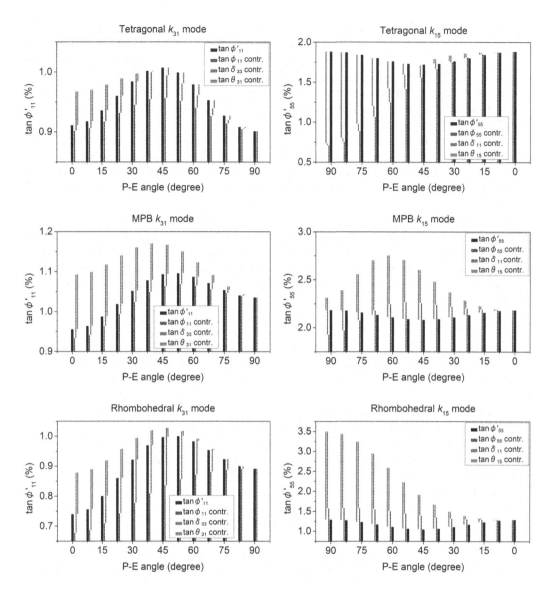

FIGURE 9.21 Contribution of extensive loss factors for the intensive elastic loss in effective k_{31} and k_{15} vibration modes.

the piezoelectric loss adds more phase lags and increases the overall intensive elastic loss. With diminishing electromechanical coupling, the intensive and extensive elastic loss becomes the same.

A possible model of the *negative extensive loss* in tetragonal k_{31} vibration mode is proposed in Figure 9.22. When the polarization canted angle γ is <45°, the positive strain will facilitate 90° domain wall to move leftward. Thus, a transient positive field is generated locally. Note the extensive loss parameters are considered under constant D condition which is electrically open-circuited. When the polarization canted angle is 45°, the positive strain will not facilitate any domain wall motion. While, when the polarization canted angle is more than 45°, the positive strain will facilitate 90° domain wall to move rightward, leading to generate transient negative electric field locally. In other words, positive and negative extensive piezoelectric loss "tan θ" are anticipated for the lower and higher angled PNZT samples, as measured in the tetragonal k_{31} mode in tan $\theta_{31}(\gamma)$ of Figure 9.19 top-left.

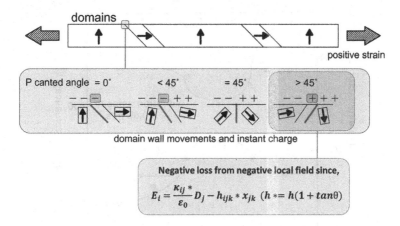

FIGURE 9.22 A possible model of negative extensive piezoelectric loss in effective k_{31} vibration mode.

9.4 COMPOSITION DEPENDENCE OF PIEZOELECTRIC LOSSES

"High Power" in the piezoelectrics stands for "high power density" in mechanical output energy converted from the maximum input electrical energy under the drive condition with 20°C temperature rise from room temperature at the maximum temperature (i.e., node) point. For an off-resonance drive condition, the figure of merit of piezo-actuators is given by the piezoelectric d constant ($\Delta L = dEL$). Heat generation can be evaluated primarily by the intensive dielectric loss tan δ' (i.e., P–E hysteresis). In contrast, for a resonance drive condition, the figure of merit is primarily the vibration velocity v_0, which is roughly proportional to $Q_m \cdot dEL$. Q_m can be considered as an amplification factor of the vibration amplitude and velocity. Heat generation is originated from the intensive elastic loss tan ϕ' (inverse value of Q_m). The mechanical power density can be evaluated by the square of the *maximum vibration velocity* (v_0^2) (as long as the density is similar), which is a sort of material's constant. Remember that there exists the maximum mechanical energy density, above which level, the piezoelectric material becomes a mere ceramic heater. High-vibration velocity piezo-materials are suitable for actuator applications such as ultrasonic motors. Our primary target for the high-power density PZT materials is set around $v_0 = 0.8$ m/s or 40 W/cm³ (of course, the higher is the better in the future), in comparison with the commercially available $v_0 = 0.3$ m/s or 5 W/cm³. Further, when we consider transformers and transducers, where both transmitting and receiving functions are required, the figure of merit will be the product of $v_0 \cdot k$ (k: electromechanical coupling factor). The power density here is the "handling" energy of a piezo-device, and primarily the "input" energy, but should be close to the "output" energy, as long as the pure piezo-materials loss is just around 1%.

9.4.1 PZT-BASED CERAMICS

9.4.1.1 PZT Binary System

Let us discuss high-vibration velocity materials based on PZTs first. Figure 9.23 shows the mechanical Q_m vs. basic composition x at two effective (rms) vibration velocities $v_0 = 0.05$ and 0.5 m/s for Pb(Zr$_x$Ti$_{1-x}$)O$_3$ doped with 2.1 at.% of Fe.[2] The decrease in mechanical Q_m with an increase of vibration level is minimum around the rhombohedral-tetragonal morphotropic phase boundary (MPB 52/48). In other words, the smallest Q_m material under a small vibration level becomes the highest Q_m material under a large vibration level, which is very suggestive. The data obtained by a conventional impedance analyzer with a small voltage/power do not provide any information relevant to high-power performances. The reader should notice that most of the materials with $Q_m > 1,200$ in a company catalog are degraded dramatically at an elevated power measurement. This is the major reason why our ICAT/Penn State group has been putting

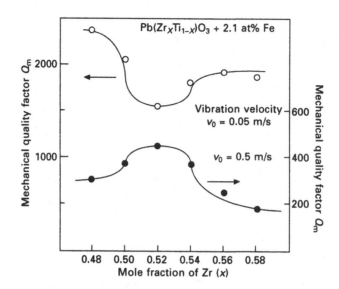

FIGURE 9.23 Mechanical Q_m vs. basic composition x at two effective vibration velocities $v_0 = 0.05$ and $0.5\,\text{m/s}$ for $Pb(Zr_x Ti_{1-x})O_3$ doped with 2.1 at.% of Fe.[2]

significant efforts on the High-Power Piezoelectric Characterization System (HiPoCS™) developments in these 35 years.

9.4.1.2 Improved High-Power PZTs

The typical rms value of v_{max} for commercially available materials, defined by the temperature rise of 20°C from room temperature, is around 0.3 m/s for rectangular samples operating in the k_{31} mode (like a Rosen-type transformer).[19] Pb(Mn, Sb)O$_3$ (PMS)–PZT ceramics with the v_{max} of 0.62 m/s are currently used for NEC transformers.[2] By doping the PMS–PZT or Pb(Mn, Nb)O$_3$–PZT with rare-earth ions such as Yb, Eu, and Ce, we further improved high-power performance, operational with v_{max} up to 1.0 m/s.[20,21] Compared with commercially available piezoelectrics, ten times (square of 3.3 times of v_0) higher input electrical and output mechanical energy can be expected from these new materials without generating significant temperature rise, which corresponds to 50 W/cm². Figure 9.24 shows the dependence of the maximum vibration velocity v_0 (20°C temperature rise) on the atomic % of rare-earth ion, Yb, Eu, or Ce in the PMS–PZT-based ceramics. Enhancement in the v_0 value is significant by the addition of a small amount of the rare-earth ion.[22] The high-power performance improvement by doping will be discussed in Section 9.5 from the mechanism's viewpoint.

9.4.1.3 Semi-hard PZT

Section 6.4 introduced a diesel fuel injection valve with a piezoelectric ML actuator.[23] Because of very-high responsivity of the ML actuator, additional vibration ringing was initially observed after the trapezoidal voltage application. It is noteworthy that the vibration ringing increases the heat generation significantly, in addition to the fuel injection operation problem. When we drive a piezoelectric device under a pulse drive with some vibration overshoot and ringing, in addition to the normal "intensive" dielectric loss under a high-electric field drive, the resonating ringing generates more heat via the elastic loss. Figure 9.25 shows the correlation among mechanical quality factor Q_m and piezoelectric constant d_{33} for various PZT ceramics. You can find two major categories in this figure: (1) hard PZT with high $Q_m > 1,000$ and low piezoelectric constant $d_{33} < 300$ pC/N, and (2) soft PZT with low $Q_m < 100$ and high piezoelectric constant $400 < d_{33} < 700$ pC/N. Though soft PZTs are preferably used for off-resonance actuator applications in general, because of their small Q_m (<100), additional heat generation associated with the vibration ringing could not be neglected. To the contrary, though hard PZTs with high Q_m are good from the heat generation viewpoint,

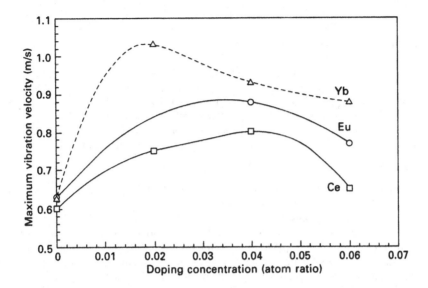

FIGURE 9.24 Dependence of the maximum vibration velocity v_0 (20°C temperature rise) on the atomic % of rare-earth ion, Yb, Eu, or Ce in the PMS–PZT-based ceramics.[21]

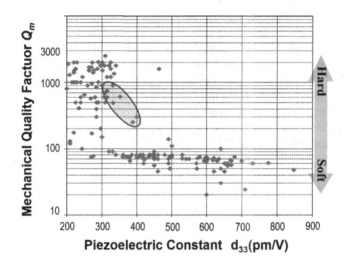

FIGURE 9.25 Correlation among mechanical quality factor and piezoelectric constant for various PZT ceramics.

the displacement amplitude is small (small d_{33}) and vibration ringing is larger due to the vibration amplification by the factor Q_m. In conclusion, compromised semi-hard PZT compositions with intermediate Q_m (300–800) and intermediate d_{33} (300–400 pC/N) were chosen for this practical application (circled area on Figure 9.25).

Example Problem 9.2

Provide the unit of the piezoelectric constant d_{33}. When we use the equation: $x_3 = d_{33}E_3$, since the unit of electric field is [V/m] and strain is unit-less, the unit of the piezoelectric constant should be [m/V]. On the contrary, when we use the equation: $P_3 = d_{33}X_3$, since the unit of stress is [N/m²] and that of polarization is [C/m²], the unit of the piezoelectric constant should be [C/N]. Verify the unit equivalency, [m/V] = [C/N].

Solution

Let us start from the Gibbs energy description:

$$G = -(1/2)s^E X^2 - dXE - (1/2)\varepsilon_0 \varepsilon^X E^2 \tag{P9.2.1}$$

The energy unit per unit volume in Eq. (P9.2.1) is given by [J/m^3]. In the mechanical system, since the work is provided by the product of force and displacement ($W = F \times \Delta L$), the energy unit should be the product of force [N] and displacement [m], that is, [J] = [Nm]. In the electrical system, since the force of a charge q [C] is given by $q \times E$, and integrating this force along the distance, we obtain the potential energy $\int qEdL = qV$.

Thus, the energy unit [J] should be also equivalent to [CV]. In conclusion, [J] = [Nm] = [CV]. The relation, [m/V] = [C/N] can be derived easily.

Let us also reconsider the units for all above constants. The unit of elastic compliance s^E ($\sim G/X^2$) is given by [J/m^3]/[N/m^2]^2 = [Jm/N^2]. Using [J] = [Nm], we obtain [m^2/N]. The unit of permittivity $\varepsilon_0 \cdot \varepsilon^X$ ($\sim G/E^2$) is given by [J/m^3]/[V/m]^2 = [J/V^{2}m]. Using [J] = [CV], we obtain [C/Vm] = [F/m]. Finally, the unit of the piezoelectric constant d is given by [J/m^3]/[N/m^2][V/m] = [J/NV]. Depending on the usage of [Nm] or [CV] for [J], we can obtain either [m/V] or [C/N].

9.4.2 PB-FREE PIEZOELECTRICS

The 21st century is called "The Century of Environmental Management". In 2006, European Community started *RoHS* (*Restrictions on the use of certain Hazardous Substances*), which explicitly limits the usage of lead (Pb) in electronic equipment. We may face to the regulation of PZT, most famous piezoelectric ceramics in the consumer usage, initiated by the government in these several years. Pb (lead)-free piezo-ceramics have been developed after 1999, which are classified into three: (Bi, Na)TiO$_3$ (BNT),[24] (Na, K)NbO$_3$ (NKN),[25] and tungsten bronze (TB).[26] *NKN* systems exhibit the highest performance because of the MPB usage, and in particular, in a structured ceramic manufacturing with flaky powders. *TB* types are another alternative choice for resonance applications, because of high Curie temperature and low loss. A sophisticated preparation technology for fabricating oriented ceramics was developed with an ML configuration: that is, preparation under strong magnetic field.[26] Refer to Section 2.7.5. The author does not think Bi^{3+} is much safer than Pb^{2+} (the ionic structure is similar), there will not be a regulation for the Bi usage, because Bi is not familiar to politicians.

Though there is a lot of research related with lead-free piezoelectric materials, there is quite few research on high-power characterization of the lead-free piezoelectric materials. The study by ICAT/Penn State enlightened the high-power characteristics of Pb-free piezoelectric ceramic compared to hard PZT. Though the RoHS regulation is the trigger of the Pb-free piezoelectric development, if we find superior performances in Pb-free piezoelectrics, this new category material may contribute significantly to the industrial and military applications.

9.4.2.1 High-Power Performance in Pb-Free Piezoelectrics

High-power characteristics were investigated with our high-power piezoelectric characterization system (HiPoCS).[27,28] Tested samples were prepared in collaboration with Toyota R&D, Honda Electronics, Korea University, and Rutgers University, including:

1. NKN – (Na$_{0.5}$K$_{0.5}$)(Nb$_{0.97}$Sb$_{0.03}$)O$_3$ prepared with 1.5% mol CuO addition
2. BNT-BT-BNMN-0.82(Bi$_{0.5}$Na$_{0.5}$)TiO$_3$-0.15BaTiO$_3$-0.03(Bi$_{0.5}$Na$_{0.5}$)(Mn$_{1/3}$Nb$_{2/3}$)O$_3$
3. BNKLT-0.88(Bi$_{0.5}$Na$_{0.5}$)TiO$_3$-0.08(Bi$_{0.5}$K$_{0.5}$)TiO$_3$-0.04(Bi$_{0.5}$Li$_{0.5}$)TiO$_3$ with Mn-doped and -undoped.

Figure 9.26 shows the high-power characterization results for a disk sample. The mechanical quality factors at resonance (Q_A) and at antiresonance (Q_B) do not decrease with increasing vibration level [*mechanical energy density* ($(1/2)\rho v_{rms}^2$)] in Pb-free piezoelectrics, in comparison with

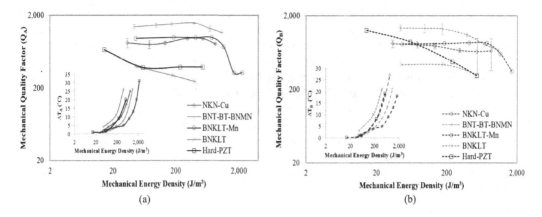

FIGURE 9.26 Temperature rise and mechanical quality factors for (a) resonance (ΔT_A and Q_A) and (b) anti-resonance (ΔT_B and Q_B) as functions of mechanical energy density ($u_{mech} \sim \rho v_{rms}^2$) for lead-free and hard PZT piezoelectric ceramics.

PZT's trend. It is worth noting that the *vibration velocity* (v_{rms}) reaches 0.8 m/s in all three Pb-free compositions but that a low mass density (30%–40% lower than the PZT density) can easily generate large vibration velocity. Therefore, *mechanical energy density* (($1/2)\rho v_{rms}^2$) should be used in order to compare the performance superiority in terms of the maximum vibration level of the materials. When we compare the high-power performance among the same species such as PZTs, we may use the vibration velocity (and the maximum vibration velocity) as the figure-of-merit parameter, since the density is almost the same. The maximum mechanical energy density defined with 20°C increase of the temperature on the nodal point in NKN and BNT-BT-BNMN is superior, when compared to hard PZTs with their sharp decrease in Q_A with increasing vibration level. At antiresonance, the high-power behavior trend for this material did not change. The mechanical quality factor at the antiresonance (Q_B) also remained constant up to the maximum vibration level. Regarding the figure-of-merit (actual vibration level, $Q_m \cdot k_p$) change with the vibration level, the PZT shows better performance in the low-power vibration range, because of higher k_p value in PZT. However, in the high-power range (>1,000 J/m³), only Pb-free piezoelectrics can practically be adopted.

As discussed in Section 6.3, much higher maximum vibration velocity in Pb-free piezo-ceramics such as (Na, K)NbO₃-based materials, than in the PZTs, seems to be partially originated from much larger thermal conductivity in the NKN-based material than the PZTs, as shown in Table 6.2.

9.4.2.2 Piezoelectric Loss Contribution in Pb-Free Piezoelectrics

The Q_A and Q_B values in Pb-free are almost the same up to the maximum vibration velocity, v_{max}. This trend is also distinctly different from the hard PZT, where Q_B is always greater than Q_A, experimentally (depending on the piezoelectric loss). Comparison of mechanical quality factors at resonance (Q_A) and antiresonance (Q_B) provides an aspect to the material losses (i.e., dielectric (tan δ'), mechanical (tan ϕ'), and piezoelectric (tan θ')) behavior for each different composition. Both soft and hard PZTs result in $Q_B > Q_A$, as we already discussed. While, regarding the Pb-free piezoelectrics, hard Bi-perovskite ceramics (i.e., BNT-BKT-BLT-Mn and BNT-BT-BNMN) seem to have the minimum influence from the piezoelectric loss (tan θ') when excited at high vibration levels since they showed the $Q_A > Q_B$ relationship. Cu-doped NKN ceramics seem to have more influence from the piezoelectric loss (tan θ') ($Q_A \cong Q_B$). Softer BNT-BKT-BLT ceramics seem to have the largest influence ($Q_B > Q_A$ for BNT-BKT-BLT) similar to the hard PZT ($Q_B > Q_A$ for hard PZT) at high-power levels.

We present here an intuitive domain reorientation atomic model for explaining the difference of the piezoelectric loss contribution among Pb-based and Pb-free piezo-ceramics, though the model is composed of too simplified assumptions. As Figure 9.27 shows the crystal structure of $PbTiO_3$ and $BaTiO_3$, the spontaneous polarization is largely contributed by perovskite A-ion (Pb) shift in $PbTiO_3$, while B-ion (Ti) shift is the largest contribution in $BaTiO_3$, due to anisotropic large ionic polarizability of Pb^{2+} ion (Ba^{2+} has a spherical and closed electron shell). If we illustrate 90° domain wall models for both A-ion shift and B-ion shift spontaneous polarization materials, Figure 9.28 may be obtained. As we discussed in Figure 9.12, the piezoelectric loss may be originated from *Gauss Law*, div $(D) = \sigma$ (charge). That means, the piezoelectric loss is a measure of the polarization alignment easiness from the "head-to-head" or "tail-to-tail" to the "head-to-tail". In the perovskite A-ion model piezoelectric, large size A-ions need to move largely with distorting the crystal frame during this reorientation process, which seems to require higher energy, leading to a large piezoelectric

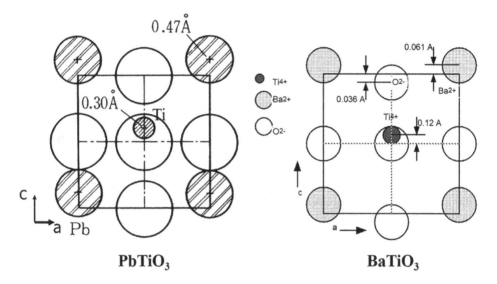

FIGURE 9.27   Crystal structures of $PbTiO_3$ and $BaTiO_3$.

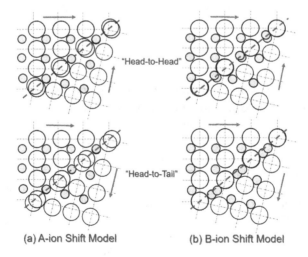

FIGURE 9.28   Domain reorientation atomic models for explaining the piezoelectric loss. (a) A-ion shift model and (b) B-ion shift model.

loss tan θ'; while in the B-ion model piezoelectric, small size B-ion can easily move in a relatively large "rattling" space in a solid frame created by A-ions. This explains large piezoelectric loss factor contribution in PZT and small contribution in Pb-free BT piezoelectrics.

9.5 DOPING EFFECT ON PIEZOELECTRIC LOSSES

9.5.1 "HARD" AND "SOFT" PZTs

We have already used the terminologies "Hard" and "Soft" on the PZT piezo-ceramics without detailing the performance or mechanisms. The "Hard" and "Soft" piezoelectrics are distinguished by the coercive field; "Hard" higher than 1 kV/mm (such as PZT 8), and "Soft" lower than 100 V/mm (such as PZT 5). The piezoelectrics with the coercive field in-between are called "Semi-soft" or "Semi-hard" (such as PZT 4). This section will clarify the difference. Small amounts of dopants sometimes dramatically change the dielectric and electromechanical properties of ceramics. Donor doping tends to facilitate domain wall motion, leading to enhanced piezoelectric charge coefficients, d, and electromechanical coupling factors, k, producing what is referred to as a "soft piezoelectric". Acceptor doping, on the other hand, tends to pin domain walls and impeding their motion, leading to an enhanced mechanical quality factor, Q_m, producing what is called a "hard piezoelectric". Table 9.4 summarizes the advantages and disadvantages of "soft" and "hard" piezoelectrics and compares their characteristics with an electrostrictive material, $Pb(Mg_{1/3}Nb_{2/3})O_3$ (PMN)-based ceramics. The electrostrictive ceramic is commonly used for positioning devices where hysteresis-free performance is a primary concern. However, due to their high permittivity, the electrostrictive devices are generally used only for applications that require slower response times. On the other hand, the soft piezoelectric materials with their relatively low permittivity and high-piezoelectric charge coefficients, d, can be used for applications requiring a quick response time, such as pulse drive devices like inkjet printers and fuel injection valves. However, soft piezoelectrics generate a significant amount of heat when driven at resonance, due to their small mechanical quality factor, Q_m. Thus, for ultrasonic motor applications, hard piezoelectrics with a larger mechanical quality factor are preferred despite the slight sacrifices incurred with respect to their smaller piezoelectric strain coefficients, d, and the electromechanical coupling factors, k.

Let us consider the crystallographic defects produced on the perovskite lattice due to doping. *Acceptor* ions, such as Fe^{3+}, lead to the formation of oxygen deficiencies ($\square$) in the PZT lattice, and the resulting defect structure is described by:

$$Pb\left(Zr_y Ti_{1-y-x} Fe_x\right)\left(O_{3-x/2} \square_{x/2}\right)$$

Acceptor doping allows for the easy reorientation of deficiency-related dipoles. These dipoles are comprised of an Fe^{3+} ion (effectively the negative charge because it is situated in the 4+ Ti site) and an oxygen vacancy (effectively the positive charge). The oxygen deficiencies are produced at

---

TABLE 9.4
Advantages (+) and Disadvantages (−) of Soft and Hard PZT Piezoelectrics, Compared with the Features of an Electrostrictive Material, $Pb(Mg_{1/3}Nb_{2/3})O_3$ (PMN)

Material	d	k	Q_m	Off-Resonance Applications	Resonance Applications
PMN	High$^+$ (DC bias)$^-$	High$^+$ (DC bias)$^-$	Low$^-$ (DC bias)$^-$	High displacement$^+$ No hysteresis$^+$	Broad bandwidth$^+$
Soft PZT-5H	High$^+$	High$^+$	Low$^-$	High displacement$^+$	Heat generation$^-$
Hard PZT-8	Low$^-$	Low$^-$	High$^+$	Low strain$^-$	High AC displacement$^+$

high temperature (>1,000°C) during sintering, but the oxygen ions are still able to migrate at temperatures well below the Curie temperature (even at room temperature), because the oxygen and its associated vacancy are only 2.8 Å apart and the oxygen can readily move into the vacant site as depicted in Figure 9.29a.

In the case of donor dopant ions, such as Nb^{5+}, a Pb deficiency is produced and the resulting defect structure is designated by the following:

$$\left(Pb_{1-x/2}\square_{x/2}\right)\left(Zr_y Ti_{1-y-x}Nb_x\right)O_3$$

Donor doping is not very effective in generating movable dipoles, since the Pb ion cannot easily move to an adjacent A-site vacancy due to the proximity of the surrounding oxygen ions as depicted in Figure 9.29b. "Soft" characteristics are, therefore, observed for donor-doped materials. Another factor that should be considered here is that lead-based perovskites, such as PZT, tend to be p-type semiconductors due to the evaporation of lead during sintering and are thus already "hardened" to some extent by the lead vacancies that are produced. Hence, donor doping will compensate the p-type, then exhibit large piezoelectric charge coefficients, d, but will also exhibit pronounced aging due to their soft characteristics.

There are three types of the impurity dipole (originated from a pair of the acceptor ion and oxygen vacancy) alignment possibility as schematically visualized in Figure 9.30: (a) random alignment, (b) unidirectionally fixed alignment, and (c) unidirectionally reversible alignment. Accordingly, the expected polarization vs. electric hysteresis curve will be (a) high coercive field P–E loop, in

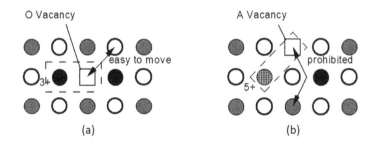

FIGURE 9.29 Lattice vacancies in PZT containing: (a) acceptor and (b) donor dopants.

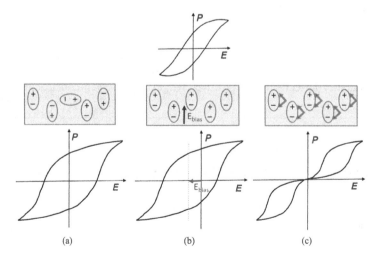

FIGURE 9.30 Impurity dipole alignment possibility and the expected P vs. E hysteresis curves: (a) random alignment, (b) unidirectionally fixed alignment, and (c) unidirectionally reversible alignment.

comparison with that of the original undoped ferroelectric, (b) DC bias field P–E loop, and (c) double hysteresis loop. We discuss more details in the following sections.

9.5.2 Dipole Random Alignment

9.5.2.1 Pseudo-DC Drive

The "soft" and "hard" characteristics are reflected in the *coercive field E_C* or more precisely in the stability of the domain walls. A piezoelectric is classified as "hard" if it has a coercive field >1 kV/mm and "soft" if the coercive field is <100 V/mm. Consider the transient state of a 180° domain reversal that occurs at a domain wall associated with a configuration of head-to-head polar domains (Figure 9.31). We know from *Gauss' law* that:

$$\text{div } D = \rho, \tag{9.13}$$

where D is the electric displacement and ρ is the charge density. The domain wall is very unstable in a highly insulating material (i.e., $\rho = 0$) and, therefore, readily reoriented and the coercive field for such a material is found to be low. However, this head-to-head configuration is stabilized in a slightly conductive (some movable charges) material, and thus, a higher coercive field is required for polarization reversal and the associated domain wall movement to occur. These two cases are illustrated schematically in Figure 9.31. As we explained above, the free charges associated with defect structures present in a doped PZT material. The presence of acceptor dopants, such as Fe^{3+}, in the perovskite structure is found to produce oxygen deficiencies (that is movable), while donor dopants, such as Nb^{5+}, produce A-site deficiencies (non-movable). Only the acceptor doping generates movable dipoles, which contribute to ρ and stabilize the domain walls.

These simple defect models help us to understand and explain various changes in the properties of a perovskite ferroelectric of this type that occur with doping. The effect of donor doping in PZT on the field-induced strain response of the material was examined for the soft piezoelectric composition $(Pb_{0.73}Ba_{0.27})(Zr_{0.75}Ti_{0.25})O_3$.[29] The parameters *maximum strain*, x_{max}, and the *degree of hysteresis*, $\Delta x/x_{max}$, are defined in terms of the hysteresis response depicted in Figure 9.32a. The maximum strain, x_{max}, is induced under the maximum applied electric field (1 kV/mm in Japanese Industrial Standard). The degree of hysteresis, $\Delta x/x_{max}$, is just the ratio of the strain induced by half the maximum applied electric field to the maximum strain, x_{max} under unipolar drive (i.e., $0 - E_{max}$). The effect of acceptor and donor dopants (2 atm% concentration) on the induced strain and degree of hysteresis is shown in Figure 9.32b. It is seen that materials incorporating high valence donor-type ions on the B-site (such as Ta^{5+}, Nb^{5+}, W^{6+}) exhibit excellent characteristics as positioning actuators, namely, enhanced induced strains and reduced hysteresis. On the other hand, the low valence acceptor-type ions (+1, +2, +3) tend to suppress the strain and increase the hysteresis and the coercive field under large field drive. Although acceptor-type dopants are not desirable when

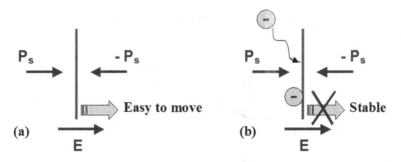

FIGURE 9.31 Stability of 180° domain wall motion in: (a) an insulating material and (b) a material with free charges.

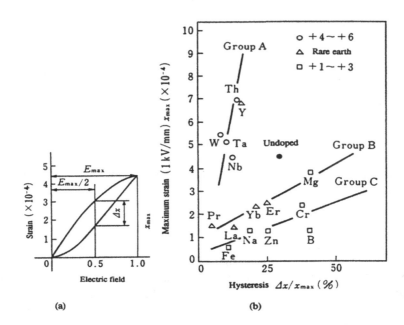

FIGURE 9.32 Maximum strain and hysteresis in $Pb_{0.73}Ba_{0.27}(Zr_{0.75}Ti_{0.25})O_3$-based ceramics: (a) hysteresis curve showing the parameters needed for defining the maximum strain, x_{max}, and the degree of hysteresis, $\Delta x/x_{max}$, and (b) the dopant effect on actuator parameters.[29]

designing actuator ceramics for positioner applications, acceptor doping is important in producing "hard" piezoelectric ceramics under low-field drive, which are preferred for ultrasonic motor and transformer applications. In this case, the acceptor dopant acts to pin domain walls, resulting in the high-mechanical quality factor characteristic of a hard piezoelectric.

When we focus on rare-earth ion doping, though all ions' valence is the same 3+ and they are located primarily in the perovskite A site, we find an interesting trend, as shown in Figure 9.33. The horizontal axis is the periodic table sequence of rare-earth atom, and the maximum strain x_{max}, and the ionic radius are plotted on this figure. The maximum strain increases with the periodic table sequence, which may be inversely related with the ionic radius. Smaller rare-earth ions may not be sited only on the A site (donor effect) but also on the B site (acceptor effect). If this supposition is accepted, small rare-earth ions exhibit the compensated effect of donor and acceptor; more facilitating the domain wall motion like "miracle ions".

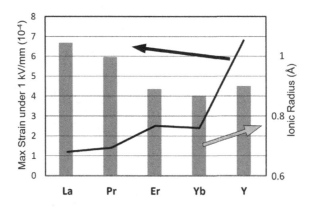

FIGURE 9.33 Maximum strain x_{max} and the ionic radius plotted in the period sequence of rare-earth ions.

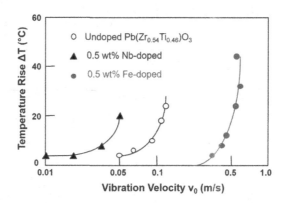

FIGURE 9.34 Temperature rise, ΔT, plotted as a function of effective vibration velocity, v_0, for undoped, Nb-doped, and Fe-doped PZT samples.[30]

9.5.2.2 Resonance Drive

Let us now consider high-power piezoelectric ceramics for ultrasonic (AC drive) applications. When the ceramic is driven at a high vibration rate [i.e., under a relatively large AC electric field (of course, much smaller than the coercive field) at the resonance], heat will be generated in the material resulting in significant degradation of its piezoelectric properties. A high-power device such as an ultrasonic motor or transformer therefore requires a very "hard" piezoelectric with a high-mechanical quality factor, Q_m, to reduce the amount of heat generated. The temperature rise at the nodal point (i.e., plate center in the k_{31} type) in undoped, Nb-doped, and Fe-doped PZT plate specimens is plotted as a function of vibration velocity (rms value measured at the plate end) in Figure 9.34.[30] A significant reduction in the generation of heat is apparent for the Fe-doped (acceptor-doped) ceramic. Commercially available hard PZT ceramic plates tend to generate the maximum vibration velocity around 0.3 m/s (rms) when operated in a k_{31} mode. Even when operated under the higher applied electric field strengths, the vibration velocity will not increase for these devices, but the additional energy is merely converted into heat (like ceramic heater!).

Higher maximum vibration velocities have been realized in PZT-based materials modified by dopants that effectively reduce the amount of heat generated in the material and thus allow for the higher rates of vibration. NEC, Japan, developed ML piezoelectric transformers with the $(1-z)\mathrm{Pb}(\mathrm{Zr}_x\mathrm{Ti}_{1-x})\mathrm{O}_3 - (z)\mathrm{Pb}(\mathrm{Mn}_{1/3}\mathrm{Sb}_{2/3})\mathrm{O}_3$ composition.[31] The maximum vibration velocity of 0.62 m/s occurs at the $x = 0.47$ and $z = 0.05$ composition and is accompanied by a 20°C temperature rise from room temperature. Incorporation of additional rare-earth dopants to this optimum base composition results in an increase in the maximum vibration velocity to 0.9 m/s at 20°C temperature rise, already introduced in Figure 9.24.[22] This increased vibration rate represents a threefold enhancement over that typically achieved by commercially produced hard PZT devices and corresponds to an increase in the vibration energy density by an order of magnitude with minimal heat generated in the device. The mechanism of this stable high-power performance is explained in the next section.

9.5.3 Unidirectionally Fixed Dipole Alignment

"Hard" PZT is usually used for high-power piezoelectric applications, because of its high coercive field, in other words, the stability of the domain walls. Acceptor ions, such as Fe^{3+}, introduce oxygen deficiencies in the PZT crystal (in the case of donor ions, such as Nb^{5+}, Pb deficiency is introduced). Thus, in the conventional model, the acceptor doping causes "domain wall pinning" through the easy reorientation of deficiency-related dipoles, leading to "hard" characteristics (Domain Wall Pinning Model[32]). ICAT/Penn State University explored the origin of our high-power piezoelectric ceramics and found that the *internal bias field* model seems to be better for explaining our material's characteristics, rather than the conventional domain wall pinning model.

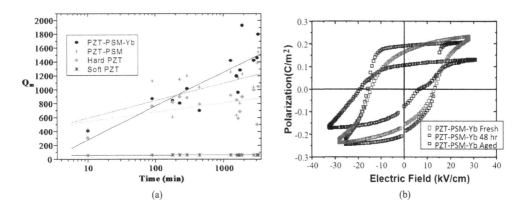

(a)   (b)

FIGURE 9.35 (a) Change in the mechanical Q_m with time lapse (minute) just after the electric poling, measured for various commercial soft and hard PZTs, PSM–PZT, and PSM–PZT doped with Yb.[33] (b) Polarization vs. electric field hysteresis curves measured for the Yb-doped PMS–PZT sample just after poling (fresh), 48 hours after, and a week after (aged).

High mechanical Q_m is essential in order to obtain a high-power material with a large maximum vibration velocity. Figure 9.35a exhibits suggestive results in the mechanical Q_m increase with time lapse (minute) after the electric poling, measured for various commercial soft and hard PZTs, PSM–PZT, and PSM–PZT doped with Yb.[22,33] It is worth noting that the Q_m values for commercial hard PZT and our high-power piezoelectrics were almost the same, slightly higher than soft PZTs, and around 200~300 immediately after the poling. After a couple of hours passed, the Q_m values increased more than 1,000 for the "hard" materials, while no change was observed in the "soft" material. The increasing slope is the maximum for the Yb-doped PSM–PZT. We also found a contradiction that this gradual increase (in a couple of hours) in the Q_m cannot be explained by the abovementioned "domain wall pinning" model, which is hypothesized that the oxygen-deficit-related dipole should move rather quickly in millisecond scale.

Figure 9.35b shows the polarization vs. electric field hysteresis curves measured for the Yb-doped PMS-PZT sample immediately after poling (fresh), 48 hours after, and a week after (aged).[34,35] Remarkable aging effects could be observed: (a) in the decrease in the magnitude of the remnant polarization, and (b) in the *positive internal bias* electric field growth (i.e., the hysteresis curve shifts leftward with respect to the external electric field axis). The phenomenon (a) can be explained by the local domain wall pinning effect, but the large internal bias (close to 1 kV/mm) growth (b) seems to be the origin of the high-power characteristics. Suppose that the vertical axis shifts rightward (according to 1 kV/mm positive internal bias field) as in Figure 9.36, in comparison with the Uchida-Ikeda's domain reorientation model in Figure 9.10, the larger negative electric field is required for realizing the 180° polarization reversal, leading to the resistance enhancement against generating the hysteresis or heat with increasing the applied AC voltage.

Finally, let us propose the origin of this "internal bias field" growth. Based on the presence of the oxygen deficiencies and the relatively slow (a couple of hours) growth rate, we assume here the oxygen deficiency diffusion model, which is illustrated in Figure 9.37.[34,35] Under the electric poling process, the defect dipole P_{defect} (a pair of acceptor ion and oxygen deficiency) will be arranged parallel to the external electric field. After removing the field, the oxygen diffusion occurs, which can be estimated in a scale of hour at room temperature. Taking into account slightly different atomic distances between the A and B ions in the perovskite crystal in a ferroelectric (asymmetric) phase, the oxygen diffusion probability will be slightly higher for the downward, as shown in the figure schematically, leading to the increase in the defect dipole with time. This may be the origin of the *internal bias electric field*. You can easily understand that the dipole alignment is unidirectionally fixed along the polarization direction.

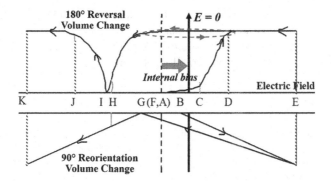

FIGURE 9.36   High-electric field drive on a piezoelectric with the internal bias field.

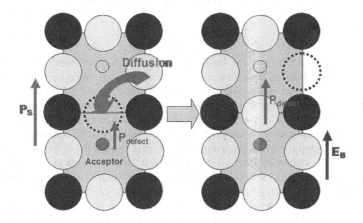

FIGURE 9.37   Oxygen deficiency diffusion model for explaining the internal bias electric field growth.[34,35]

9.5.4   Unidirectionally Reversible Dipole Alignment

Tan and Viehland reported a double hysteresis in K-doped PZT ceramics (PKZT).[36] Figure 9.38a shows a $P–E$ hysteresis curve observed in an "aged" sample of 4 at% K-doped $Pb(Zr_{0.65}T_{0.35})O_3$. Probably due to the unidirectionally (along the spontaneous polarization direction) switchable

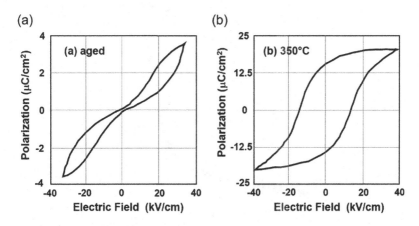

FIGURE 9.38   (a) Room temperature $P–E$ hysteresis curve observed in an "aged" 4 at% K-doped $Pb(Zr_{0.65}T_{0.35})O_3$. (b) $P–E$ hysteresis at 350°C.[36]

impurity dipole, though the hysteresis curve is symmetric (no internal DC bias field), the polarization is pinched around $E = 0$. However, with increasing the temperature, this pinching curve was released to the normal high-coercive field hysteresis. A 150°C is reported to be the temperature near which mobile defects begin to move to be a random alignment.

Example Problem 9.3

Researcher's common misconception include the confusion among voltage and electric field. Supposing a PZT disk with top and bottom surfaces electroded, when we apply high DC voltage on the top by keeping the bottom electrode ground, is this disk poled upward or downward? The correct answer is "downward". Now, next question: The polarization P vs. electric field E relation in a PZT specimen shown in Figure 9.39 (solid line) is cited from a journal paper written by a famous professor. He argued that "the positive electric field bias was observed in the P–E hysteresis". Consider what is wrong about his claim, and why this mistake happens.

Solution

If we believe his P–E hysteresis curve measurement is correct, we need to conclude that the "negative electric field bias" is observed in the hysteresis curve, because this specimen needs just a small negative field to realize the polarization reversal to $-P_S$, while a much larger positive field to realize the another polarization reversal to $+P_S$. However, this paper's content needs the conclusion of "positive field bias". So, what is wrong with his measured P–E curve? During a practical measurement using a Sawyer-Tower circuit, we usually obtain the voltage V vs. charge Q (from the integration of the current) relation first, then most of the researcher will take positive voltage in the rightward direction. The problem usually exists in the next step; that is, during changing voltage to electric field. Many researchers usually change the horizontal scale by dividing the voltage V by the specimen thickness t, without changing the hysteresis curve shape. If you remember that field is given by $E = -\text{grad} (V)$, you can easily understand that electric field is obtained by $(-V/t)$ or you need to plot E in the opposite direction to the voltage axis. In practice, the hysteresis curve should be rotated by 180° with respect to the origin (0,0), which is shown by the dashed line on Figure 9.39, where you can find now the "positive" bias field correctly. Probably his research assistant made this primitive mistake on the axis direction exchange during $V \rightarrow -E$.

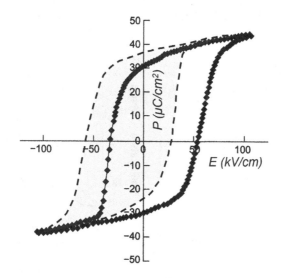

FIGURE 9.39   Biased P–E hysteresis curve.

9.6 GRAIN SIZE EFFECT ON HYSTERESIS AND LOSSES

To understand the grain size dependence of the piezoelectric properties, we must consider two size regions: The micrometer range in which multiple domain walls exist, and sub-micrometer range in which the ferroelectricity becomes destabilized. We will primarily discuss the former region in this section.

9.6.1 PSEUDO-DC DRIVE

Figure 9.40a shows the transverse field-induced strains of 0.8 at% Dy-doped fine-grain ceramic BaTiO$_3$ (grain diameter around 1.5 μm) and of the undoped coarse-grain ceramic (50 μm), as reported by Yamaji et al.[37] As the grains become finer, under the same electric field, the absolute value of the strain decreases and the hysteresis becomes smaller. This is explained by the increase in coercive field for 90° domain rotation with decreasing grain size (i.e., "hardening"). The grain boundaries (with many dislocations on the grain boundary) "pin" the domain walls and do not allow them to move easily. The decrease of grain size seems also to make the phase transition of the crystal much more diffused. Although the effective value of d_{33} decreases in the Dy-doped sample, the temperature dependence is remarkably improved for practical applications. It should be noted that Yamaji's experiment cannot separate the effect due to intrinsic grain size from that due to dopants.

Uchino and Takasu studied the effects of grain size on PLZT.[38] We obtained PLZT (9/65/35) powders by coprecipitation. Various grain sizes were prepared by hot pressing and by changing sintering periods, without using any dopants. PLZT (9/65/35) shows significant *dielectric relaxation* (frequency dependence of the permittivity) below the Curie point of about 80°C, and the dielectric constant tends to be higher at lower frequency. We prepared various grain size samples in the range of 1–5 μm. For grain size larger than 1.7 μm, the dielectric constant decreases with decreasing grain size. Below 1.7 μm, the dielectric constant increases rapidly. Figure 9.40b shows the dependence of the longitudinal field-induced strain on the grain size. As the grain size becomes smaller, the maximum strain decreases monotonically. However, when the grain size becomes <1.7 μm, the hysteresis is reduced. This behavior can be explained as follows: with decreasing grain size t_{crys}, ferroelectric (ferroelastic) domain size d decreases as $d \propto \sqrt{t_{crys}}$ [Refer to Equation (9.39)]. Since the domain wall energy also

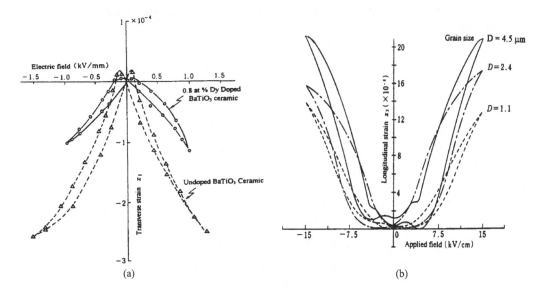

FIGURE 9.40 (a) Electric field-induced strain curves in Dy doped and undoped BaTiO$_3$ ceramic samples.[37] (b) Grain size dependence of the induced strain in PLZT ceramics.[38]

decreases, walls can be pinned more easily by the grain boundaries. Thus the strain becomes smaller. However, below a sort of critical size about 1.7 μm, the surface tension hydrostatic pressure (due to the size reduction) may facilitate the phase transition to an antiferroelectric in this particular PLZT composition, which may release the domain wall pinning, leading to the permittivity increase.

9.6.2 Resonance Drive

Sakaki et al. studied on the grain size effect on high-power performances, from practical device application viewpoint.[39] The vibration velocity vs. the temperature rise was investigated for various grain size soft Nb-doped PZTs from 0.9 to 3 μm. Among two different grain sizes, 0.9 and 3.0 μm, a higher maximum vibration velocity 0.40 m/s is observed for the 0.9 μm-grain ceramic, which is larger than 0.30 m/s of the 3.0 μm-grain ceramic. Furthermore, the temperature rise for the fine grain ceramic is observed to be about 40% lower than the coarse-grain ceramic near the maximum vibration velocity (0.30 m/s). This trend has suggested that a higher vibration velocity and lower heat generation can be achieved by reducing the grain size, which has been confirmed by investigating the heat generation phenomenon as a function of the grain size at various vibration velocities, and the observed trend is shown in Figure 9.41. The reason of this "hardening" with a decrease in the grain size may be the same as in Section 9.6.1, that is, smaller domain wall size and energy increase the pinning of walls. A linear trend in heat generation can been found with grain size. In addition, the slope of heat generation is found to increase with an increment of the vibration velocity. This in turn suggested effective control over heat generation with lowering grain size of ceramic.

It is known that an Nb-doped PZT with a molecular formula $\left(Pb_{1-y/2}\square_{y/2}\right)\left(Zr_x Ti_{1-x-y} Nb_y\right)O_3$ is free of oxygen vacancies, hence, the movable space charge or impurity dipole may not be observed in this material. Thus, it is apparent that the increment of the mechanical quality factor (Q_m) with reducing the grain size is not caused by space charge effect of oxygen vacancies but by the grain size effect. The origin of such effect is speculated due to the following reasons:

1. The change in the configuration of the domain structure,
2. The pinning effect by the grain boundaries with reduction in the grain size (i.e., grain boundaries contribute as additional pining points).

Similar to the discussion in Section 9.6.1, smaller domain size and domain wall energy may increase the pinning of walls.

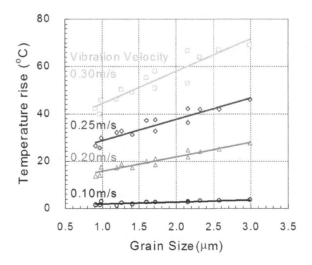

FIGURE 9.41 Grain size dependence of the temperature rise of the specimen for various vibration velocity.[39]

9.7 DC BIAS FIELD AND STRESS EFFECT ON HIGH-POWER PERFORMANCE

High-power piezoelectric devices, such as Langevin transducers and medical ultrasound transducers, are usually working under external stress/electric field biases. Although the performance of these materials transducers has been thoroughly studied, there is lack of study on the fundamental piezoelectric loss parameters at high-power conditions, which is important in the computer simulation in designing the optimum transducer configuration. One of the main figures of merit of piezoelectric materials at high-power conditions is the mechanical quality factor, which represents the loss/heat generation in these materials. A comprehensive study of loss mechanism under external biases can be both useful for enhancement of piezoelectric high-power applications and also it is beneficial for achieving a better understanding of loss mechanisms in general. In this section, our studies on the material property changes, in particular loss factors (dielectric, elastic and piezoelectric losses) of soft piezoelectric PZTs, are introduced under external electric field bias and compressive stress.

9.7.1 DC ELECTRIC FIELD BIAS EFFECT

9.7.1.1 DC Bias Field vs. Q_m

Internal field, which was attributed from acceptor defect clusters, plays important role in domain kinetics. In Section 9.5.3, we indicated that the internal bias field is observed in the high-power piezoelectrics, which seems to stabilize domain configuration.[35]

For both "soft" and "hard" PZTs, a relatively first "fast aging stage" exists which is associated with the release of grain boundary stress and charge imbalance accumulated during poling process. The aging due to elastic relaxation can't shift the electric hysteretic curve, that is, $E_i = 0$. For acceptor doped PZTs an additional "longer aging process" associated with the reorientation of the local defect clusters exists. During longer aging, the oxygen vacancy will diffuse into energetically favored site through thermal activation which results in the buildup of the internal bias field.[33] Dipole reorientation is a much slower process than polarization switching. The relaxation time is mainly governed by the thermally activated oxygen diffusion. The aging behavior in acceptor-doped ceramics is mainly attributed to the reduction of domain wall contribution to the piezoelectric, elastic, and dielectric properties. Internal bias field can suppress the amount of switchable polarization, leading to dramatical increase in the mechanical quality factor Q_m. The decrease of materials properties such as remnant polarization with time corresponds to an increasing shift of P–E hysteresis along the E axis (negative direction).

Expecting a high-power performance improvement similar to the "internal bias field", we examined the "external DC bias field" effect on hard and soft PZT k_{31}-type rectangular plate specimens. Figure 9.42a top shows the Q_m value changes as a function of external positive DC bias for aged "hard" specimen PZT-PSM-Lu. After aging process, the Q_m value was developed as high as 1,660 already, by applying external positive bias, the Q_m value increased by 14%. At the bias field level of 150 V/mm, the Q_m value reached to 1,900. The simultaneous k_{31} value measurement of the "hard" specimen is given in Figure 9.42a bottom. With increasing positive bias field, the k_{31} value of PZT-PSM-Lu sample decreases only slightly, and no significant change was observed.

Figure 9.42b top and bottom show the Q_m and k_{31} value change under various external DC bias level for "soft" PZT. Similar to the "freshly" poled specimens, the Q_m value of the PZT sample held unchanged at room temperature for 1 month after poling, which increases with increasing positive bias level; a significant increase by 35% was observed in the Q_m under DC bias field, while, k_{31} value change is very small with DC bias levels. In the donor-modified "soft" PZT, Pb vacancy is generated for the charge compensation, which is relatively large and separated; therefore, no transport would happen at relatively low temperature. After the relatively "fast aging" in the first stage, which is associated with the release of the stress and charge imbalance caused during poling, no "long-term aging" stage related to the internal bias field development exists for the "soft" PZT. Thus, the Q_m and k_{31} values did not change for the "soft" PZTs even 1 month after poling.

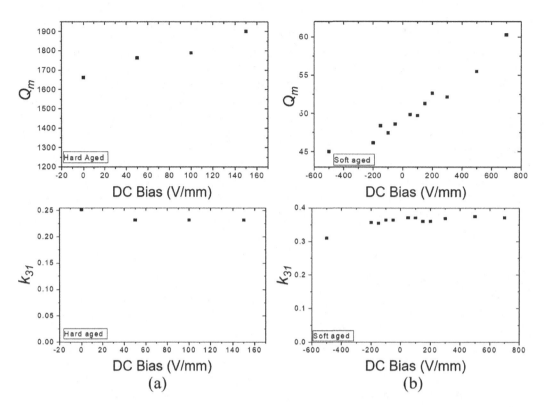

FIGURE 9.42 DC bias electric field effect on the mechanical quality factor Q_m and electromechanical coupling factor k_{31} for (a) hard PZT (aged) and (b) soft PZT.[35]

9.7.1.2   DC Bias Electric Field Effect on the Piezoelectric Performance

Our report[40] further clarified the stabilization mechanism of the ferroelectric performance under the DC bias electric field with respect to the three dielectric, elastic, and piezoelectric losses. We mainly focused on the dependence of properties of the PZT k_{31}-type plates under different vibration velocity conditions as a function of external DC bias field, by comparing the results among hard and soft PZTs. Though we[41] and Wang et al.[42] reported "Effect of DC bias field on the complex materials coefficients of piezoelectric resonators" previously, these papers did not identify three losses separately.

Bansal et al. performed continuous admittance spectrum method under DC bias electric field on the hard (PIC 144) and soft (PIC 255, PI Ceramic GmbH, Germany) PZTs.[40] Since the vibration velocity was <0.030 m/s, we did not observe the heat generation more than 1°C even at the resonance frequency range for both hard and soft PZTs. Figure 9.43a and b shows the dielectric permittivity (at off-resonance frequency), dielectric loss; elastic compliance, elastic loss; and piezoelectric constant, piezoelectric loss for hard and soft PZTs, respectively. As it can be seen from Top of Figure 9.43, the dielectric constant ε_{33}^X change was −0.8% for the hard PZT and −1.7% per 100 V/mm for the soft PZT for a positive bias field and 0.5% for hard PZT and 1.2% for soft PZT in the case of a negative bias field. This may indicate a sort of threshold of the domain wall dynamics, suggesting a two-step mechanism. To the contrary, the intensive dielectric loss tan δ' (note one order of magnitude difference between the hard and soft PZTs) shows a decreasing tendency with a positive DC bias field with a similar bend around 200 V/mm and an increasing tendency with a negative DC bias field. Though both samples did not exhibit significant dielectric loss change, the change rate (−0.4% per 100 V/mm) of the hard PZT is less than the rate (−1.5% per 100 V/mm) of the soft PZT for a positive DC bias field.

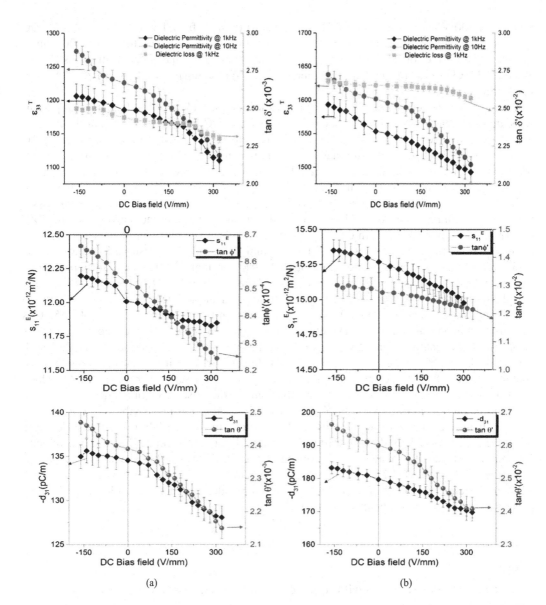

FIGURE 9.43 DC bias electric field effect on permittivity, elastic compliance, piezoelectric constant, and their respective losses for (a) hard PZT (aged) and (b) soft PZT.[39]

The % change in elastic compliance s_{11}^E was −0.7% per 100 V/mm for the hard PZT, in comparison to −1.7% for the soft PZT for positive bias field and 0.5% for hard PZT and 1.4% for soft PZT in the case of negative bias field. The elastic compliance decrease with the positive DC bias field indicates that the PZT ceramics gets hardened as the DC bias field is applied whereas effect of negative DC bias field is completely opposite. The elastic loss tan ϕ' (inverse of the mechanical quality factor at resonance) showed a shift of −1.1% per 100 V/mm for the hard PZT and −2.4% for soft PZT with increasing positive DC bias field. Also, it showed a shift of 0.8% per 100 V/mm for the hard PZT and 1.9% for the soft PZT under negative DC bias field.

Regarding the piezoelectric constant $|d_{31}|$, the decrease for the hard PZT (−1% per 100 V/mm) was less than that for the soft PZT (−1.5% per 100 V/mm) under a positive DC bias field. Also, apart from the real parameter, piezoelectric loss change of the hard PZT was again smaller

(−1.9% per 100 V/mm) for the hard PZT as compared to the soft PZT (−3.1%). For negative DC bias field, the piezoelectric constant magnitude increase for the hard PZT (0.7% per 100 V/mm) was less han for the soft PZT (1.2% per 100 V/mm). Apart from that, piezoelectric loss change of the hard PZT was again smaller (1.5% per 100 V/mm) as compared to the soft PZT (2.8% per 100 V/mm).

We provided the dependency of dielectric, elastic, and piezoelectric coefficients (real parameters and imaginary losses) on the vibration velocity and DC bias field comprehensibly.[40] Note that as we drive the piezoelectric ceramic at a high-vibration velocity, we are introducing AC stress in the material. Table 9.5 summarizes the change in material's coefficients with the vibration velocity and DC electric field. Using a burst mode, our results excluded any temperature rise during the measurements. It has been found that with the vibration velocity increase, the dielectric constant, elastic compliance, piezoelectric coefficient, and their corresponding losses increase for both hard and soft PZTs. However, the change is more pronounced in the soft PZT as compared to the hard PZT. In contrast, the influence of a DC bias field depends strongly of the direction of the field with respect to the original poling field. A positive DC bias electric field affects in a way, that is, decrease in the dielectric constant, elastic compliance, piezoelectric coefficient, and their corresponding losses, whereas a negative DC bias field affects in a completely opposite way. The decrease rate of these physical parameters under small vibration level is enhanced under large vibration level, as shown in the third column in Table 9.5. This situation can be visualized in the 3D plot in Figure 9.44, showing the dependence of elastic loss tan ϕ' on the externally applied DC bias field and the vibration velocity at the k_{31} sample length edge in the hard (PIC 144) and soft (PIC 255) PZT. In comparison with a relatively smooth plane contour for the soft PZT, a clear bend on the contour plane is observed for the hard PZT, which suggests two-stage domain wall dynamic mechanism (further discussion in Section 9.8). We can find that the intensive elastic loss tan ϕ' shows two different trends under vibration velocity and DC bias field: increase with the vibration velocity (domain wall destabilization) and decrease with positive DC bias field (domain wall stabilization). Roughly speaking, the vibration velocity 100 mm/s (+3% and 5% change in s_{11}^E for the hard and soft PZTs) exhibits the almost equivalent "opposite" change rate (−1.7% × 2 and −2.3% × 2 change in s_{11}^E for the hard and soft PZTs) of 200 V/m DC electric field. That is, the heat generation under a high-vibration level can be suppressed by the bias electric field application.

TABLE 9.5
Change in Material Properties with Applied Positive DC Bias Field, in Comparison with Vibration Velocity[40]

% Change in Material Properties	Vibration Velocity (per 0.1 m/s)		Electric Field (per 100 V/mm at 0.01 m/s)		Electric Field (per 100 V/mm at 0.3 m/s)	
	Hard PZT (%)	Soft PZT (%)	Hard PZT (%)	Soft PZT (%)	Hard PZT (%)	Soft PZT (%)
Dielectric constant (ε_{33}^T)	+2.9	+5	−0.8	−1.7	–	–
tan δ_{33}'	–	–	−0.4	−1.5	–	–
Elastic compliance (s_{11}^E)	+3	+5	−0.7	−1.7	−1.7	−2.3
tan φ_{11}'	+75	(+)~0 & 75	−1.1	−2.4	−3.5	−4.2
Piezoelectric constant (d_{31})	+2.7	+5.1	−1	−1.5	–	–
tan θ_{11}'	+70	+20	−1.9	−3.1	–	–
Electromechanical coupling factor (k_{31})	+3.5	+5.3	Change within error limits	Change within error limits	–	–

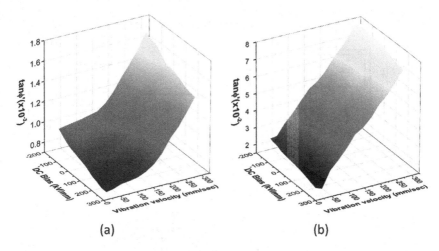

FIGURE 9.44 3D plot showing the variation of elastic loss for PZTs as a function of DC bias field and vibration velocity for (a) hard PZT and (b) soft PZT.[40]

9.7.2 DC STRESS BIAS EFFECT

High-power piezoelectric devices are often subjected to external mechanical constraints or biases, in many applications such as underwater transducers. While their performance and property changes under these external biases have been widely investigated by different researchers[43–49] from a practical application viewpoint, there is a significant lack of fundamental research on the loss mechanism in terms of three existing losses (dielectric, elastic, and piezoelectric) under external stress bias. Introduced below is based on our preliminary studies.[50]

9.7.2.1 Bolt-Clamped Langevin Transducer

In order to investigate the effect of uniaxial stress, two different configurations can be used; quasi-static and resonant.[51] Although proposed methodologies based on these configurations were able to capture the piezoelectric material's real part property changes under compressive stress, they have suffered from the difficulty in studying the piezoelectric loss mechanism (i.e., imaginary performance) at the resonance condition, due to the experimental complications. In studying the compressive mechanical bias stress dependence of the loss mechanism, we adopted a modified bolt-clamped Langevin transducer.[50] Figure 9.45 shows schematic experimental diagram for measuring DC bias stress effect on the Langevin transducer, which consists of mass/load, active PZT, and inactive PZT parts. A semi-symmetric configuration was chosen to provide a symmetrical and uniform AC-stress distribution over the active PZT samples (center six-ring specimens). The inactive/dummy PZT samples (side 2×2 inactive rings) were introduced to increase the length/thickness ratio of structure and reduce the clamping effect. The structure was designed in a way that the fundamental longitudinal resonance is far from higher resonance frequencies, and therefore, no interference or spurious modes would be expected by applying external load on the samples, and that a uniform AC-stress ($X_{33}^{\text{AC,rms}}$) and DC-stress (X_{33}^{DC}) distribution along the longitudinal direction on the active six PZT rings (less than $\pm10\%$ for AC stress and $<0.1\%$ for DC stress). A soft piezoelectric Pb(Zr, Ti)O$_3$ (PIC-255, PI Ceramic GbmH, Germany) was examined from a scientific viewpoint, and an equivalent circuit methodology with three losses based on the fundamental longitudinal mode was developed with symmetrical head and tail masses to facilitate the analysis on a modified bolt-clamped Langevin transducer. The measurement was conducted between 20 and 42 MPa range, because too low compressive stress does not provide stable/reliable results under repeating cycles, and too high generates the PZT ceramic cracks. Refer to Ref. [50] for the experimental details.

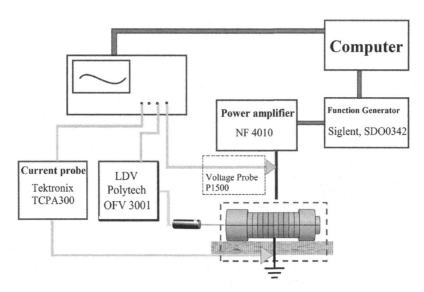

FIGURE 9.45 Schematic experimental setup for measuring DC bias stress effect on the Langevin transducer.

9.7.2.2 Six-Terminal Equivalent Circuit Approach

Figure 9.46 shows the Langevin transducer equivalent circuit (EC) with short-circuited inactive PZT configuration (including six real and six imaginary elements). This EC shows a left-half of the transducer in Figure 9.45, composed of three parts: (1) the right-side circuit for an active PZT ring part, (2) the center circuit for a dummy PZT ring part (under short-circuit condition), and (3) the left-side circuit for an aluminum head mass. In order to include loss parameters, material properties were introduced into the equivalent circuit using complex forms. Furthermore, the load parts (inactive PZT and aluminum loads) were also added to the mechanical branch of equivalent circuit replacing edge forces. The vibration velocity $\dot{u}$ at the head mass end corresponds to the EC mechanical branch current, and the end is short-circuited because of force $F = 0$. Voltage and current at the input electric branch port are measured for obtaining the admittance frequency spectrum.

9.7.2.3 DC Stress Bias Dependence of Physical Parameters and Loss Factors

Figure 9.47 summarizes the dependence on the compressive stress along the polarization direction of (a) intensive dielectric constant ε_{33}^X and dielectric loss $\tan \delta'$; (b) intensive elastic compliance (ε_{33}^X) and elastic loss ($\tan \phi_{33}'$); (c) piezoelectric constant (d_{33}) and intensive piezoelectric loss ($\tan \theta_{33}'$);

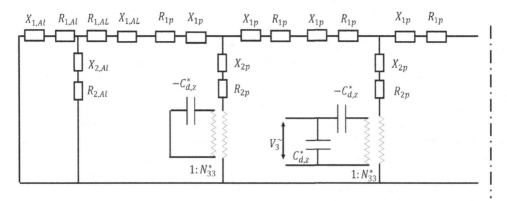

FIGURE 9.46 A Langevin transducer symmetrical equivalent circuit with short-circuited inactive PZT configuration (including six real and six imaginary elements).

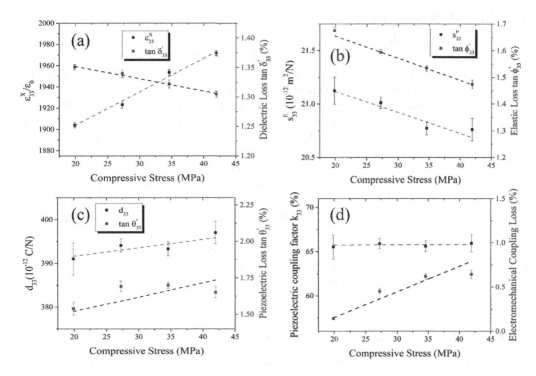

FIGURE 9.47 Dependence on the compressive stress along the polarization direction of (a) intensive dielectric constant ε_{33}^X and dielectric loss $\tan \delta'$; (b) intensive elastic compliance (s_{33}^E) and elastic loss ($\tan \phi_{33}'$); (c) piezoelectric constant (d_{33}) and intensive piezoelectric loss ($\tan \theta_{33}'$); and (d) piezoelectric coupling factor (k_{33}) and its corresponding k_{33}^2 loss.

and (d) piezoelectric coupling factor (k_{33}) and its corresponding loss. The permittivity shows a monotonic increase under compressive stress (+3.67% increase at maximum stress level 42 MPa), while the dielectric loss decreases slightly with an increase of compressive stress along polarization direction (–3.6% decrease at maximum stress level 42 MPa). The elastic constant showed a monotonous stiffening behavior (elastic compliance decrease) under compressive stress (+2% increase at maximum stress level 42 MPa). Also, as can be seen in Figure 9.47, similar to the dielectric loss, the elastic loss decreases remarkably under compressive stress (–12% at maximum stress level). While the piezoelectric constant showed a small increase under compressive stress (1% at highest stress level), the intensive piezoelectric loss showed considerable increase (~12% at highest stress level). Also, as can be seen in Figure 9.47d, the piezoelectric coupling factor (k_{33}) was almost constant over small compressive stress range. While, the electromechanical coupling loss factor ($2 \tan \theta_{33}' - \tan \phi_{33}' - \tan \delta_{33}'$) is increasing with an increase of compressive stress.

While the piezoelectric constant and coupling factor of soft PZT do not show a major change, the loss parameters, especially elastic loss, show a considerable improvement. The decrease in elastic loss under compressive stress (which indicates lower heat generation) in conjunction with stable piezoelectric constant over low stress levels can be beneficial for piezoelectric resonance/high-power applications. Also, the increase of the electromechanical coupling loss leads to more benefit of the antiresonance mode usage.

In a brief summary, the positive DC electric field and compressive (negative) stress exhibit a similar decreasing trend in real physical parameters such as permittivity and elastic compliance (except for piezoelectric d constant), which can be roughly explained by the equilibrium phenomenology. However, the loss dependence trend cannot be understood easily at present. We need to wait for a detailed loss mechanism model.

9.8   EXTENDED RAYLEIGH LAW APPROACH

The introduction of the complex parameters into dielectric permittivity, elastic compliance, and piezo-electric constant, described as $\varepsilon_3^{X*} = \varepsilon_3^X \left(1 - jtan\delta_{33}'\right), s_{11}^{E*} = s_{11}^E \left(1 - jtan\phi_{11}'\right), d_{31}^* = d_{31}\left(1 - jtan\theta_{31}'\right)$, means merely slight time lag of the extensive parameters D or x from the intensive E or X, with no particular insight into domain reorientations. The introduction of "j" generates the flat *elliptical shape hysteresis* automatically with a rounded curve at the maximum (or minimum) electric field or stress range, which shows a discrepancy from the actual hysteresis curve with sharp kink at the end points (see Figure 4.2). We will construct domain-dynamic models to explain high-power piezoelec-tric performances in this section and Section 9.9. The Rayleigh law approach provides a connection concept of the losses with microscopic domain wall dynamics described in Section 9.9.

Ferromagnetic materials have domains of uniform spin/dipole orientation and behave similarly to ferroelectric materials under application of external magnetic field. Under external fields, the domain in ferromagnetic materials can switch polarity. These materials also have intrinsic and extrinsic contributions to their properties. Based on principles of domain wall motion under an external field, the change in material properties due to reorientation of domain walls with increas-ing amplitude driving yields the relationship known as the *Rayleigh law.* Though originally derived for ferromagnetic materials, researchers have applied its premises to ferroelectric materials under external stresses and electrical fields with success.[52,53]

9.8.1   THE PRANDTL–TOMLINSON MODEL FOR DRY FRICTION

We introduce the "*Prandtl–Tomlinson Model*" here which explains plastic deformations in a crys-tal from the *dry friction* mechanism viewpoint on atomic scales.[54,55] This model is a hybrid type between the Coulomb (friction), solid and viscous damping, introduced in Chapter 3. Let us con-sider the simplest 1D model of a mass m in a periodic potential with the wave number k under an external force F with a viscous damping (proportional to velocity) factor ξ, as illustrated in Figure 9.48a. Using the coordinate u of the mass, and the period force amplitude N, we obtain the following dynamic equation:

$$m\ddot{u} = F - \xi\dot{u} - N\sin(ku) \tag{9.14}$$

When the mass is at rest and a force is applied to it, its equilibrium position should satisfy the equation,

$$F = N\sin(ku). \tag{9.15}$$

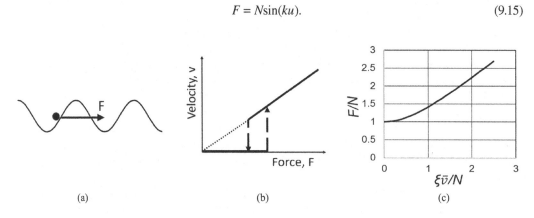

(a)   (b)   (c)

FIGURE 9.48   (a) Prandtl–Tomlinson model: A point mass in a periodic potential. (b) Macroscopic law of friction: dependence of the average velocity on the force. (c) Friction force of the Prandtl–Tomlinson model vs. average velocity in the overdamped case.

Only when $F \leq N$, the above equation gives a solution. When u is small, this force proportional to u corresponds to the *solid damping*. Thus, the static friction force is expressed by

$$F_S = N. \tag{9.16}$$

When the force is larger, $F \geq N$, the mass will enter macroscopic motion without equilibrium position. The movement of the mass can be described as a quasi-static friction process: the dependence of the average velocity on the applied force is explained as *macroscopic law of friction*, as illustrated in Figure 9.48b. When the force is small, the mass does not move; that is, there is some dead region to overcome the friction force.

9.8.1.1 Under-Damping Case

When the damping factor ξ is small, we discuss the case in small force ($F \leq N$). Since the motion is periodic, we assume that the work performed by an external force F over a period of $a = 2\pi/k$ is given by the energy loss, $\int_0^T \xi v^2(t)dt$:

$$\frac{2\pi F}{k} = \int_0^T \xi v^2(t)dt = \int_0^a \xi v(u)du = \xi \int_0^a \sqrt{\frac{2}{m}\left(E_0 + \frac{N}{k}\cos(ku)\right)}du. \tag{9.17}$$

Here, we assumed $E_0 = \frac{mv^2}{2} - \frac{N}{k}\cos(ku) \approx$ constant; energy conservation during a short period by neglecting the energy loss via the damping. The small force F' is given by Eq. (9.17) with $E_0 = N/k$:

$$\frac{F'}{N} = \frac{4}{\pi}\frac{\xi}{\sqrt{mkN}} \tag{9.18}$$

By equating the kinetic friction force F' to the static friction force N, we can estimate the critical threshold of the damping factor ξ between the under- and over-damping conditions.

9.8.1.2 Over-Damping Case

In the case of the over-damping, we can neglect the inertia term $\ddot{u}$ in Eq. (9.14):

$$0 = F - \xi\dot{u} - N\sin(ku), \text{ or}$$

$$\dot{u} = \frac{du}{dt} = \frac{F}{\xi} - \frac{N}{\xi}\sin(ku) \tag{9.19}$$

One special period a is traversed in the time period T:

$$T = \int_0^{2\pi/k} \frac{du}{\frac{F}{\xi} - \frac{N}{\xi}\sin(ku)} = \frac{\xi}{kN}\int_0^{2\pi} \frac{dz}{\frac{F}{N} - \sin(z)} = \frac{\xi}{kN}\frac{2\pi}{\sqrt{\left(\frac{F}{N}\right)^2 - 1}}. \tag{9.20}$$

Thus, the average speed is given by

$$\bar{v} = \frac{a}{T} = \frac{\sqrt{F^2 - N^2}}{\xi}, \tag{9.21}$$

and the force is expressed as follows as a function of the average speed $\bar{v}$:

$$F = \sqrt{N^2 - (\xi\bar{v})^2}. \tag{9.22}$$

Figure 9.48c shows the dependence of the force on the product of the damping factor ξ and the average velocity $\bar{v}$ (both parameters are normalized by the period force amplitude N). When the velocity is small, the friction force seems to be constant. With increasing the velocity, the force will increase in proportion to the velocity above the threshold velocity.

When we interpret the above mass, the periodic potential and the mass velocity as the effective domain wall mass, the crystal lattice periodic potential and the domain wall velocity, respectively, a similar domain dynamic response is anticipated to the phenomena illustrated in Figure 9.48. A hysteresis is expected in the relation between the force and the domain displacement (or stress vs. strain), and the velocity vs. force may exhibit the nonlinear behavior as in Figure 9.48c.

9.8.2 Conventional Rayleigh Law

The *Rayleigh law* has become a successful analytical tool to describe the change of material properties in ferroelectric materials with driving amplitude. However, its application has solely been toward off-resonance measurements, mostly at low power. Our team developed the Rayleigh law into a tool to analyze high-power resonance properties of piezoelectric materials.[56] Preliminary analysis on the elastic properties (s_{11}^E and $\tan\phi_{11}'$) has been done on the k_{31} mode plate, taking into account both *reversible* and *irreversible domain wall motions*, which correspond to the statuses of $F \leq N$ and $F \geq N$, respectively, and the *Rayliegh coefficent* α_X was introduced accordingly.

In order to apply the Rayleigh law to resonance measurements, a new derivation must be performed to account for the energy distribution in resonating k_{31} plate samples. The derivation is summarized here. First, the elastic energy u_e of a resonating sample (k_{31}) with length L is derived as

$$u_e = \frac{1}{4s_{11}^E}\left(u_{0,1}\frac{\pi}{L}\right)^2, \quad (u_{0,1}: \text{displacement at the end}) \tag{9.23}$$

under a supposition of sinusoidal distribution of displacement on the k_{31} plate. Then the energy lost due to elastic losses is defined according to the loss tangent and stored energy:

$$w_e = 2\pi u_e \tan\phi_{11}'. \tag{9.24}$$

Next, the Rayleigh law is defined in terms of RMS stress, due to the stress variation in the sample:

$$s_{11}^E = s_{11,i}^E + \alpha_X X_{1,\mathrm{rms}}, \tag{9.25}$$

$$\frac{\partial u_1}{\partial x} = \left(s_{11,i}^E + \alpha_X X_{1,\mathrm{rms}}\right)X_{1,\mathrm{rms}} \pm \frac{\alpha_X}{2}\left(X_{1,\mathrm{rms}}^2 - X_1^2\right), \tag{9.26}$$

where α_X is the *Rayleigh coefficient*. Note that the strain $\left(\dfrac{\partial u_1}{\partial x}\right)$ curve is composed of two parabolic curves with sharp kinks at the maximum stress points, closer to the stress–strain loop actually observed. Integrating the area in the parabolic hysteresis defined by the Rayleigh law (Figure 9.49), the energy lost according to α_X is

$$w_e = \frac{\alpha_x}{3\sqrt{2}}\left(u_{0,1}\frac{\pi}{Ls_{11}^E}\right)^3 \tag{9.27}$$

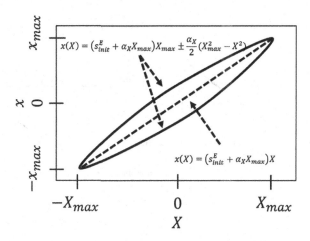

FIGURE 9.49   Rayleigh law for elastic response of a ferroelectric material.

Combining these equations, the relationship between the Rayliegh coefficent α_X and the equivalent elastic loss tangent is derived

$$\tan\phi'_{11} = \frac{2\alpha_x}{3L\sqrt{2}} u_{0,1} \left(\frac{1}{s^E_{11}}\right)^2 \tag{9.28}$$

Using this formulation, it is possible to verify the relationship between Rayliegh coefficent α_X and the loss tangent, thus verifying the applicability of the Rayleigh law.

9.8.3   Application of Hyperbolic Rayleigh Law

The *hyperbolic Rayleigh law* introduces nonlinearity in the Rayleigh law formulation based on a domain wall dynamics perspective. This model was introduced by Borderon,[57] but here it is extended toward elastic measurement in high-power conditions. Additionally, the significance of the model's parameters is elucidated by the work of ICAT (see Figure 7.20a, sample PIC 144). The hyperbolic Rayleigh law introduces nonlinearity in the prethreshold stress region; this is essential to describe the behavior seen in the hard PZT material [Remember the clear bend (threshold) in Figure 9.44a]. The previous representation of the Rayleigh law is only true for a larger stress level than the threshold stress X_{th}. If the stress is lower than this, domain wall vibrations will provide the majority of the extrinsic response, and thus, the elastic compliance remains relatively constant as a function of applied stress. The threshold stress X_{th} is a measure of the degree of pinning in the material; this stress does not show a clearly defined feature, where sometimes only the linear and the high-field region can be realized under electric fields which are experimentally considered. Without the hyperbolic Rayleigh law, the threshold level is usually determined by inspection, which may lead to a large error in its reported value and, therefore, large error in analysis of microstructure properties. This threshold stress level may be related with the threshold vibration velocity in Figure 7.19a, below which the Q_m value is rather constant and above which the Q_m starts dramatic reduction.

The hyperbolic is schematically represented in Figure 9.50a. In large stress conditions, the threshold stress is passed and the traditional Rayleigh law can be applied. However, at low stress levels, apparent nonlinearity exists which cannot be modeled by the Rayleigh law; thus, the hyperbolic Rayleigh law is introduced. Figure 9.50b shows the change in elastic compliance with effective applied RMS stress for a hard PZT material. From the calculation of the threshold stress from the hyperbolic Rayleigh parameters, the threshold stress is larger than the largest stress level measured. When we consider the "blocking" stress ≈ 35–40 MPa, which may be slightly larger than the

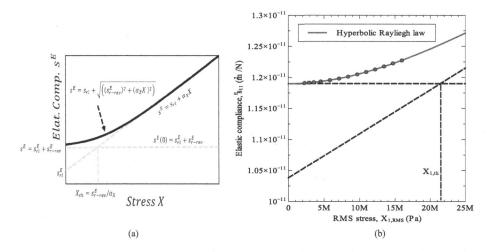

FIGURE 9.50 (a) Hyperbolic Rayleigh law – illustrated parameter relationships and (b) hyperbolic fitting to change in elastic compliance with stress in hard PZT.[56]

coercive stress for the domain reorientation in PZTs, the experimentally obtained threshold stress 22 MPa as RMS value is in reasonable agreement with our speculation. The meaning of this is that hard PZT can practically only be driven in its prethreshold region (i.e., reversible domain wall motion). This explains the discrepancies found by previous researchers in applying the Rayleigh law to nonlinear behavior below the threshold field.

9.9 FERROELECTRIC DOMAIN WALL DYNAMICS – PHENOMENOLOGY

One of the key items in the loss mechanism of piezoelectrics is to elucidate the loss factors phenomenologically from the domain wall dynamics viewpoint. We describe first on the static domain formation, and the domain wall models (no loss integrated), then discuss the domain wall dynamics without dopants (delay time and loss integrated). Finally, we will consider the dopant effect on the domain wall mobility, which is directly related with the loss factor difference between the "hard" and "soft" PZTs. This section is largely devoted to a comprehensive domain structure textbook, *Domain Structure in Ferroelectrics and Related Materials* by Sidorkin. Though the author tried to derive at least the dielectric loss formula from the domain wall dynamic model, the discussion is still immature. The reader is strongly requested to expand the theoretical approach so as to include the elastic and piezoelectric losses in the future.[58]

9.9.1 DOMAIN FORMATION – STATIC STRUCTURE

It has been assumed in the fundamental phenomenological descriptions presented so far that the materials in question are *monodomain single crystals* and the application of the field does not change their state of polarization so easily. We provided the macroscopic coercive field in Example Problem 2.4, in which whole the crystal changes from the $+P_S$ state to the $-P_S$ state instantaneously, which gives a huge value higher than 10 kV/mm. When considering actual ferroelectric ceramics, however, this assumption does not hold. Typical ferroelectric material may have a multiple-domain structure even in a single crystal form, and a more complicated domain structure in a polycrystalline ceramic, as schematically illustrated in Figure 9.2. In practice, the apparent coercive field (to switch the polarization) is 1 kV/mm or less, 1/10 smaller than the value calculated phenomenologically above, because the polarization reversal is realized through the domain wall movement under relatively small electric field. The domain pattern change was observed during the 90° polarization reorientation process in a BaTiO$_3$ single crystal (4 mm tetragonal crystal symmetry at room temperature) as shown in Figure 9.4, where electric field was applied in the up/down $\updownarrow$ directions on that page. Except under the maximum field, multiple 90° domain walls are observed. Figure 9.51a shows 180° domain walls observed in the BaTiO$_3$ (001) plate without electrode. Why do multiple domains appear? The simplest answer is "because the multiple-domain state exhibits lower energy than the monodomain state".

9.9.1.1 Depolarization Field

Let us start from a multidomain model on a ferroelectric plate, as shown schematically in Figure 9.51b, where only the up/down spontaneous polarization (i.e., 180° domain walls) arrangement is considered just from the simplicity viewpoint. Two key energies, in addition to the polarization energy in a monodomain ferroelectric, to be discussed on the domain structure formulation are (1) *depolarization energy* and (2) *domain wall energy*.

We consider first the depolarization energy. Taking into account Gauss'Law: div $D = \rho$, where $D = \varepsilon_0 E + P = \varepsilon\varepsilon_0 E + P_S$, we obtain:

$$\text{div } E = \frac{1}{\varepsilon\varepsilon_0}\left(\rho - \text{div } P_S\right). \tag{9.29}$$

Polarization P is composed of the induced polarization $\varepsilon_0(\varepsilon - 1)E$ and the spontaneous polarization P_S. If free charge ρ is provided on the surface via crystal conductivity (like p- or n-type semiconductor) or surrounding medium (humid air), in particular, when the surface has an electrode, internal field E can be zero as an equilibrium status (imagine the electrode on the top surface in Figure 9.51b). However, if $\rho = 0$ due to a highly resistive crystal status without electrodes, as in Figure 9.51b, so-called *depolarization field*:

$$E = -\left(P_S/\varepsilon\varepsilon_0\right) \tag{9.30}$$

should be generated in a monodomain crystal (so as to keep $D = 0$), originated from the surface *bound charge* (i.e., originated from P_S). The dipole charges ±q inside the crystal are canceled out in the adjacent area. This discussion is similar to the content in Section 4.3.2.2, E-constant and D-constant.

Thus, the depolarization energy can be calculated as

$$W_e = (1/2)\int dV\varepsilon\varepsilon_0 E^2 = (1/2\varepsilon\varepsilon_0)\int dV \cdot P_{S^2} \tag{9.31}$$

When the crystal surface is structured by twin up and down polarization domains with equal volumes, the plus and minus bound charges (P_S) can be compensated by the free charges (ρ) drift on the surface electrode, so that the depolarization energy can be diminished significantly and a monodomain status is maintained, however without the surface electrode, the multidomain configuration is preferably induced for reducing this energy.

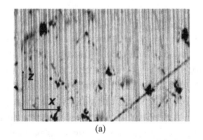
(a)

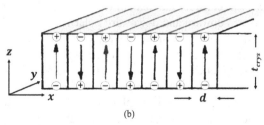
(b)

FIGURE 9.51 (a) A 180° domain walls in BaTiO$_3$ (001) plate (top view, no electrode). (b) Multidomain model of 180° domain walls on a ferroelectric plate (no electrode).

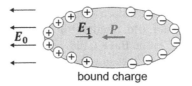

bound charge

FIGURE 9.52 Ellipsoid dielectric model for calculating depolarization factor.

The above discussion can be modified by considering the domain shape as below. Figure 9.52 shows a hypothetical ellipsoid dielectric material for calculating the depolarization field.[59] Macroscopic field in a body can be obtained by the sum of $E = E_0 + E_1$, where E_0 is the external field and E_1 is the "depolarization field". Then, the depolarization field is expressed by:

$$E_{1x} = -\frac{L_x P_x}{\varepsilon_0}, E_{1y} = -\frac{L_y P_y}{\varepsilon_0} \; E_{1z} = -\frac{L_z P_z}{\varepsilon_0}, \tag{9.32}$$

where L_x, L_y, and L_z are *depolarization factors*, which depend on the ratios of the ellipsoid principal axes. And $L_x + L_y + L_z = 1$. Depolarization field can be calculated as a field generated by *fictitious charges* on the ellipsoid surface (i.e., inside ellipsoid is cavity), as $L = \frac{abc}{2} A$, where A is an ellipti-

cal integral $\left(K(k) = \int_0^{\pi/2} \frac{d\theta}{\sqrt{1 - k^2 \sin^2 \theta}} \right)$. $L = 1/3$ for sphere, $L = 1$ for thin slab normal direction, and

$L = 0$ for thin slab in-plane direction.

$$W_e = (1/2) \int dV \varepsilon \varepsilon_0 E^2 = (\varepsilon/\varepsilon_0) \int dv L^2 P^2{}_s \tag{9.33}$$

Mitsui and Furuichi provided a simpler description of the depolarization energy, by taking an aspect ratio of the domain shape (d/t_{crys}) instead of L, where d is the domain width, and t_{crys} is the ferro-electric plate thickness:[60]

$$W_e = \varepsilon^* P^2 V \left(\frac{d}{t_{crys}} \right) \tag{9.34}$$

where V is the crystal volume. In short summary, the depolarization energy promotes the multidomain status: the thinner the domain width d and the thicker crystal plate t_{crys}, the more preferable to reduce the depolarization energy.

9.9.1.2 Domain Wall Energy

Now, we will consider the domain wall energy. The 180° *domain wall* in Figure 9.51b is the interface between upward and downward polarization domains. Though it is rather thin (similar to the unit cell dimension ≪ domain size d), it should have some width (90 degree domain wall has much larger thickness). In order to switch the polarization vector in an opposite direction, two possible domain wall models are proposed in Figure 9.53: (a) *Neel wall*, in which keeping the polarization magnitude constant, the vector arrow rotates, and (b) *Bloch wall*, in which without changing the direction, the vector arrow length changes and the direction reverses (i.e., zero polarization at the wall center). In a ferromagnetic domain wall's case, the spin exchange energy suppresses a drastic magnetization change, leading to an increase in the domain wall thickness, while the anisotropy

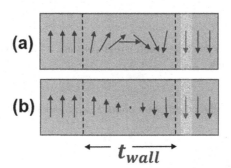

FIGURE 9.53 Ferroelectric domain wall models: (a) Neel wall and (b) Bloch wall.

(elastic) energy reduces the domain wall thickness. Since the exchange energy exceeds to the anisotropy energy in magnetic materials, wall thickness is rather thick (wider than hundreds of unit cells). On the contrary, in a ferroelectric domain wall's case, compared with the dipole interaction energy, the anisotropy elastic energy contributes significantly; leading to a narrow domain wall thickness, t_{wall}, around a couple of unit cell size.[61] The domain wall energy can be expressed as

$$W_{\text{wall}} = \left(A/t_{\text{wall}}\right) + Bt_{\text{wall}} \tag{9.35}$$

where A and B are *dipole interaction energy* and *anisotropy elastic energy*, respectively, and $B \gg A$ in most of ferroelectric materials. Taking the energy minimization condition, $\dfrac{\partial W_{\text{wall}}}{\partial t_{\text{wall}}} = 0$, we can obtain the optimum t_{wall} and the wall energy W_{wall}:

$$t_{\text{wall}} = \sqrt{A/B} \tag{9.36}$$

$$W_{\text{wall}} = 2\sqrt{AB} \tag{9.37}$$

Now, we can finally obtain the stabilized domain size d on the ferroelectric crystal plate with volume V and thickness t_{crys}. We consider the sum of the depolarization energy Eq. (9.34) and the domain wall energy Eq. (9.37) × (number of walls V/d):

$$W_e + W_{\text{wall}}\left(\frac{V}{d}\right) = \varepsilon^* P^2 V\left(\frac{d}{t_{\text{crys}}}\right) + W_{\text{wall}}\left(\frac{V}{d}\right) \tag{9.38}$$

Minimizing again the total energy, the optimized domain width d can be related with the crystal plate thickness t_{crys} as follows:

$$d = \sqrt{\frac{W_{\text{wall}}\, t_{\text{crys}}}{\varepsilon^* P^2}}. \tag{9.39}$$

In summary, the depolarization energy reduces the domain width d, while the domain wall energy increases the domain width, leading to a compromised domain width related with the square root of the crystal thickness. Accordingly, ferroelectric thin films (even in a single crystal-like epitaxial) may exhibit very small domain width, while we need some techniques to obtain monodomain crystals, such as *surface screening* with electrode semiconductor (charge carriers). The domain wall energy introduced above can phenomenologically be calculated again in Section 9.9.2.

Example Problem 9.4

Explain the permittivity difference between stress constant ε^X and strain constant ε^x from the depolarization field viewpoint.

Solution

- Under free condition, the polarization under E_0 is provided by

$$\varepsilon_0 \varepsilon^X E_0.$$

(P9.4.1)

- Under clamped condition ($x = 0$), from

$$x = s^E X + dE_0 \equiv 0.$$

(P9.4.2)

we need stress equal to $X = -\dfrac{d}{s^E} E_0$. Accordingly, this stress generates the polarization

$P = d\left(-\dfrac{d}{s^E} E_0\right)$ under open-circuit condition, which corresponds to the *depolarization field*

$$E_{\text{dep}} = \frac{P}{\varepsilon_0 \varepsilon^X} = -\frac{d^2}{\varepsilon_0 \varepsilon^X s^E} E_0.$$

(P9.4.3)

If we denote $k^2 = \dfrac{d^2}{\varepsilon_0 \varepsilon^X s^E}$, $E_{\text{dep}} = -k^2 E_0$. Thus, the effective field is to be $\left(1 - k^2\right)E_0$, thus the induced polarization should be reduced by this amount or $\varepsilon^x = \left(1 - k^2\right)\varepsilon^X$.

9.9.2　Ginzburg–Landau Functional – Domain Wall Structure Phenomenology

This section handles the discussion conducted in Section 9.9.1 from the advanced phenomenological viewpoint; that is, the inter-coupling between the polar clusters.

9.9.2.1　Formulation of Ginzburg–Landau Functional

Landau phenomenological theory was introduced in Section 2.5.2, where we assume that the Landau free energy F in 1D is represented in terms of polarization P as:

$$F(P,T) = (1/2)\alpha P^2 + (1/4)\beta P^4,$$

(9.40)

in the simplest second-order phase transition's case. That is, only the intra-coupling in the uniform polar cluster distribution (i.e., monodomain status) is taken into account. Only the coefficient α is temperature dependent,

$$\alpha = (T - T_0)/\varepsilon_0 C,$$

(9.41)

where C is taken as a positive constant called the *Curie–Weiss constant* and T_0 is equal to or lower than the actual transition temperature T_C (*Curie temperature*), in general (the first-order transition). β is supposed to be constant and positive in this second-order transition case. The phenomenological formulation should be applied for the entire temperature range over which the infinitely large crystal is uniform in its paraelectric and ferroelectric states. We discuss the form of this expansion in the case of multidomain status in a finite-size specimen shown in Figure 9.51b, without taking into account the depolarization energy initially (i.e., a rather thick single crystal specimen). Taking the Cartesian coordinate, z: polarization direction, x: perpendicular to the 180° domain wall planes, we consider the $P(x)$ distribution function in a narrow slab x to $x + \Delta x$ (Figure 9.51b). If as the first approximation we assume that the cells (or slabs), x, do not influence each other, the free energy can be decomposed into a sum (or an integral in a continuum approximation) of contributions of each cell. Since $+P_S$ and $-P_S$ do not change the energy in Eq. (9.40) at all, the energy form is exactly the same. In the second step, we take the inter-coupling between neighboring cells into account by a term describing an increase of free energy if the polarization $P(x)$ in neighboring cells differ from each other. $x = 0$ is taken at the domain boundary center point, and $P(x) = +P_S$ at $x = \infty$, and $P(x) = -P_S$ at $x = -\infty$, without losing generality. This is achieved by a term $\kappa(\nabla P(x))^2$; that is, "second power" (symmetry with respect to the crystal up-side-down) of grad $(P(x))$. Thus, we represent F in the form of the famous *Ginzburg–Landau functional*:

$$F(P(x),T) = \int dx \left[\frac{\alpha}{2} P(x)^2 + \frac{\beta}{4} P(x)^4 + \frac{\kappa}{2}(\nabla P(x))^2 \right] \tag{9.42}$$

If we adopt the *equation for the relaxation of $P(x)$*, that is, if we assume that the polarization change ΔP with time Δt is proportional to the free energy decrease with respect to the polarization change ΔP, we obtain *the time-dependent Ginzburg–Landau equation*:

$$\frac{dP}{dt} = -\frac{\partial F}{\partial P} = -\alpha P - \beta P^3 + \kappa \Delta P \tag{9.43}$$

where the proportional constant is taken as (-1) from the simplicity viewpoint, and P is now treated as function of space, x, and time t. The time-dependent equations are handled in Section 9.9.4. The typical features of Eq. (9.43) in the static condition are:

- A linear term, $-\alpha P$, where the coefficient α changes its sign at the Curie–Weiss temperature $T = T_0$
- A nonlinear term, $-\beta P^3$, which serves for a stabilization of the system
- A "diffusion" term, $\kappa \Delta P$, where Δ is the Laplacian operator.

Example Problem 9.5

In order to derive the following second-derivative equation, which is equivalent to Eq. (9.43) under steady-state condition (i.e., $\frac{dP}{dt} = 0$):

$$\kappa \left(\frac{d^2 P}{dx^2} \right) = \alpha P + \beta P^3, \tag{P9.5.1}$$

we consider the minimization of the time-independent Ginzburg–Landau functional with respect of P,

$$F(P(x),T) = \int_{-\infty}^{\infty} \left[\frac{\alpha}{2} P(x)^2 + \frac{\beta}{4} P(x)^4 + \frac{\kappa}{2} \left(\frac{dP}{dx} \right)^2 \right] dx. \tag{P9.5.2}$$

Using the minimization condition, $(\partial F / \partial P) = 0$, derive Eq. (P9.5.1).[58]

Solution

We denote $f(P) = f(P) = \frac{\alpha}{2} P(x)^2 + \frac{\beta}{4} P(x)^4$. Then, we consider slight change of polarization δP around the energy minimum point P. Let us take the expansion series:

$$f(P + \delta P) = f(P) + \frac{f'(P)}{1!} \delta P + \frac{f''(P)}{2!}(\delta P)^2 \tag{P9.5.3}$$

$$\frac{\kappa}{2}\left[\frac{d(P+\delta P)}{dx}\right]^2 = \frac{\kappa}{2}\left[\frac{dP}{dx} + \frac{d(\delta P)}{dx}\right]^2 = \frac{\kappa}{2}\left(\frac{dP}{dx}\right)^2 + \kappa\left(\frac{dP}{dx}\right)\frac{d(\delta P)}{dx} + \frac{\kappa}{2}\left[\frac{d(\delta P)}{dx}\right]^2 \quad \text{(P9.5.4)}$$

Then, we obtain

$$F(P+) = \int_{-\infty}^{\infty}\left[f(P+\delta P) + \frac{\kappa}{2}\left(\frac{d(P+\delta P)}{dx}\right)^2\right]dx$$

$$= \int\left[f(P) + \frac{\kappa}{2}\left(\frac{dP}{dx}\right)^2\right]dx + \int\left[f'(P)\delta P + \kappa\left(\frac{dP}{dx}\right)\frac{d(\delta P)}{dx}\right]dx + \cdots \quad \text{(P9.5.5)}$$

$$= F(P) + \delta F + \delta^2 F + \cdots$$

The first term in the right-hand side of Eq. (P9.5.5) describes the thermodynamic potential of the optimum distribution $P(x)$, with respect to which the variation is performed. It coincides with the original Eq. (P9.5.2). The following terms represent, respectively, the first and second variations of the free energy.

The equality to zero of the first variation $\delta F = 0$ enables us to find the distribution $P(x)$, corresponding to the minimum F. Taking into account integration by parts

$$\int\left[f'(P)\delta P + \kappa\left(\frac{dP}{dx}\right)\frac{d(\delta P)}{dx}\right]dx = \int\left[f'(P) - \kappa\left(d^2P/dx^2\right)(\delta P)\right]dx \quad \text{(P9.5.6)}$$

The derivation of the above Eq. (P9.5.6) is from the two facts that $\frac{d}{dx}\left(\frac{dP}{dx}\delta P\right) = \frac{d^2P}{dx^2}\delta P +$ $\frac{dP}{dx}\frac{d(\delta P)}{dx}$, and that $\int_{-\infty}^{\infty}\frac{d}{dx}\left(\frac{dP}{dx}\delta P\right)dx = \frac{dP}{dx}\delta P\Big|_{-\infty}^{\infty} = 0$ (note that the slope $\frac{dP}{dx}$ should be zero at a point far from $x = 0$).

Since Eq. (P9.5.6) should be zero for any δP, the expression in the braces is equal to zero:

$$\kappa\left(\frac{d^2P}{dx^2}\right) = \frac{df}{dx} = \alpha P + \beta P^3 \quad \text{(P.9.5.7)}$$

The sign of the second variation $\delta^2 F$ makes it possible to evaluate the stability (maximum or minimum) of the corresponding solution. We will skip this process here and remain it in Chapter Problem 9.1.

9.9.2.2 Domain Wall Formula

We now solve the static domain wall structure equation from Eq. (9.43):

$$\kappa\left(\frac{d^2P}{dx^2}\right) = \alpha P + \beta P^3 \quad \text{(9.44)}$$

The domain wall exists at $x = 0$ and away from the boundary where the homogeneous state is implemented, that is, the second derivative $\left(\frac{d^2P}{dx^2}\right) = 0$. The value of P at a point far from the wall can be determined from the conventional equation:

$$\alpha P_0 + \beta P_0^3 = 0, \quad \text{(9.45)}$$

which in the case of ferro-phase with the spontaneous polarization P_0 gives

$$P_0^2 = -\alpha/\beta \ (\alpha < 0) \tag{9.46}$$

To determine the structure of the 180° domain wall in the vicinity of $x = 0$, we use the following boundary conditions:

$$\begin{cases} P(+\infty) = P_0 \\ P(-\infty) = -P_0 \end{cases} \tag{9.47}$$

First, both parts of Eq. (9.44) are multiplied by (dP/dx) and integrated with respect of dx, taking into account the conditions Eq. (9.47). Consequently, we obtain

$$\frac{\kappa}{2}\left(\frac{dP}{dx}\right)^2 = f[P(x)] - f[P_0], \tag{9.48}$$

where $f(P) = \dfrac{\alpha}{2} P(x)^2 + \dfrac{\beta}{4} P(x)^4$. Separating the variables in Eq. (9.48) gives

$$\int \frac{dP}{\sqrt{f[P(x)] - f[P_0]}} = \int \frac{dx}{\sqrt{\kappa/2}}. \tag{9.49}$$

Taking into account the specific value P_0^2, the difference $f[P(x)] - f[P_0]$ is transformed to the form

$$f[P(x)] - f[P_0] = \frac{\beta}{4}\left[P_0^2 - P^2\right]^2. \tag{9.50}$$

Expanding the resultant difference of the squares into multipliers

$$\frac{1}{P_0^2 - P^2} = \frac{1}{2P_0(P_0 + P)} + \frac{1}{2P_0(P_0 - P)} \tag{9.51}$$

and integrating Eq. (9.49), taking Eqs. (9.50) and (9.51) into account, we obtain

$$\frac{1}{\sqrt{\beta}P_0} \int \left[\frac{dP}{(P_0 + P)} + \frac{dP}{(P_0 - P)}\right] = \frac{1}{\sqrt{\beta}P_0} \ln \frac{(P_0 + P)}{(P_0 - P)} = \sqrt{\frac{2}{\kappa}}x \tag{9.52}$$

Assuming that the center of the domain boundary is situated at $x = 0$, where $P = 0$, we obtain the following distribution of polarization around the boundary.

$$P(x) = P_0 \tanh\left(\frac{x}{\delta}\right) \quad \left[\text{Note that } \tanh x = \frac{e^x - e^{-x}}{e^x + e^{-x}}\right]. \tag{9.53}$$

$$\delta = \frac{1}{P_0}\sqrt{\frac{2\kappa}{\beta}} = \sqrt{\frac{2\kappa}{-\alpha}} \tag{9.54}$$

It is important to note that the polarization distribution around the domain wall follows a hyperbolic tangent function in this simple *Ginzburg–Landau Functional* case. In accordance with Eq. (9.53) and Figure 9.54, quantity δ is naturally referred to as the half of the domain boundary thickness. As indicated by Eq. (9.54), this thickness depends greatly on temperature and increases significantly

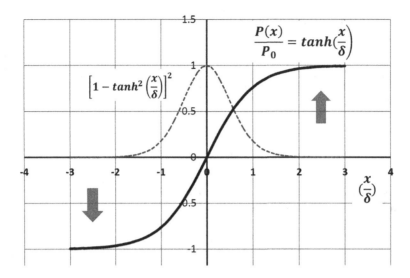

FIGURE 9.54 Ferroelectric domain wall models: polarization distribution (solid line) and domain wall energy distribution (dashed line).

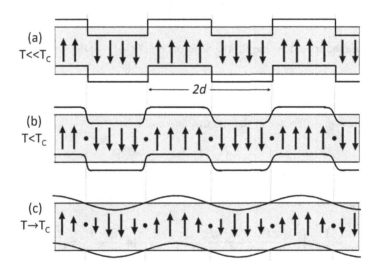

FIGURE 9.55 Polarization distribution configurations in a multidomain structure with the width d at a low temperature (a), an intermediate temperature (b), and a temperature close to Curie temperature (c).

when approaching the Curie point T_C (= T_0 in our simplest second-order transition) due to $\alpha \to 0$. Figure 9.55 illustrates expected polarization distribution configurations in a multiple domain structure with the width d and domain wall thickness δ at a low temperature (a), at an intermediate temperature (b), and a temperature close to the Curie temperature (c). A step function-like thin domain wall at a low temperature becomes diffused with increasing the temperature, with exhibiting the polarization $P(x)$ distribution in a hyperbolic tangent in terms of position coordinate x. With approaching to the Curie temperature, the domain wall becomes very vague, as though the polarization distribution is sinusoidal with the periodicity of $2d$ (Figure 9.55c). This polarization configuration seems to be the optical soft-phonon mode with the wave vector $k = 2\pi/\lambda = \pi/d$.

9.9.2.3 Domain Wall Energy

The surface density of the energy of the stationary wall γ_0 is obtained as a result of substituting the distribution Eq. (9.53) into Eq. (9.42) less the energy of the homogeneous state. Consequently, taking into account that $dP_0/dx = 0$, the first integral Eq. (9.48) and the ratio Eq. (9.50), we find

$$\gamma_0 = \int_{-\infty}^{\infty} \left[f[P(x)] - f[P_0] + \frac{\kappa}{2} \left(\frac{dP}{dx} \right)^2 \right] dx \quad \left[\text{from Eq. (9.48)} \right]$$

$$= 2 \int_{-\infty}^{\infty} \left[f[P(x)] - f[P_0] \right] dx \quad \left[\text{using Eq. (9.50)} \ f[P(x)] - f[P_0] = \frac{\beta}{4} \left[P_0^2 - P^2 \right]^2 \right]$$

$$= \frac{\beta}{2} P_0^4 \int_{-\infty}^{\infty} \left[1 - \tanh^2 \left(\frac{x}{\delta} \right) \right]^2 dx = -\frac{\alpha P_0^2 \delta}{2} \int_{-\infty}^{\infty} \frac{\frac{dx}{\delta}}{\cosh^4 \left(\frac{x}{\delta} \right)} \quad \left[\text{using Eq. (9.46)} \ P_0^2 = \alpha/\beta \right]$$

$$= -\frac{2}{3} \alpha P_0^2 \delta = \frac{4}{3} P_0^2 \left(\frac{\kappa}{\delta} \right) \quad \left[\text{from Eq. (9.54)} \right] \tag{9.55}$$

Note here that a general integral relation, $\displaystyle \int_{-\infty}^{\infty} \frac{dx}{\cosh^4(x)} = \frac{\sinh x}{3\cosh^3(x)} \bigg|_{-\infty}^{\infty} + \frac{2}{3} \int_{-\infty}^{\infty} \frac{dx}{\cosh^2(x)} = $

$\displaystyle \frac{2}{3} \tanh x \bigg|_{-\infty}^{\infty} = \frac{4}{3}.$

9.9.3 STATIC DOMAIN STRUCTURES – ADVANCED FORM

Since we obtain the domain wall energy from the phenomenology in Eq. (9.55), we reconsider the static domain structures in an advanced fashion by integrating all energy terms. We will work again on a 180° domain arrangement shown in Figure 9.51b with domain width d and domain crystal thickness t_{crys}. Following the Chenskii's approach,[62] we describe the

$$G_1 - G_1^0 = \int \left(\frac{\alpha}{2} P^2 + \frac{\beta}{4} P^4 \right) dV + W_e + W_{\text{dip}} + W_x \tag{9.56}$$

Here, W_e, W_{dip}, and W_x are *depolarization energy*, *dipole interaction energy* (i.e., the *domain wall energy*, given in Eq. (9.55) excluding the polarization energy $\left[\frac{\alpha}{2} P_0^2 + \frac{\beta}{2} P_0^4 \right]$ term), and *elastic energy* (via piezoelectric effect, $x_i = g_{ij} P_j$), which are provided, respectively, by

$$W_e = \varepsilon^* P_0^2 V \left(\frac{d}{t_{\text{crys}}} \right), \tag{9.57}$$

$$W_{\text{dip}} = -K\alpha P_0^2 \delta \left(\frac{V}{d} \right), \quad \left[K = \frac{2}{3} \text{ in Eq. (9.55), but different for the first-order transition.} \right] \tag{9.58}$$

$$W_x = \frac{1}{2} \int_V X \cdot x dV = \frac{1}{2} \int_V c_{ij} x_i x_j \, dV = \frac{1}{2} c_{ij} g_{ij} P_j g_{jk} P_k. \tag{9.59}$$

Now, Eq. (9.56) can be rewritten as

$$G_1 - G_1^0 = \frac{\alpha}{2} P_0^2 + \frac{\beta}{4} P_0^4 + \varepsilon^* P_0^2 V \left(\frac{d}{t_{\text{crys}}} \right) - K \alpha P_0^2 \delta \left(\frac{V}{d} \right) + \frac{1}{2} cg^2 P_0^2 \tag{9.60}$$

From the static condition, $\dfrac{dP}{dt} = -\dfrac{\partial G_1}{\partial P} = 0$, we obtain

$$\frac{\partial G_1}{\partial P} = 0 = \left(2 \frac{\varepsilon^* d}{t_{\text{crys}}} - 2K\delta\alpha/d + \alpha + cg^2 \right) P_0 + \beta P_0^3$$

$$P_0^2 = -\frac{1}{\beta} \left(\alpha + cg^2 + 2 \frac{\varepsilon^* d}{t_{\text{crys}}} - 2K\delta\alpha/d \right)$$

Then, the equilibrium domain width d and the polarization P_0 can be obtained from the minimization condition $\dfrac{\partial G_1}{\partial d} = 0$;

$$d = \sqrt{\frac{-K\delta\alpha t_{\text{crys}}}{\varepsilon^*}} \tag{9.61}$$

$$P_0^2 = -\frac{\alpha + cg^2}{\beta} \tag{9.62}$$

Note that the spontaneous polarization is slightly modified by the existence of the piezoelectricity under the multidomain configuration, and the elastic energy term cg^2 should be replaced, depending on the crystal symmetry, by $(c_{33}g_{33}^2 + 2c_{11}g_{31}^2)$ in the 4mm case, for example.

9.9.4 Domain Wall Dynamics in Phenomenology

We consider now to determine the parameters of the moving domain wall, referring to the *Prandtl–Tomlinson Model* introduced in Section 9.8.1. We will assume the effective mass of the domain wall, m^*, and its displacement, u, in a periodic potential modulated by the distance a (maybe the crystal unit cell). We further assume the effective charge, e^*, which links u with polarization P and a.

9.9.4.1 Domain Wall Dynamic Equation

In order to discuss the dynamic response, we introduce the kinetic energy T into the potential energy expressed in Eq. (9.42). A general expression of the kinetic energy $T = \dfrac{1}{2} \rho \left(\dfrac{\partial u}{\partial t} \right)^2$ (ρ: density of the crystal) can be rewritten as

$$T = \frac{1}{2} \mu \left(\frac{\partial P}{\partial t} \right)^2, \tag{9.63}$$

taking into account $P = e^* u / a^3$ (proportional to the displacement), and a new parameter μ is a sort of mass proportional to the material's density normalized by the effective charge, which is verified as the "mobility" later, and is provided by

$$\mu = \rho a^6 / e^{*2}. \tag{9.64}$$

The full energy expression (kinetic and potential energy) can be provided by

$$\Phi = \int dx[F + T]$$

$$= \int dx \left[\frac{1}{2}\mu \left(\frac{\partial P}{\partial t}\right)^2 + \frac{1}{2}\kappa \left(\frac{\partial P}{\partial x}\right)^2 + \frac{\alpha}{2}P^2 + \frac{\beta}{4}P^4 \right] \tag{9.65}$$

Note that we neglect the depolarization energy and elastic energy in the present discussion. Under the energy minimum condition, $\frac{\partial \Phi}{\partial P} = 0$, we derive

$$\mu \frac{\partial^2 P}{\partial t^2} - \kappa \frac{\partial^2 P}{\partial x^2} + \alpha P + \beta P^3 = 0 \quad \left[\text{Refer to Example Problem 9.5}\right] \tag{9.66}$$

When we consider the dissipation Γ and the presence of the external electric field E, which is actually essential to discuss the dielectric loss, we can expand the formula as follows:

$$\mu \frac{\partial^2 P}{\partial t^2} + \Gamma \frac{\partial P}{\partial t} - \kappa \frac{\partial^2 P}{\partial x^2} + \alpha P + \beta P^3 = E \tag{9.67}$$

9.9.4.2 Domain Wall Motion under Zero Field

In order to understand the domain wall configuration change with the motion, we initially consider the simpler Eq. (9.66) [i.e., no loss]. We will adopt an assumption that the distribution profile of polarization in the moving wall maintains $P(x, v) = P(x - vt)$, where v is the velocity of the domain wall motion; that is, the domain wall behaves as a "traveling wave". This leads to a general relation,

$$\frac{\partial P}{\partial t} = -v \frac{\partial P}{\partial x}. \tag{9.68}$$

Then, taking new parameters

$$x' = x - vt,$$

$$k' = k - \mu v^2 = k\left(1 - v^2/c_0^2\right), \tag{9.69}$$

$$c_0^2 = k/\mu, \tag{9.70}$$

we can rewrite Eq. (9.66) in the following equation:

$$\kappa' \frac{\partial^2 P}{\partial x'^2} = \alpha P + \beta P^3 \tag{9.71}$$

The solution is obvious from Eq. (9.53) and expressed as

$$P(x, v) = P_0 \tanh\left(\frac{x - vt}{\delta'}\right), \text{ where} \tag{9.72}$$

$$\delta' = \sqrt{\frac{2\kappa'}{-\alpha}} = \sqrt{\frac{2\kappa\left(1 - v^2/c_0^2\right)}{-\alpha}} = \delta\sqrt{\left(1 - v^2/c_0^2\right)} \cdot \left[\alpha < 0\right]. \tag{9.73}$$

The limiting velocity of the domain wall motion is $c_0 = \sqrt{\kappa/\mu}$, and the domain wall width seems to be reduced by the factor of $\sqrt{\left(1 - v^2/c_0^2\right)}$, a similar formula to the relativistic "Lorentz factor". Let us determine the domain wall energy under the velocity v: from $\Phi - \Phi(P_0)$, obtain as

$$\gamma(v) = -\frac{2}{3}\alpha P_0^2 \delta \cdot \frac{1}{\sqrt{\left(1 - v^2/c_0^2\right)}} = \frac{\gamma_0}{\sqrt{\left(1 - v^2/c_0^2\right)}}. \tag{9.74}$$

where $\gamma_0 = -\frac{2}{3}\alpha P_0^2 \delta$ as given in Eq. (9.55). If we adopt a similar relationship among "energy and the mass" to the relativity theory (i.e., energy = mc^2), we may introduce the effective mass m^* as follows:

$$\gamma(v) = \frac{\gamma_0}{\sqrt{\left(1 - v^2/c_0^2\right)}} = m^*(v)c_0^2. \tag{9.75}$$

The effective mass is expressed as

$$m^*(v) = \frac{\gamma_0/c_0^2}{\sqrt{\left(1 - v^2/c_0^2\right)}} = \frac{m_0^*}{\sqrt{\left(1 - v^2/c_0^2\right)}}, \quad \left[m_0^* = \gamma_0/c_0^2\right] \tag{9.76}$$

which increases with increasing the velocity. When the velocity is small, the kinetic energy is estimated by

$$T = \left(m^*(v) - m_0^*\right)c_0^2 \approx \frac{1}{2}m_0^* v^2, \tag{9.77}$$

which re-proved the validity of the effective mass description, taking into account the initial introduction of the kinetic energy $T = \frac{1}{2}\rho\left(\frac{\partial u}{\partial t}\right)^2$.

9.9.4.3 Domain Wall Motion with Dissipation under an Electric Field

Taking into account the dissipation and the electric field application, let us derive the dielectric loss from the formula expressed in Eq. (9.67):

$$\mu\frac{\partial^2 P}{\partial t^2} + \Gamma\frac{\partial P}{\partial t} - \kappa\frac{\partial^2 P}{\partial x^2} + \alpha P + \beta P^3 = E.$$

We first examine its asymptotic solution. Away from the boundary, where all the derivatives are equal to zero, the asymptotic values of the polarization are the roots of the equation:

$$\alpha P + \beta P^3 = E. \tag{9.78}$$

As the derivation process is schematically illustrated in Figure 9.56a, under $E \neq 0$, the three roots, P_1, P_2, and P_3, have no longer the symmetry: "$P_0 = P_1 = \sqrt{-\alpha/\beta}$ and $P_3 = 0$ under $E = 0$" changes to "$P_1 > P_0$, $|P_2| < P_0$ and $P_3 \neq 0$". Figure 9.56b shows the expected distribution of polarization in a domain wall under the external electric field E, in comparison with the stationary profile under $E = 0$.

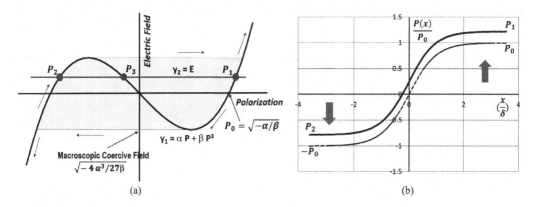

FIGURE 9.56 (a) Roots of the polynomial $\alpha P + \beta P^3 = E$ under various external field E. (b) Distribution of polarization in a stationary domain wall and in a domain wall under the external electric field.

Absolute permittivity $\varepsilon \cdot \varepsilon_0$ is obtained from $\left(\dfrac{\partial P}{\partial E}\right)$, and from Eq. (9.78),

$$\varepsilon \cdot \varepsilon_0 = \left(\frac{\partial E}{\partial P}\right) = 1/\left(\alpha + 3\beta P^2\right) \tag{9.79}$$

Thus, the relative permittivity under very small electric field E at P_1, P_2, and P_3 is given, respectively, by taking into account $P_1 = P_2 = \sqrt{-\alpha/\beta}$ and $P_3 = 0$, as follows:

$$P_1, P_2 : \varepsilon\varepsilon_0 = 1/(-2a) = C/2(T_0 - T) \tag{9.80}$$

$$P_3 : \varepsilon\varepsilon_0 = 1/(a) = -C/(T_0 - T) \; \left[\text{hypothetical negative capacitance at } T < T_0\right] \tag{9.81}$$

On the other hand, the macroscopic coercive field is obtained from the maximum/minimum point of $y_1 = \alpha P + \beta P^3$ curve: that is, from $\dfrac{\partial y_1}{\partial P} = 0$, $\alpha + 3\beta \cdot P^2 = 0$. Then, $P = \sqrt{-\alpha/3\beta}$. Since the coercive field is obtained from the maximum y_1 point,

$$y_1 = \alpha P + \beta P^3 = \sqrt{-\alpha/3\beta}\left[\alpha + \beta(-\alpha/3\beta)\right]$$

$$= \sqrt{-4\alpha^3/27\,\beta} \tag{9.82}$$

We may not need to consider the AC electric field amplitude higher than this macroscopic coercive field (usually > 10 kV/mm).

As the second step, we rewrite Eq. (9.67) by using the normalized parameters, $p = P/P_0$, $E' = E/(-\alpha)P_0$:

$$\frac{1}{2}\frac{\delta^2}{c_0^2}\frac{\partial^2 p}{\partial t^2} + \frac{\Gamma}{\alpha}\frac{\partial p}{\partial t} - \frac{1}{2}\delta^2\frac{\partial^2 p}{\partial x^2} + p + p^3 = E'. \tag{9.83}$$

Taking further a new position/time parameter ξ for the moving domain wall with the velocity v,

$$\xi = \sqrt{2}\,(x - vt)/\delta\sqrt{\left(1 - v^2/c_0^2\right)}, \tag{9.84}$$

and the relationships,

$$\frac{\partial^2 p}{\partial t^2} = \frac{\partial^2 p}{\partial \xi^2} \frac{2v^2}{\delta^2 \left(1 - v^2/c_0^2\right)}, \tag{9.85}$$

$$\frac{\partial^2 p}{\partial x^2} = \frac{\partial^2 p}{\partial \xi^2} \frac{2}{\delta^2 \left(1 - v^2/c_0^2\right)}, \tag{9.86}$$

Equation (9.83) can be transformed into

$$\frac{\partial^2 p}{\partial \xi^2} + \bar{v}\frac{\partial p}{\partial \xi} - p - p^3 + E' = 0, \tag{9.87}$$

where

$$\bar{v} = \frac{\sqrt{2}\Gamma}{\alpha\delta\sqrt{\left(1 - v^2/c_0^2\right)}} v. \quad [\text{Since } \alpha < 0, \bar{v} < 0] \tag{9.88}$$

Because of the damping factor Γ, the domain wall experiences a sort of dragging force proportional to $\bar{v}$, which is also proportional to the wall velocity v with an opposite direction (i.e., viscous damping).

When the electric field E is not large, the solution of Eq. (9.87) may be modified from the asymptotic solution for $\bar{v} = 0$ (no dissipation) and $E' = 0$: that is,

$$\frac{p}{p_0} = \tanh\left(\frac{x}{\delta}\right) = \frac{\exp\left(\dfrac{x}{\delta}\right) - \exp\left(-\dfrac{x}{\delta}\right)}{\exp\left(\dfrac{x}{\delta}\right) + \exp\left(-\dfrac{x}{\delta}\right)} = 1 - \frac{2}{\exp\left(\dfrac{2x}{\delta}\right) + 1}. \tag{9.89}$$

Denoting the three roots under E by

$$P_1/P_0 = a, P_2/P_0 = b, P_3/P_0 = c, \tag{9.90}$$

let us find the solution of Eq. (9.87) in the following form, according to Ref.[63]:

$$p(\xi) = a + \frac{b - a}{\exp\left(\dfrac{(b-a)\xi}{\sqrt{2}}\right) + 1}. \tag{9.91}$$

Then, the dimensionless roots of polynomial should satisfy the following relation:

$$p^3 + p - E' = (p - a)(p - b)(p - c). \tag{9.92}$$

We can initially obtain the following general relationship among a, b, and c:

$$a + b + c = 0, ab + bc + ca = 1, \text{ and } abc = E'. \tag{9.93}$$

Substitution of Eq. (9.91) into Eq. (9.87) verifies that the function of Eq. (9.91) is the solution of Eq. (9.87) only when

$$\bar{v} = -\frac{(a+b-2c)}{\sqrt{2}} = \frac{3c}{\sqrt{2}}. \tag{9.94}$$

By equating Eq. (9.88) and Eq. (9.94), and $c = P_3/P_0 \approx \varepsilon\varepsilon_0 E/P_0 = (1/\alpha)E/P_0$, we obtain

$$\frac{\sqrt{2}\Gamma}{\alpha\delta\sqrt{(1-v^2/c_0^2)}} v = \frac{3}{\sqrt{2}}(1/\alpha)E/P_0.$$

Thus, when $v \ll c_0$, we can write the velocity in proportion to the applied electric field E:

$$v = \frac{3}{2}\frac{E}{P_0}\frac{\delta}{\Gamma} = \mu E, \tag{9.95}$$

where the proportional constant – mobility – can be expressed by

$$\mu = \frac{3}{2}\frac{\delta}{P_0}\frac{1}{\Gamma} \tag{9.96}$$

It is interesting to note that the mobility expression, Eq. (9.96), can be directly derived from the original equation, Eq. (9.67), by assuming the following items:

1. The external field and the dissipation factor have little influence on the moving domain wall profile
2. The domain wall profile is determined by Eq. (9.66) without field nor dissipation
3. The dissipation term and electric field may be equivalent; $\Gamma\frac{\partial P}{\partial t} = E$
4. $\frac{\partial P}{\partial t} = -v\frac{\partial P}{\partial x} \approx -v(P_0/\delta)$.

Therefore, $E = \Gamma\frac{\partial P}{\partial t} = -\Gamma v(P_0/\delta)$, or $v = \frac{3}{2}\frac{E}{P_0}\frac{\delta}{\Gamma}$, exactly the same as Eq. (9.95).

When we consider the steady polarization induction $P = P_0 + p = P_0 + p_0 e^{j(\omega t+\Phi)}$ (Φ: phase lag), according to the input AC electric field $E = E_0 e^{j\omega t}$. From the following equation

$$\mu\frac{\partial^2 p}{\partial t^2} + \Gamma\frac{\partial p}{\partial t} + \alpha(P_0 + p) + \beta(P_0 + p)^3 = \mu\frac{\partial^2 p}{\partial t^2} + \Gamma\frac{\partial p}{\partial t} + (\alpha + 3\beta P_0^2)p = E, \tag{9.97}$$

we obtain

$$\varepsilon\varepsilon_0 = \frac{\partial P}{\partial E} = \frac{\partial p}{\partial E} = \frac{p_0}{E_0} = \frac{e^{-j\Phi}}{-\omega^2\mu + j\omega\Gamma - 2\alpha}. \tag{9.98}$$

If we adopt the following notations:

$$\omega_0^2 = -2\alpha/\mu, \tag{9.99}$$

$$2\zeta\omega_0 = \Gamma/\mu, \tag{9.100}$$

the permittivity and the dielectric loss $\tan\Phi$ can be obtained as

$$|\varepsilon\varepsilon_0| = \frac{1}{\sqrt{\left(-2\alpha - \mu\omega^2\right)^2 + \left(2\mu\zeta\omega_0\omega\right)^2}}, \tag{9.101}$$

$$\tan\Phi = \frac{2\mu\zeta\omega_0\omega}{-2\alpha - \mu\omega^2}. \tag{9.102}$$

For a low frequency, $\varepsilon\varepsilon_0$ approaches to $(1/(-2\alpha))$ and $\tan\Phi$ is expressed as $\Gamma\omega/(-2\alpha)$. In conclusion, the dielectric loss is expressed in proportion to the domain wall viscous damping factor Γ and the measuring frequency ω in the above single domain-wall model. It is also worthy to note that dielectric loss $\tan\Phi$ will enhance significantly with increasing the temperature to the Curie point $(-2\alpha \rightarrow 0)$. Both permittivity and the dissipation factor $\tan\Phi$ are expected to exhibit the maximum around the Curie temperature range, which is consistent with the experimentally-known fact.

9.9.5   DOMAIN WALL VIBRATIONS IN MULTIDOMAIN STRUCTURES

We discussed on the single domain wall dynamics in Section 9.9.4. We will discuss in this section on the domain wall vibrations in more realistic multidomain structures (i.e., in a finite crystal thickness specimen), as illustrated in Figures 9.51 and 9.55. As shown in Figure 9.56b, the domain wall position will translate perpendicularly to the electric field or spontaneous polarization direction. We first discuss on the 180° domain reversal from the domain nucleation model, which concludes that the translational domain boundary motion is easier under the smaller electric field, in comparison with the macroscopic coercive field level in a monodomain single crystal. Then, cooperative domain boundary vibrations are discussed for investigating further microscopic origin of the dielectric loss in piezoelectric materials. The hardening mechanism in the acceptor-doped PZTs is also explained in the last part.

9.9.5.1   180° Domain Reversal Model

The 180° domain wall does not translate parallelly by keeping its complete flatness, as schematically shown in Figure 9.56b in a more precise microscopic viewpoint. The flat-dagger domain nucleus model is introduced here, according to a paper by Craik et al.[64] Figure 9.57 shows a flat-dagger domain structure which is initiated under the external electric field E during the 180° domain reversal. This small domain nucleus may be stable or unstable depending on the nucleus

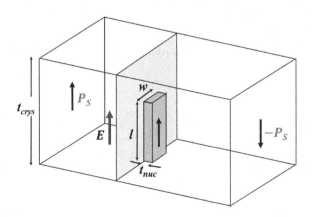

FIGURE 9.57 Flat-dagger domain structure model under the external electric field E during the 180° domain reversal.

size and applied electric field E. We assume the size relation, $t_{nuc} \ll w \ll l$, and its nucleus volume $V = t_{nuc} wl$. We can consider the domain nucleus growth or shrink (disappearance) in two directions: (1) side wall translation (i.e., motion of wl and $t_{nuc}l$ faces) and (2) front wall translation (i.e., $t_{nuc}w$ face motion). If we denote the domain wall energy density, γ_w' and γ_w'' for the side wall and front wall, respectively, the free energy increase in terms of this nucleus (volume V and area A) can be evaluated by

$$\Delta G_1 = -2EP_S V + \gamma_w A$$
$$= -2EP_S\, t_{nuc} wl + \gamma_w' wl + \gamma_w'' t_{nuc} w, \tag{9.103}$$

where the side contribution from the $t_{nuc}l$ face was neglected due to $t_{nuc}l \ll wl$. The first term $-2EP_S$ corresponds to the polarization energy (per unit volume) change from $-P_S$ to P_S under the field E. In order to maintain this nucleus, the *critical nucleus size*, below which the nucleus disappears, can be evaluated from the energy minimization in terms of size (l and t_{nuc}):

$$\frac{\partial \Delta G_1}{\partial l} = 0 \rightarrow t_{nuc}^* = \gamma_w'/(2EP_S), \tag{9.104}$$

and

$$\frac{\partial \Delta G_1}{\partial t_{nuc}} = 0 \rightarrow l^* = \gamma_w''/(2EP_S). \tag{9.105}$$

Note that the critical nucleus size becomes smaller with an increase of electric field. In this case, the critical energy of this domain nucleus is obtained as

$$\Delta G_1^* = \gamma_w' \gamma_w''/(2EP_S) \tag{9.106}$$

Sideways growth of domains plays an important role in the polarization reversal in BT.[65,66] The domain wall velocity varies as a function of applied field (verified experimentally):

$$v = v_\infty \exp(-\delta/E), \tag{9.107}$$

where δ is the *activation energy* for sideways wall motion. Similarly, if we suppose that the domain nucleation rate n is given by

$$n = n_\infty \exp(-\alpha/E), \tag{9.108}$$

because the nucleation rate varies proportionally to $\exp(-\Delta G_1/kT)$, we can obtain the relationships:

$$\alpha = w\gamma_w' \gamma_w''/(2kTP_S) \tag{9.109}$$

or

$$n = n_\infty \exp\left[-\gamma_w' \gamma_w'' w/(2kTP_S)E\right] \tag{9.110}$$

In conclusion, the velocity of the domain wall motion follows initially by the exponential law, $v = v_\infty \exp(-\delta/E)$, for a small field range, owing to the domain nucleation rate (Eq. 9.107), then this dependence is changed to follow the linear law, $v = \mu E$, for a larger field (Eq. 9.95), where the domain wall may shift parallelly by keeping its plane flatness.

9.9.5.2 Translational Vibrations of the Multidomain Structures

We consider now free translational vibrations of the multidomain structures shown in Figure 9.51b. Under such vibrations, the domain walls are displaced as a whole unit in the direction normal to the polar axis, when we ignore the domain nucleation. The factors causing the formation of bending dynamics of domain walls lead to the formation of translational vibrations of the domain structures. These domain wall displacements also lead to changes in the distribution of charges of spontaneous polarizations. Figure 9.58 illustrates the translational displacements of the domain walls in a ferroelectric with 180° domain structure in the quasi-acoustic (a) and quasi-optical branches of vibrations of the domain structure (b). Let us assume that the domain walls remain flat and the value of the translational displacement of the $2n$-th wall is denoted U_{2n}, which is much less than the average domain width d. The specimen sizes are denoted as L_x, L_y, and L_z in a Cartesian coordinate system shown in Figure 9.58.

Taking the electric potential φ, the electrostatic energy (i.e., depolarization energy) of this specimen is expressed by

$$\Phi = \frac{L_y}{2} \iint P(x,z) E_z \, dx \, dz = -\frac{L_y}{2} \iint P(x,z) \frac{d\varphi}{dz} \, dx \, dz \qquad (9.111)$$

Supposing that the depolarization energy change with the domain wall displacement U_n is small, we use an expansion series formula:

$$\Phi = \Phi_0 + \frac{1}{2} \sum_{n,n'} \frac{\partial^2 \Phi}{\partial U_n \partial U_{n'}} \bigg|_0 U_n U_{n'} = \Phi_0 + \frac{1}{2} \sum_{n,n'} \tilde{K}(n,n') U_n U_{n'} \qquad (9.112)$$

Note that the linear terms (proportional to U_n) of the expansion series should disappear because of the equivalency of energy for $+U_n$ and $-U_n$.

On the basis of Eq. (9.112), the dynamic equation of the n-th domain wall

$$\tilde{m}\ddot{U}_n = -\frac{\partial \Phi}{\partial U_n} \qquad (9.113)$$

takes the following form

$$\tilde{m}\ddot{U}_n + \tilde{K}(n,n) U_n + \sum_{n \neq n'} \tilde{K}(n,n') U_{n'} = 0 \qquad (9.114)$$

Here, $\tilde{m} = m L_y L_z$, m is the density of the local effective mass of the domain wall and $\tilde{K}(n, n')$ is the first derivative of depolarization energy in terms of the wall displacement.

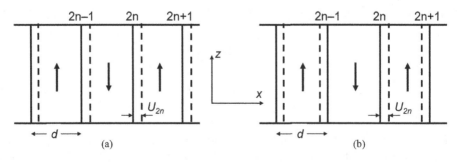

FIGURE 9.58 Translational displacements of the domain walls in a ferroelectric with 180° domain structure in the quasi-acoustic (a) and quasi-optical branches of vibrations of the domain structure (b).

In order to conduct the calculation of effective stiffness coefficient $\tilde{K}(n, n')$ in Eq. (9.114), acting on the domain walls displaced from the equilibrium positions, the electrostatic equation should be supplemented by the dynamic charge distribution and its related depolarization energy. Though Figure 9.51b illustrates no surface free-charge status, observed immediately after the spontaneous polarization generated by decreasing the temperature passing through the Curie point, ferroelectric plate surfaces with the spontaneous polarization normal to the plate are usually covered by the migrating charges from the air-atmosphere gradually after minute to hour time lapse. As discussed in Section 4.3.2.2, from the Gauss law, $\mathrm{div}\ \boldsymbol{E} = \dfrac{1}{\varepsilon_0 \varepsilon}\left(\sigma - \mathrm{div}\ \mathrm{P_S}\right)$, when no charge the depolarization field exists anti-parallel to the spontaneous polarization; while the surface charge will seal the

depolarization field. Figure 9.59 illustrates the equivalence of the resultant charge for equal displacement of the domain boundaries in the following extreme cases: (a) non-compensated charges (immediately after the P_S generation), and (b) completely compensated charges of the spontaneous polarization on the surface (after minute to hour time lapse). Note the same pulse-like charge distribution generated at the domain boundary under dynamic range (1 kHz or higher) vibration condition, regardless of whether the electric field in the initial state has or has not been compensated by the free carriers. For the calculation simplicity, we assume a periodic step-function polarization distribution along x direction with a periodicity $2d$, as in Figure 9.55a by neglecting the domain wall thickness. This means that the surface free charge can be presented as the sum of a period distribution, corresponding to the equilibrium domain structure, on the top surface ($z = L_z/2$):

$$\sigma_0(x) = \begin{cases} -P_0, & (2n-1)d < x < 2nd \\ P_0, & 2nd < x < (2n+1)d \end{cases} , \tag{9.115}$$

and antisymmetric to it on the bottom surface ($z = -L_z/2$). For a small displacement of the domain boundaries $U_n \ll d$, the variation of the charge $\sigma_1(x)$ under the wall displacement U_n from the equilibrium status is represented with delta (or impulse) function by

$$\sigma_1(x) = \sum_n \gamma_n \delta(x - nd) \tag{9.116}$$

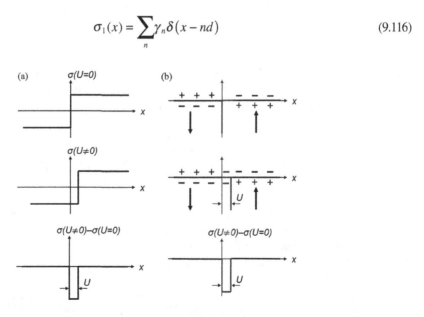

FIGURE 9.59 Equivalence of the resultant charge for equal displacement of the domain boundaries in the cases: (a) non-compensated charges and (b) completely compensated charges of the spontaneous polarization on the surface.

where $\gamma_n = 2P_0 U_n$ or $-2P_0 U_n$ at $z = L_z/2$ or $-L_z/2$, respectively. Or

$$\sigma_1(x,z) = \sum_n \gamma_n \delta(x - nd) \left[\delta\left(z - \frac{L_z}{2}\right) - \delta\left(z + \frac{L_z}{2}\right) \right] \tag{9.117}$$

From the constitutive equations:

$$X_i = c_{ij}^E x_j - e_{ij} E_j,$$
$$D_i = e_{ij} x_j + \varepsilon_0 \varepsilon_{ij}^x E_j, \tag{9.118}$$

taking the first derivative of the above equations $\left[E = -\dfrac{d\varphi}{dz}, x_j = \dfrac{du_j}{dx} \right]$ in terms of coordinate in order to introduce the impulse charge $\sigma_1(x, z)$, we obtain the following equations:

$$\rho \ddot{u}_i = c_{ijkl}^E \frac{\partial^2 u_k}{\partial x_i \partial x_j} + e_{kij} \frac{\partial^2 \varphi_k}{\partial x_i \partial x_j} \tag{9.119}$$

$$\sum_n \gamma_n \delta(x - nd) \left[\delta\left(z - \frac{L_z}{2}\right) - \delta\left(z + \frac{L_z}{2}\right) \right] = e_{ijk} \frac{\partial^2 u_j}{\partial x_k \partial x_i} - \varepsilon_0 \varepsilon_{ij}^x \frac{\partial^2 \varphi_k}{\partial x_i \partial x_j} \tag{9.120}$$

Taking into account the periodicity in terms of the x direction, we adopt the Fourier transform with wave vector k (based on $2\pi/2d$) for the potential and displacement[58]:

$$\varphi(x,z) = \iint \varphi_k e^{j\omega t} e^{-jk_x x} e^{-jk_z z} \frac{dk_x dk_z}{(2\pi)^2} \tag{9.121}$$

$$u_i(x,z) = \iint u_{ik} e^{j\omega t} e^{-jk_x x} e^{-jk_z z} \frac{dk_x dk_z}{(2\pi)^2} \tag{9.122}$$

where we adopted the time dependence of φ_k and u_{ik} in proportion to $e^{j\omega t}$. Assuming the elastic isotropy of the piezo-material, we introduce the normalized Fourier coefficient $\tilde{\varphi}_k = \varphi_k / \gamma$, which can be obtained as follows, taking the piezoelectric effect into account:

$$\tilde{\varphi}_k = 4\pi \left\{ \varepsilon_0 \varepsilon_{ij}^x k_i k_j + \frac{4\pi e_{ijk} k_k k_i}{\rho\left(c_t^2 \tilde{k}^2 - \omega^2\right)} \left[e_{pmj} k_p k_j - \frac{c_l^2 - c_t^2}{\left(c_l^2 \tilde{k}^2 - \omega^2\right)} e_{plj} k_p k_j k_l k_m \right] \right\}^{-1} \tag{9.123}$$

where $\tilde{k}^2 = k_x^2 + k_z^2$, and $c_t^2 = \mu/\rho$ and $c_l^2 = (\lambda + 2\mu)/\rho$ are the transverse and longitudinal sound velocities, with material's density ρ and *Lamé parameters* (first parameter λ and *shear modulus μ*). [Refer to Chapter Problem 9.3 to familiarize yourself with the *Lamé parameters*.] From the charge distribution Eq. (9.117), potential Eq. (9.121) is calculated as

$$\varphi_1(x,z) = \int \sigma_k \tilde{\varphi}_k e^{-jk\sigma} \frac{dk}{(2\pi)^2} = \sum_{n'} \gamma n' \int \tilde{\varphi}_k e^{-jk_x(x - dn')} \left[e^{-jk_z\left(z - \frac{L_z}{2}\right)} - e^{-jk_z\left(z + \frac{L_z}{2}\right)} \right] \frac{dk_x dk_z}{(2\pi)^2}. \tag{9.124}$$

Thus, the energy difference Eq. (9.112) becomes

$$
\Phi - \Phi_0 = \frac{1}{2} \sum_{n,n'} \tilde{K}(n,n') U_n U_{n'}
$$

$$
= L_y \int \sigma_1\left(x, z = \frac{L_z}{2}\right) \varphi_1\left(x, z = \frac{L_z}{2}\right) dx = L_y \sum_n \sum_{n'} \gamma_n \gamma_{n'} \tilde{\varphi}_k e^{-jk_x d(n-n')} \left[1 - e^{-jk_z L_z}\right] \frac{dk_x dk_z}{(2\pi)^2}
$$

$$(9.125)$$

Thus,

$$
\tilde{K}(n,n') = \tilde{K}(n-n') = (-1)^{n-n'} 8 P_0^2 L_y \int \tilde{\varphi}_k e^{-jk_x d(n-n')} \left[1 - e^{-jk_z L_z}\right] \frac{dk_x dk_z}{(2\pi)^2} \tag{9.126}
$$

We are now ready to solve the dynamic Eq. (9.114) under zero damping. As imagined, the adjacent walls (in the notations used, these are the walls with even $2n$ and odd $2n - 1$ numbers) differ by the alternation of the signs of the domains separated by them from negative (with a negative direction of P_0) to positive and vice versa for the displacement in the positive direction along the x axis. Therefore, the motions of these walls should be examined separately. We describe the displacements of the even and odd numbered walls in the form of flat waves

$$
U_{2n} = U_0(2n) \cdot e^{j[kd2n - \omega t]},
$$

$$
U_{2n-1} = U_0(2n-1) \cdot e^{j[kd(2n-1) - \omega t]}, \tag{9.127}
$$

where $U_0(2n)$ and $U_0(2n-1)$ are the amplitudes of the corresponding vibrations and k is the wave vector. Equation (9.114) can be transformed as

$$
-\tilde{m}\omega^2 U_0(2n) + \tilde{K} U_0(2n) + 2 \sum_{n=1}^{\infty} \tilde{K}(2n-1) \cos kd(2n-1) \cdot U_0(2n-1)
$$

$$
+2 \sum_{n=1}^{\infty} \tilde{K}(2n) \cos kd(2n) \cdot U_0(2n),
$$

$$(9.128)$$

$$
-\tilde{m}\omega^2 U_0(2n-1) + \tilde{K} U_0(2n-1) + 2 \sum_{n=1}^{\infty} \tilde{K}(2n-1) \cos kd(2n-1) \cdot U_0(2n)
$$

$$
+2 \sum_{n=1}^{\infty} \tilde{K}(2n) \cos kd(2n) \cdot U_0(2n-1),
$$

Grouping the coefficients of amplitudes $U_0(2n)$ and $U_0(2n-1)$ and equating the determinant consisting of these coefficients to zero, we obtain the eigen frequencies of the domain wall vibrations:

$$
\begin{vmatrix} -\tilde{m}\omega^2 + \tilde{K}_1 & \tilde{K}_2 \\ \tilde{K}_2 & -\tilde{m}\omega^2 + \tilde{K}_1 \end{vmatrix} = 0,
$$

where

$$\tilde{K}_1 = \tilde{K} + 2\sum_{n=1}^{\infty} \tilde{K}(2n)\cos kd(2n),$$

$$\tilde{K}_2 = \sum_{n=1}^{\infty} \tilde{K}(2n-1)\cos kd(2n-1). \tag{9.129}$$

We can now find two eigen frequencies:

$$\omega_-^2 = \left(\tilde{K}_1 - \tilde{K}_2\right)\big/\tilde{m}, \text{ and } \omega_+^2 = \left(\tilde{K}_1 + \tilde{K}_2\right)\big/\tilde{m}, \tag{9.130}$$

and corresponding displacement relations, respectively:

$$\left(\frac{U_0(2n)}{U_0(2n-1)}\right) = 1, \text{ and } \left(\frac{U_0(2n)}{U_0(2n-1)}\right) = -1. \tag{9.131}$$

According to Eqs. (9.130) and (9.131), we understand that there are two branches of eigen vibrations of domain walls. The first branch corresponds to vibrations in Figure 9.58a which do not induce the polarization, and may be called as a *quasi-acoustic mode*; while the second branch induces the polarization (because the $+P_0$ and $-P_0$ domains expand or shrink out-of-phase) as in Figure 9.58b and be called as a *quasi-optic mode*. Under the external field excitation, the quasi-optic mode is preferably excited, if the operating frequency is high enough (MHz or higher).

9.9.5.3 Domain Motion Relaxation versus Ionic Doping

It is well known that the acceptor ion doping such as Fe^{3+} increases the mechanical quality factor significantly in PZTs or decreases elastic loss tangent. We will consider in this section ion doping effect in terms of the domain wall dynamics viewpoints. We discussed in Section 9.9.5.2 natural undamped translational vibrations of the domain walls in terms of quasi-elastic term $\bar{K}U_n$. Depending on the mobility of the defects created by the ion doping, we expect the modification of this $\bar{K}$ value. Their influence on the mobility of the domain walls can be implemented in two cases, which are macroscopically categorized as "dry" and "viscous" friction, respectively. In the first case, the motion of the domain wall is in a system of stationary obstacles consisting of successive acts of detachment of the wall from stationary "stoppers" with its further capture by other defects. The second case takes place if the domain wall interacts with a system of mobile defects accompanying its motion.[67]

9.9.5.3.1 Stationary Stopper Case

We will determine the force acting on the moving walls from the direction of the defects interacting with the wall. The power (per unit area) of the domain wall, required by the external electric field to overcome the resistance of defects to displacement of the arbitrary domain wall, is equal to

$$F \cdot \dot{U} = \int \frac{\partial \mathcal{U}}{\partial t} n(x,t)dx = \int \frac{\partial \mathcal{U}}{\partial U} \cdot \dot{U} \cdot n(x,t)dx = -\dot{U}\int \frac{\partial \mathcal{U}}{\partial x} \cdot n(x,t)dx, \tag{9.132}$$

where $\mathcal{U} = \mathcal{U}(x - U(t))$ is introduced to give the energy increase associated with the wall bending from its equilibrium position (again symmetric with respect to x) in the system of points of its pinning by defects, that is, linked with extra bending of the wall. U is the wall coordinate of the plane of the average orientation of the domain wall interacting with the defects, $\dot{U}$ its velocity, and $n(x,t)$ is the volume concentration of points of the wall pinning by the defects.

We assume that the time dependence of the $n(x, t)$ is described by kinetic equation with the single relaxation time τ:

$$\frac{dn}{dt} = -\frac{n - n_\infty}{\tau} \tag{9.133}$$

where $n_\infty(x)$ is the equilibrium distribution of pinning points in the given region of the crystal, which may be regarded as having a Heaviside Step function Θ:

$$n_\infty = n \cdot \Theta\left(\tilde{U} - |x|\right). \tag{9.134}$$

Here n is the volume concentration of the defects, and the displacement $\tilde{U}$ is introduced to characterize the maximum distance of the defect from the plane of the average orientation of the wall at which the wall is still captured by the defect (i.e., pinning). Since the solution of Eq. (9.133) has the form

$$n(x,t) = \int_{-\infty}^{t} \exp\left(-\frac{t-\xi}{\tau}\right) n_\infty(\xi) \frac{d\xi}{\tau}, \tag{9.135}$$

substituting it in the following equation

$$F = -\int \frac{\partial \mathcal{U}}{\partial x} \cdot n(x,t) dx, \tag{9.136}$$

we obtain

$$
\begin{aligned}
F &= -\int \frac{\partial \mathcal{U}\left(x - U(t)\right)}{\partial x} \cdot \int_{-\infty}^{t} \exp\left(-\frac{t-\xi}{\tau}\right) n_\infty\left(x - U(\xi)\right) \frac{d\xi}{\tau} dx \\
&= -\int_{-\infty}^{t} \exp\left(-\frac{t-\xi}{\tau}\right) \cdot \int_{-\infty}^{\infty} \frac{\partial \mathcal{U}(x)}{\partial x} n_\infty\left(U(t) - U(\xi)\right) dx \frac{d\xi}{\tau} \\
&= -\int_{-\infty}^{t} \exp\left(-\frac{t-\xi}{\tau}\right) \cdot \mathcal{E}\left[U(t) - U(\xi)\right] \frac{d\xi}{\tau}.
\end{aligned} \tag{9.137}
$$

Here, we denote

$$\mathcal{E}\left[U(t) - U(\xi)\right] = \int_{-\infty}^{\infty} \frac{\partial \mathcal{U}(x)}{\partial x} n_\infty\left(U(t) - U(\xi)\right) dx. \tag{9.138}$$

Using a harmonic approximation and taking $\Delta = \left(U(t) - U(\xi)\right)$ small,

$$\mathcal{E}\left[U(t) - U(\xi)\right] = \mathcal{E}[\Delta] = \mathcal{E}[0] + \left.\frac{\partial \mathcal{E}}{\partial \Delta}\right|_0 \cdot \Delta + \cdots = K\left[U(t) - U(\xi)\right]. \tag{9.139}$$

Expanding Eq. (9.114) by introducing the external electric field E and the above relaxation analysis, we obtain the following equation for an arbitrary domain wall:

$$\tilde{m}\ddot{U} + KU(t) + F(U) = 2P_0 E(t). \tag{9.140}$$

Replacing $F(U)$ in Eq. (9.140) with Eq. (9.137), we obtain

$$\tilde{m}\ddot{U} + KU(t) + \bar{K}U(t) - \bar{K}\int_{-\infty}^{t}\exp\left(-\frac{t-\xi}{\tau}\right)\cdot U(\xi)\frac{d\xi}{\tau} = 2P_0E(t). \tag{9.141}$$

By taking the harmonic oscillation $E = E_0 e^{j\omega t}$, we assume $U(t) = U_0 e^{j(\omega t + \alpha)}$, then we can obtain the following relation:

$$U(t) = \frac{2P_0E(t)}{-\tilde{m}\omega^2 + (K + \bar{K}) - \bar{K}/(1 + j\omega\tau)}. \tag{9.142}$$

In the practically important case of relatively low frequency ω, we can approximate Eq. (9.142) as

$$U(t) \approx \frac{2P_0E(t)}{(K + \bar{K})}\left\{\left[1 + \frac{\dfrac{\bar{K}}{K}}{(1 + \omega^2\tau_c^2)}\right] - j\frac{\omega\tau_c\dfrac{\bar{K}}{K}}{(1 + \omega^2\tau_c^2)}\right\} \tag{9.143}$$

where

$$\tau_c = \tau \cdot \frac{(k + \bar{K})}{k} \tag{9.144}$$

9.9.5.3.2   Mobile Defect Case

The motion of the domain wall, interaction with a system of mobile defects, is described using a similar approach in the above described, with the only difference that the role of $n(x, t)$ in all expression here is performed directly by the concentration of defects in the given location in the crystal, and therefore, τ here is the relaxation time of the defective atmosphere. Thus, in contrast to the above discussion, devoted to consideration of the passage of the domain wall through the system of stationary stoppers, where time τ characterizes the relaxation properties of the domain wall, the time τ characterizes the mobility of defects, that is, depends mainly on the activation energy of its motion. When the crystal contains relatively mobile defects, the domain wall in the initial states is flat but enriched (or depleted, depending on the sign of energy $\mathcal{U}_0$ of the boundary interaction with the defects) by the atmosphere of the defects. Substituting Eq. (138) in the case of interaction of the wall with the atmosphere of mobile defects of

$$n_\infty(x) = n + n \cdot \left[\exp\left(\frac{\mathcal{U}_0}{T}\right) - 1\right]\delta(x - U(t)), \tag{9.145}$$

As the equilibrium distribution shows that the equation of motion of the domain wall and its solution here are also described by Eqs. (9.141)–(9.144).

Example Problem 9.6

Figure 9.60 shows the temperature dependence of inverse permittivity (bottom) and dielectric loss (top) of a soft PZT ceramic specimen, measured along the poling direction in a poled state, and an unpoled state, after annealing at 500°C for 30 minutes. Explain the following facts from the phenomenological viewpoints:

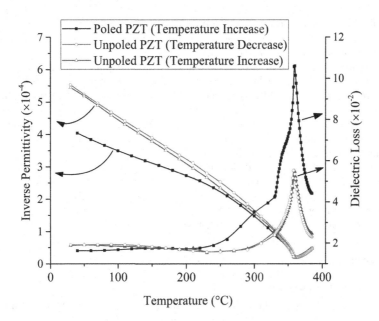

FIGURE 9.60 Inverse permittivity and dielectric loss changes with temperature in poled and unpoled soft PZT specimen.

1. The permittivity at room temperature in the poled sample is larger than that in the unpoled sample.
2. The permittivity values merge around 200°C for both poled and unpoled samples.
3. The dielectric loss $\tan\delta'_{33}$ shows the maximum at the Curie temperature.
4. The dielectric loss at room temperature in the poled sample is smaller than that in the unpoled sample.
5. The dielectric loss values cross around 200°C for both poled and unpoled samples, and that in the poled sample is higher than that in the unpoled sample above 200°C.

Solution

Poling process on a piezo-ceramic specimen provides

- Spontaneous polarization alignment
- Crystallographic anisotropy
- Reduction of domain walls
- Residual tensile stress along the spontaneous polarization direction.

1. PZT ceramic specimens exhibit either higher or lower permittivity, ε_{33}, along the spontaneous polarization direction, in comparison with that along the perpendicular orientation, depending on the doped ions. This particular soft PZT composition seems to show higher permittivity along the P_S direction, because the average permittivity between ε_{33} and ε_{11} in the unpoled specimen shows lower value.
2. The permittivity values merge around 200°C for both poled and unpoled samples. This may be explained by the poled state release with an increase in temperature. Even if the polarization alignment is roughly maintained upward or downward, once non-180° domain reorientation starts, the permittivity approaches to the non-poled state.
3. At low frequency ω, such as 1 kHz, the dielectric loss $\tan\delta = \Gamma\omega/(-2\alpha)$ (Eq. 9.102). Since $-\alpha = \dfrac{T_C - T}{\varepsilon_0 C}$, with increasing the temperature around the Curie temperature T_C, we can expect the maximum of loss $\tan\delta$.

4. The dielectric loss at room temperature in the poled sample is smaller than that in the unpoled sample. The loss along the constrained direction is smaller than the transverse direction, in general. Also, if the poling process reduces the domain wall numbers to create more monodomain-like status in a grain, we can expect lower dielectric loss.

5. We may adopt a similar "poled state release" model with an increase in temperature to the above item (2). Even if the polarization alignment is roughly maintained upward or downward, once non-180° domain reorientation starts, the dielectric loss approaches to the non-poled state. Further interesting point is higher dielectric loss in the poled sample than that in the unpoled sample above 200°C. This may be originated from the residual tensile stress generated during the poling process. As we discussed in Figure 9.47, we demonstrated the decrease in dielectric loss tan δ with the compressive stress along the polarization direction. Thus, when the sample is under the tensile stress, we may expect the increase in dielectric loss tan δ. This higher dielectric loss is observed only in the first temperature rise process, and it decreases significantly after annealing (i.e., after removing the residual stress) at 500°C for 30 minutes.

CHAPTER ESSENTIALS

1. The Uchida-Ikeda Model:

   The saturation values of the polarization and the strain of a tetragonal and rhombohedral ceramic under high electric field are summarized as follows:

$$\text{Tetragonal: } P_3 \rightarrow (0.831)P_S \qquad x_3 \rightarrow (0.368)S_S$$

$$\text{Rhombohedral: } P_3 \rightarrow (0.861)P_S \quad x_3 \rightarrow (0.424)S_S$$

2. Loss in PZTs:
   1. The antiresonance Q_B is always larger than the resonance Q_A in PZTs: This is a significant contradiction with the IEEE Standard assumption, leading to the necessity of the "Standard" revision.
   2. The intensive (prime) losses are larger than the corresponding extensive (non-prime) losses: This is understood by the boundary condition difference between "intensive" and "extensive"; that is, Free or Clamped/Constrained conditions.
   3. The intensive piezoelectric losses are significantly larger than the intensive dielectric or elastic losses in PZTs. That is, $\tan\theta' > (\tan\delta' + \tan\phi')/2$: This is not true for Pb-free piezoelectrics.
   4. There is apparent loss anisotropy in dielectric, elastic, and piezoelectric losses, indicating the anisotropy in domain wall mobility in the crystal.

3. Electrostrictor, hard and soft PZTs:

Material	d	k	Q_m	Off-Resonance Applications	Resonance Applications
PMN	High$^+$ (DC bias)$^-$	High$^+$ (DC bias)$^-$	Low$^-$ (DC bias)$^-$	High displacement$^+$ No hysteresis$^+$	Broad bandwidth$^+$
Soft PZT-5H	High$^+$	High$^+$	Low$^-$	High displacement$^+$	Heat generation$^-$
Hard PZT-8	Low$^-$	Low$^-$	High$^+$	Low strain$^-$	High AC displacement$^+$

4. Impurity dipole alignment categories:
   a. Random alignment – higher coercive field – "domain wall pinning" model
   b. Unidirectionally fixed alignment – internal bias field
   c. Unidirectionally reversible alignment – pinched/double hysteresis

5. Ginzburg–Landau functional:

$$F\left(P(x),T\right) = \int dx \left[\frac{\alpha}{2} P(x)^2 + \frac{\beta}{4} P(x)^4 + \frac{\kappa}{2} (\nabla P(x))^2 \right]$$

Domain wall structure:

$$P(x) = P_0 \tanh\left(\frac{x}{\delta}\right), \quad \delta = \frac{1}{P_0}\sqrt{\frac{2\kappa}{\beta}} = \sqrt{\frac{2\kappa}{-\alpha}},$$

Domain wall energy:

$$\gamma_0 = -\frac{2}{3}\alpha P_0^2 \delta = \frac{4}{3} P_0^2 \left(\frac{\kappa}{\delta}\right)$$

6. Static domain structure is determined by:

$$G_1 - G_1^0 = \int \left(\frac{\alpha}{2} P^2 + \frac{\beta}{4} P^4 \right) dV + W_e + W_{\text{dip}} + W_x,$$

where W_e, W_{dip}, and W_x are depolarization energies, dipole interaction energy (domain wall energy).

7. Dynamic domain wall motion is analyzed by:

$$\mu \frac{\partial^2 P}{\partial t^2} + \Gamma \frac{\partial P}{\partial t} - \kappa \frac{\partial^2 P}{\partial x^2} + \alpha P + \beta P^3 = E,$$

where Γ is the dissipation factor and E is the applied external electric field E.

Permittivity and dielectric loss $\tan \Phi$ can be obtained as

$$|\varepsilon\varepsilon_0| = \frac{1}{\sqrt{\left(-2\alpha - \mu\omega^2\right)^2 + \left(2\mu\zeta\omega_0\omega\right)^2}},$$

$$\tan \Phi = \frac{2\mu\zeta\omega_0\omega}{-2\alpha - \mu\omega^2}.$$

Here, $\omega_0^2 = -2\alpha/\mu$, and $2\zeta\omega_0 = \Gamma/\mu$. For a low frequency,

$$\varepsilon\varepsilon_0 \approx 1/(-2\alpha)$$

$$\tan \Phi \approx \Gamma\omega/(-2\alpha).$$

CHECK POINT

1. Which is a "Soft" PZT result, left- or right-hand side figure?

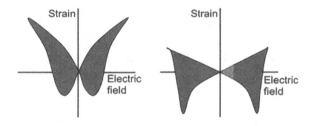

2. In a polycrystalline PZT, we observed $P_S = 27$ μC/cm². Estimate the single crystal value roughly with the Uchida-Ikeda model.

3. (T/F) The tetragonal symmetry Fe-doped PZT ceramics show the largest mechanical quality factor under vibration velocity 0.05 m/s. Thus, it is the best specimen to be used for high-power density (vibration velocity 0.5 m/s) applications. True or False?

4. (T/F) For the diesel injection valve application, the largest displacement or the largest piezoelectric d_{33} material is the most suitable candidate. Thus, very soft PZT composition should be utilized for this application. True or False?

5. (T/F) The origin of the multidomain structure in a thin ferroelectric crystal without electrodes is to reduce the depolarization energy in the specimen. True or False?

6. (T/F) The domain wall thickness in a ferroelectric crystal increases with increasing temperature up to the Curie temperature. True or False?

7. (T/F) Ferroelectrics exhibit the maximum permittivity and maximum dielectric loss just around the Curie temperature. True or False?

8. (T/F) The P–E hysteresis loop is observed in a PZT specimen as below. We can conclude that this sample possesses a "positive" internal bias field. True or False?

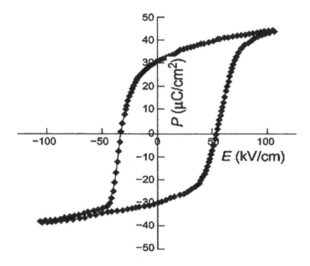

9. (T/F) The domain wall thickness in a ferroelectric barium titanate is much thicker than the domain wall thickness in a ferromagnetic Ni. True or False?

CHAPTER PROBLEMS

9.1 We consider the minimization of the time-independent Ginzburg–Landau functional with respect of P

$$F(P(x),T) = \int_{-\infty}^{\infty} \left[\frac{\alpha}{2} P(x)^2 + \frac{\beta}{4} P(x)^4 + \frac{\kappa}{2} \left(\frac{dP}{dx} \right)^2 \right] dx.$$

By changing the polarization P by δP, we calculate the free energy change

$$F(P+\delta P) = \int_{-\infty}^{\infty} \left[f(P+\delta P) + \frac{\kappa}{2}\left(\frac{d(P+\delta P)}{dx}\right)^2 \right] dx$$

$$= \int \left[f(P) + \frac{\kappa}{2}\left(\frac{dP}{dx}\right)^2 \right] dx + \int \left[f'(P)\delta P + \kappa\left(\frac{dP}{dx}\right)\frac{d(\delta P)}{dx} \right] dx + \cdots$$

$$= F(P) + \delta P + \delta^2 P + \cdots$$

From the first derivation $\delta F = 0$, we can derive

$$\kappa\left(\frac{d^2 P}{dx^2}\right) = \alpha P + \beta P^3.$$

The sign of the second variation $\delta^2 F$ makes it possible to evaluate the stability (maximum or minimum) of the corresponding solution. Calculate $\delta^2 F$ practically, then discuss the system stability from the sign of $\delta^2 F$.

Hint

$$\delta^2 F = \int \left[\frac{f''(P)}{2!}(\delta P)^2 + \frac{\kappa}{2}\left[\frac{d(\delta P)}{dx}\right]^2 \right] dx + \cdots$$

$$= \int \delta P \left[\frac{\kappa}{2}\frac{d^2}{dx^2} + \frac{1}{2}\frac{d^2 f}{dP^2} \right] \delta P dx$$

Thus, you need to solve the dynamic equation in a potential field $V(x) = \frac{1}{2}\frac{d^2 f}{dP^2} = \frac{1}{2}\left(\alpha + 3\beta P(x)^2\right)$. Verify maximum $V(x) = -\alpha > 0$ away from the boundary and $\delta^2 F > 0$. Thus, we can conclude that $\delta F = 0$ point corresponds to the *minimum* point.

9.2   In Section 9.9.2, we adopted the second-order transition case, from the static domain wall structure equation:

$$\kappa\left(\frac{d^2 P}{dx^2}\right) = \alpha P + \beta P^3.$$

When we expand the discussion to the first-order transition, $f(P) = (1/2)\alpha P^2 + (1/4)\beta P^4 + (1/6)\gamma P^6$, discuss the differences from the second-order type in terms of the domain wall structure formula and the domain wall energy.

Hint

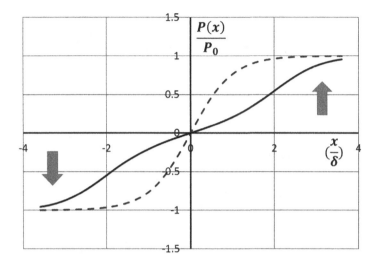

We start from the equation given by $\kappa\left(\dfrac{d^2P}{dx^2}\right) = \alpha P + \beta P^3 + \gamma P^5$, where $\beta < 0$, $\gamma > 0$. Away from the domain boundary, the spontaneous polarization is now given by

$$P_0 = \frac{\beta}{\gamma}\left(\sqrt{1 - \frac{\alpha\gamma}{\beta^2}} - 1\right).$$

Then, similar handling to Eq. (9.48), we obtain the following equation:

$$\frac{\kappa}{2}\left(\frac{dP}{dx}\right)^2 = f[P(x)] - f[P_0] = \left[P_0^2 - P^2\right]^2\left[\frac{\gamma}{6}P^2 + \frac{\beta}{4} + \frac{\gamma}{3}P_0^2\right].$$

We can obtain the following formulas:

$$P(x) = P_0 \frac{\sinh\left(\dfrac{x}{\delta}\right)}{\sqrt{\cosh^2\left(\dfrac{x}{\delta}\right) + \varepsilon}},$$

$$\delta = \sqrt{\frac{\kappa}{\gamma P_0^4 + \dfrac{\beta}{2}P_0^2}},$$

$$\varepsilon = \frac{2\gamma P_0^2}{4\gamma P_0^2 + 3\beta}.$$

Note the presence of a metastable (or flat) state at $P = 0$ as an intermediate state in the first-order transition case, in comparison with a simple $\tanh(x/\delta)$ curve. The domain wall energy density can be calculated as follows:

$$\gamma_0 = \int_{-\infty}^{\infty} \left[f[P(x)] - f[P_0] + \frac{\kappa}{2} \left(\frac{dP}{dx} \right)^2 \right] dx$$

$$= 2 \int_{-\infty}^{\infty} \left[f[P(x)] - f[P_0] \right] dx$$

$$= 2 \int_{-\infty}^{\infty} \left[P_0^2 - P^2 \right]^2 \left[\frac{\gamma}{6} P^2 + \frac{\beta}{4} + \frac{\gamma}{3} P_0^2 \right] dx$$

$$= \frac{\alpha P_0^2}{\delta} (1+\varepsilon)^2 \int_{-\infty}^{\infty} \frac{\cosh^2 t}{\varepsilon + \cosh^2 t} \, dt$$

$$= K P_0^2 \left(\frac{\kappa}{\delta} \right) \quad \left[K = \int_{-\infty}^{\infty} \frac{\cosh^2 t}{\varepsilon + \cosh^2 t} \, dt \quad \text{is a numerical factor} \right]$$

9.3 Though a poled piezoelectric ceramic has crystallographic (∞mm) symmetry (aniso-tropic), merely from the mathematical/analytical simplicity viewpoint, a treatment with "piezoelectrically anisotropic", but "elastically isotropic" assumption is often utilized. Because of its simplicity in an isotropic elastic material, *Lamé parameters* (first parameter λ and *shear modulus* μ) are often utilized. The first and second parameters, λ and μ, are defined by both shear parameters as

$$\lambda = c_{12}, \text{ and } \mu = c_{66} = \frac{1}{2}(c_{11} - c_{12}).$$

The elastic stiffness matrix is represented as follows in a symmetry:

$$(c_{ij}) = \begin{pmatrix} (\lambda + 2\mu) & \lambda & \lambda & 0 & 0 & 0 \\ \lambda & (\lambda + 2\mu) & \lambda & 0 & 0 & 0 \\ \lambda & \lambda & (\lambda + 2\mu) & 0 & 0 & 0 \\ 0 & 0 & 0 & \mu & 0 & 0 \\ 0 & 0 & 0 & 0 & \mu & 0 \\ 0 & 0 & 0 & 0 & 0 & \mu \end{pmatrix}$$

a. From the above Lamé parameters' definitions, derive the most important relation: $c_{11} = \lambda + 2\mu$

b. Derive the Poisson's ratio: $\sigma = \dfrac{c_{12}}{c_{11} + c_{12}} = \dfrac{\lambda}{2(\lambda + \mu)}$

c. Taking the material's mass density ρ, derive the sound velocity for the longitudinal and transverse waves in an isotropic material: $c_l^2 = (\lambda + 2\mu)/\rho$ and $c_t^2 = \mu/\rho$.

An alternative set of elastic parameters in an isotropic material is composed of Young's modulus E and Poisson's ration σ, where $E = 1/s_{11}$ and $\sigma = -s_{12}/s_{11}$:

$$s_{ij} = \frac{1}{E} \begin{pmatrix} 1 & -\sigma & -\sigma & 0 & 0 & 0 \\ -\sigma & 1 & -\sigma & 0 & 0 & 0 \\ -\sigma & -\sigma & 1 & 0 & 0 & 0 \\ 0 & 0 & 0 & 2(1+\sigma) & 0 & 0 \\ 0 & 0 & 0 & 0 & 2(1+\sigma) & 0 \\ 0 & 0 & 0 & 0 & 0 & 2(1+\sigma) \end{pmatrix}$$

     d.   Derive the following relations: $c_{11} = \dfrac{1-\sigma}{(1+\sigma)(1-2\sigma)} E$, $c_{12} = \dfrac{\sigma}{(1+\sigma)(1-2\sigma)} E$

REFERENCES

1. K. Uchino, *Micromechatronics*, (CRC/Dekker Press, New York, 2003).
2. S. Takahashi, S. Hirose, *Journal of Applied Physics.* **32**, 2422–2425 (1993).
3. A.F. Devonshire, *Philosophical Magazine.* **40**, 1040 (1949).
4. A.F. Devonshire, *Advances in Physics.* **3**, 85 (1954).
5. N. Uchida, T. Ikeda, *Japanese Journal of Applied Physics.* **6**, 1079 (1967).
6. N. Uchida, *Review of the Electrical Communication Laboratory.* **16**, 403 (1968).
7. N. Uchida, T. Ikeda, *Japanese Journal of Applied Physics.* **4**, 867 (1965).
8. N.A. Schmidt, *Ferroelectrics.* **31**, 105 (1981).
9. J. Kuwata, K. Uchino, S. Nomura, *Japanese Journal of Applied Physics.* **19**, 2099 (1980).
10. A. Rajapurkar, S.O. Ural, Y. Zhuang, H.-Y. Lee, A. Amin, K. Uchino, *Japanese Journal of Applied Physics.* **49**, 071502 (2010).
11. Y. Zhuang, S.O. Ural, R. Gosain, S. Tuncdemir, A. Amin, K. Uchino, *Applied Physics Express.* **2**, 121402 (2009).
12. K. Uchino, Y. Zhuang, S. Ural, A. Amin, High Power Single Crystal Piezoelectric Transformer, US Patent, US 8,395,301 B1, March 12 (2013).
13. P. Sun, Q. Zhou, B. Zhu, D. Wu, C. Hu, J.M. Cannata, J. Tian, P. Han, G. Wang, K.K. Shung, *IEEE Transactions on Ultrasonics, Ferroelectrics, and Frequency Control.* **56**(12), (2009).
14. S. Zhang, N. Sherlock, R. Meyer, Jr., T. Shrout, *Applied Physics Letters.* **94**, 162906 (2009).
15. Y. Zhuang, S.O. Ural, S. Tuncdemir, A. Amin, K. Uchino, Analysis on loss anisotropy of piezoelectrics with ∞mm crystal symmetry, *Japanese Journal of Applied Physics.* **49**, 021503 (2010).
16. M. Choi, K. Uchino, E. Hennig, T. Scholehwar, Polarization orientation dependence of piezoelectric losses in soft lead zirconate-titanate ceramics, *Journal of Electroceramics.* **40**, 16–22 (2018). doi: 10.1007/s10832-017-0085-y.
17. M. Choi, T. Scholehwar, E. Hennig, K. Uchino, Crystallographic approach to obtain intensive elastic parameters of k33 mode piezoelectric ceramics, *Journal of the European Ceramic Society.* **37**, 5109–5112 (2017).
18. X.H. Du, Q.M. Wang, U. Belegundu, K. Uchino, Piezoelectric property enhancement in polycrystalline lead zirconate titanate by changing cutting angle, *Journal of the Ceramic Society of Japan.* **107**(2), 190–191 (1999).
19. S. Hirose, M. Aoyagi, Y. Tomikawa, S. Takahashi, K. Uchino, High-Power Characteristics at Antiresonance Frequency of Piezoelectric Transducers, *Proceedings of 1995 Ultrasonics International Conference*, Edinburgh, pp. 184–187 (1995).
20. J. Ryu, H.W. Kim, K. Uchino, J. Lee, Effect of Yb addition on the sintering behavior and high power piezoelectric properties of Pb(Zr, Ti)O₃-Pb(Mn, Nb)O₃, *Japanese Journal of Applied Physics.* **42**(3), 1307–1310 (2003).
21. Y. Gao, K. Uchino, D. Viehland, Rare earth metal doping effects on the piezoelectic and polarization properties of Pb(Zr, Ti)O₃-Pb(Sb, Mn)O₃ ceramics, *Journal of Applied Physics.* **92**, 2094–2099 (2002).

22. Y. Gao, K. Uchino, Development of high power piezoelectrics with enhanced vibration velocity, *Journal of Materials Technology*. **19**(2), 90–98 (2004).

23. A. Fujii, "Piezoelectric MLs for the diesel injection valve application", Proceedings of Smart Actuators/Sensors Study Committee, JTTAS, Tokyo, December 2 (2005).

24. T. Tou, Y. Hamaguchi, Y. Maida, H. Yamamori, K. Takahashi, Y. Terashima, *Japanese Journal of Applied Physics*. **48**, 07GM03 (2009).

25. Y. Saito, *Japanese Journal of Applied Physics*. **35**, 5168–5173 (1996).

26. Y. Doshida, "Development of Lead-Free Multilayer Piezoelectric Ceramics and Study on Miniature Ultrasonic Motor using their Ceramics", *Proceedings of 81st Smart Actuators/Sensors Study Committee, JTTAS*, Tokyo, December 11 (2009).

27. E.A. Gurdal, S.O. Ural, H.Y. Park, S. Nahm, K. Uchino, High power characterization of $(Na_{0.5}K_{0.5})NbO_3$ based lead-free piezoelectric ceramics, *Sensors and Actuators A: Physical*. **200**, 44 (2013).

28. M. Hejazi, E. Taghaddos, E. Gurdal, K. Uchino, A. Safari, High power performance of manganese-doped BNT-based Pb-free piezoelectric ceramics, *Journal of the American Ceramic Society*. **97**(10), 3192–3196 (2014).

29. A. Hagimura, K. Uchino, *Ferroelectrics*. **93**, 373 (1989).

30. K. Uchino, J. Zheng, A. Joshi, Y.H. Chen, S. Yoshikawa, S. Hirose, S. Takahashi, J.W.C. de Vries, High power characterization of piezoelectric materials, *Journal of Electroceramics*. **2**, 33–40 (1998).

31. S. Takahashi, Y. Sasaki, S. Hirose, K. Uchino, *Proceedings of Materials Research Society Symposium*. **360**, 305 (1995).

32. K. Uchino, *Ferroelectric Devices*, 2nd Edition, (CRC Press, Boca Raton, FL, 2010).

33. Y. Gao, K. Uchino, D. Viehland, Time dependence of the mechanical quality factor in 'hard' lead zirconate titanate ceramics: Development of an internal dipolar field and high power origin, *Japanese Journal of Applied Physics*. **45**(12), 9119–9124 (2006).

34. Y. Gao, K. Uchino, D. Viehland, Domain wall release in 'hard' piezoelectrics under continuous large amplitude AC excitation, *Journal of Applied Physics*. **101**(11), 114110 (2007).

35. Y. Gao, K. Uchino, D. Viehland, Effects of thermal and electrical history on 'hard' piezoelectrics: A comparison of internal dipolar fields and external DC bias, *Journal of Applied Physics*. **101**(5), 054109 (2007).

36. Q. Tan, D. Viehland, *Philosophical Magazine*. (1997).

37. A. Yamaji, Y. Enomoto, E. Kinoshita, T. Tanaka, *Proceedings of 1st Mtg. Ferroelectrics Materials Application*, Kyoto, p. 269 (1977).

38. K. Uchino, T. Takasu, *Inspection*. **10**, 29 (1986).

39. C. Sakaki, B.L. Newarkar, S. Komarneni, K. Uchino, Grain size dependence of high power piezoelectric characteristics in Nb doped lead zirconate titanate oxide ceramics, *Japanese Journal of Applied Physics*. **40**, 6907–6910 (2001).

40. A. Bansal, H. Shekhani, M. Majzoubi, E. Hennig, T. Scholehwar, K. Uchino, Improving high power properties of PZT ceramics by external DC bias field, *Journal of the American Ceramic Society*. (2018). doi: 10.111/jace.15437.

41. K. Uchino, H. Negishi, T. Hirose, Drive voltage dependence of electromechanical resonance in PLZT piezoelectric ceramics, *Japanese Journal of Applied Physics*. **28**, 47–49 (1989).

42. Q.M. Wang, T. Zhang, Q. Chen, X.H. Du, Effect of DC bias field on the complex materials coefficients of piezoelectric resonators, *Sensors and Actuators A: Physical - Journal*. **109**(1–2), 149–155 (2003).

43. Q.M. Zhang, J. Zhao, *IEEE Transactions on Ultrasonics, Ferroelectrics, and Frequency Control*. **46**(6), 1518–1526 (1999).

44. D. Zhou, M. Kamlah, D. Munz, *Journal of Materials Research*. **19**(3), 834–842 (2004).

45. H.H.A. Krueger, *Journal of the Acoustical Society of America*. **42**(3), 636–645 (1967).

46. Q.M. Zhang, J. Zhao, K. Uchino, J. Zheng, *Journal of Materials Research*. **12**(1), 226–234 (1997).

47. R. Yimnirun, Y. Laosiritaworn, S. Wongsaenmai, *Journal of Physics D: Applied Physics*. **39**(4), 759 (2006).

48. M. Unruan, S. Ananta, Y. Laosiritaworn, A. Ngamjarurojana, R. Guo, A. Bhalla, R. Yimnirun, *Ferroelectrics*. **400**(1), 144–154 (2010).

49. G. Yang, S.F. Liu, W. Ren, B.K. Mukherjee, *Proceedings of the 12th IEEE International Symposium on Applications of Ferroelectrics (ISAF)*, Honolulu, HI, IEEE Cat. No. 00CH37076, Vol. 1, pp. 431–434 (2000).

50. H. Daneshpajooh, M. Choi, Y. Park, T. Scholehwar, E. Hennig, K. Uchino, Compressive stress effect on the loss mechanism in a soft piezoelectric $Pb(Zr, Ti)O_3$, *Review of Scientific Instruments*. **90**, 075001 (2019). doi: 10.1063/1.5096905.

51. C.S. Lynch, The performance of piezoelectric materials under stress, in *Advanced Piezoelectric Materials*, Ed. By K. Uchino (Woodhead Publishing, Sawston, 2010), pp. 628–659.
52. D. Damjanovic, Stress and frequency dependence of the direct piezoelectric effect in ferroelectric ceramics, *Journal of Applied Physics*. **82**, 1788–1797 (1997).
53. D.A. Hall, Review nonlinearity in piezoelectric ceramics, *Journal of Materials Science*. **36**, 4575 (2001).
54. L. Prandtl, Ein Gedankenmodell zur Kinetischen Theorie der Festen Körper, *ZAMM*. **8**, 85 (1928).
55. G.A. Tomlinson, A molecular theory of friction, *The London, Edinburgh and Dublin Philosophical Magazine and Journal of Science*. **7**(Suppl. 46), 905 (1929).
56. H.N. Shekhani, Characterization of the High Power Properties of Piezoelectric Ceramics Using the Burst Method: Methodology, Analysis, and Experimental Approach, Ph.D. Thesis, The Pennsylvania State University, August (2016).
57. C. Borderon, R. Renoud, M. Ragheb, H. Gundel, Description of the low field nonlinear ielectric properties of ferroelectric and multiferroic materials, *Applied Physics Letters*. **98**, 112903 (2011).
58. A.S. Sidorkin, *Domain Structure in Ferroelectrics and Related Materials*, (Visa Books Private Ltd., New Delhi, India, 2017). ISBN: 978-81-309-3154-8.
59. J.A. Osborn, *Physical Review*. **67**, 351 (1945); E.C. Stoner, *Philosophical Magazine*. **36**, 803 (1945).
60. T. Mitsui, J. Furuichi, *Physical Review*. **90**, 193 (1953).
61. W.J. Merz, *Physical Review*. **95**, 690 (1954).
62. E.V. Chenskii, *Physics of the Solid State*. **14**, 1940 (1973).
63. A.R. Bishop, E. Domani, J.A. Krumhansl, Quantum correction to domain walls in a model (1D) ferroelectric, *Physical Review B*. **14**(7), 2966 (1976).
64. D.J. Craik, R.S. Tebble, *Ferromagnetism and Ferromagneti Domains*, Selected Topics in Solid State Physics Vol. **4**, (North-Holland Publishing, Amsterdam, 1965).
65. R.C. Miller, A. Savege, *Physical Review*. **115**, 1176 (1959).
66. T. Mitsui, J. Furuichi, *Physical Review*. **90**, 193 (1953).
67. B.M. Darinskii, A.S. Sidorkin, A.M. Kostsov, Oscillations of the domain boundaries in real ferroelectrics and ferroelectrics-ferroelastics, *Ferroelectrics*. **160**, 35–45 (1994).

10 Practical High-Power Piezoelectric Devices

ABSTRACT

"High-Power Piezoelectrics for Practical Applications" is devoted particularly to actual high-power piezo-device developers, since I disclosed a sort of know-how from the company executive viewpoint. We will discuss on the subsidiary (but sometimes more important) items in this chapter in order to commercialize practical high-power piezoelectric components from the device designing viewpoints, in particular multilayer (ML) piezoelectric devices. Three items are described below: (1) cracking in long ML actuators, (2) surface area/volume ratio vs. heat generation, and (3) internal electrode materials.

10.1 CRACKING IN LONG ML ACTUATORS

Relatively long piezoelectric multilayer (ML) actuators are often requested for particular applications such as diesel engine injection valve control in order to generate a large displacement under large electric field at off-resonance drive. Required actuator specifications for diesel injection valve applications are summarized as:

- Displacement: 50 µm/length 50 mm (including inactive part)
  - Strain level: 0.14% at 2 kV/mm (effective $d_{33} \approx 700$ pm/V)
- Temperature stability: ±5% for −40°C to 150°C
  - Curie temperature higher than 335°C
- Life time: longer than 10^9 cycles
  - Under a severe drive condition: switching time = 50 µs
  - Electric field = 2 kV/mm
- Heat generation
- Manufacturing price
  - Tape casting cofiring technique is required.

Though the detailed compositions are the "trade secret" in each company, they are PZT-based with the Curie temperature around 300°C, the low-voltage $d_{33} \approx 350$ pm/V, and mechanical quality factor ≈ 500. Future research directions of the actuator materials will include: (1) lower-temperature sinterable PZTs, aiming at Cu internal electrode ink usage and manufacturing energy saving (refer to Section 10.3), and (2) Pb-free piezoelectric ceramics, for overcoming the social regulations such as RoHS (Restrictions on the use of certain Hazardous Substances).

The internal electrode design is not a conventional "interdigital" type for the automobile applications but still keeps the unelectroded part at the corner point of each electrode layer in order to make the external side electrode inexpensively. With increasing the actuator length up to 50 mm, these unelectroded inactive parts start to accumulate significant tensile stress under the electric field application. Figure 10.1a shows experimentally obtained results by Siemens.[1] With increasing the applied electric field on the long ML actuator, the first crack occurred at 960 V/mm at the length center of this actuator stack corner. Then, still an increase in the field, the cracks two and three occurred at almost the same fields 1,130 and 1,140 V/mm, now at the length quarter of the stack. Since the length center part already released the stress concentration once the first crack happened, the second crack happened again at the center of the segmented 1/2 length. More cracks can be

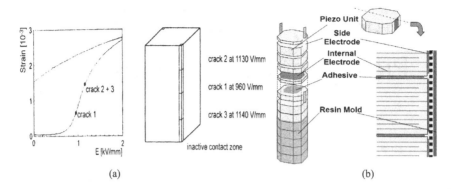

FIGURE 10.1 (a) Side crack initiation processes in an ML piezoelectric actuator.[1] (b) Multi-stacking of ML actuators for diesel injection valves.[2]

expected to occur when we apply the electric field up to the maximum 2 kV/mm. For short MLs, this tensile stress problem at the unelectroded part is usually overcome by using the external DC bias compressive stress (i.e., encapsulated with bias stress).

According to the above results, Denso Corporation decided to laminate 2 mm thick piezo-ML units by using epoxy resin to create a 50-mm long piezo-ML actuator, as shown in Figure 10.1b. Each 2-mm piezo unit consists of 80 μm PZT thin films (25 layers), which does not generate any cracks during operation up to 2 kV/mm. An 80 μm thickness was selected by the economic constraint (the thinner PZT layer demands the more Ag-Pd electrode layers, leading to more expensive production cost), not by the technological restriction. The resin mold design can reduce the unnecessary tensile stress accumulation from the stack actuators. Refer to Chapter 4 Ref. [9] for further ML electrode configurations.

10.2 SURFACE AREA/VOLUME RATIO VS. HEAT GENERATION

The temperature rise in a piezoelectric device is determined by the subtraction of heat dissipation from heat generation. The heat generation is primarily originated from the hysteresis in polarization-electric field relation (dielectric loss) under an off-resonance frequency, while from the hysteresis in strain–stress relation (elastic loss) under a resonance frequency. On the contrary, the heat dissipation is dependent on the device design/configuration.

Zheng et al. reported the heat generation at an off-resonance frequency from various configurations of ML-type piezoelectric ceramic (soft PZT) actuators.[3] Figure 10.2 shows a structure of the ML piezoelectric actuators. The temperature change with time in the actuators was monitored when driven at 3 kV/mm (high electric field) and 300 Hz (much lower frequency than the resonance

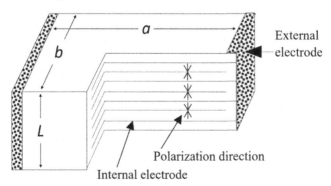

FIGURE 10.2 Structure of an ML piezo-actuator.

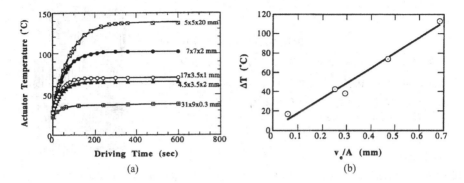

FIGURE 10.3 (a) Device temperature change with driving time for ML actuators of various sizes. (b) Temperature rise at off-resonance vs. V_e/A in various size soft PZT ML actuators, where V_e and A are the effective volume and the surface area, respectively.

frequency) (Figure 10.3a). The specimen temperature reached up to 140°C, depending on the size, showing an exponential increase with the operation time lapse. Figure 10.3b plots the saturated temperature as a function of V_e/A, where V_e is the effective volume (electrode overlapped part, abL in the figure) and A is the all surface area. Suppose that the temperature was uniformly generated in a bulk sample (no significant stress distribution, except for the small inactive portion of the external electrode sides), this linear relation is reasonable because the volume V_e generates the heat and this heat is dissipated through the surface area A. Thus, if we need to suppress the temperature rise, a small V_e/A design is preferred.

In a hexagon configuration, a flat plate design is more suitable than a cubical shape, while in a cylindrical configuration, a tube design is better than a solid rod shape, as you can easily imagine. When you need a large-area ML actuator for generating large force, you had better use four-arrayed MLs with 1/4-area ML units to increase the surface area.

10.3 INTERNAL ELECTRODE MATERIALS

Though we have developed "high-power density" piezoelectric ceramics, ML and/or cofiring are the key to develop actual "high-power" components from the device designing viewpoint. However, the present Ag-Pd electrode structure includes twofold problems: (1) Pd is expensive, and (2) although Ag migration during sintering and under electric field applied can be suppressed by Pd, the electrode conductance is significantly decreased with Pd content. The latter is the major problem in designing ML ultrasonic motors and transformers, because the electrode Joule loss appears to be large, leading to heat generation and low efficiency. In order to solve the problems, pure Cu or pure Ag electrode is a key. However, the ML samples need to be sintered at a relatively low temperature (900°C or lower) in a reduced atmosphere, when utilizing Cu-embedded electrodes. Thus, low-temperature sintering of "hard"-type PZTs is a necessary technology to be developed. Unlike soft PZTs, most of the conventional dopants to decrease the sintering temperature have not been used, because these dopants also degrade the Q_m value significantly.

10.3.1 ELECTRODE METAL PERFORMANCE

Ural et al. measured the electric conductivities of various internal electrodes commercially available, after firing on PZT green sheets.[6] Figure 10.4 shows the test electrode strip design and obtained electrical conductivities of silver, silver/platinum, and copper electrodes. The conductivities of the electrodes were 4.4, 1.6, and 1.6 mΩ/sq for Ag/Pt, Ag, and Cu, respectively. Ag/Pt (or Ag/Pd) electrode, exhibiting the highest resistance, is not favorable for cofired ML piezoelectric ultrasonic

	Data Sheet	Experimental
DuPont QS171 (Ag/Pt)	4.5 mΩ/sq	4.4 mΩ/sq
DuPont 6145 (Ag)	< 3 mΩ/sq	1.7 mΩ/sq
DuPont 9922 (Cu *) * N_2 atmosphere fired	1.6 mΩ/sq	1.7 mΩ/sq

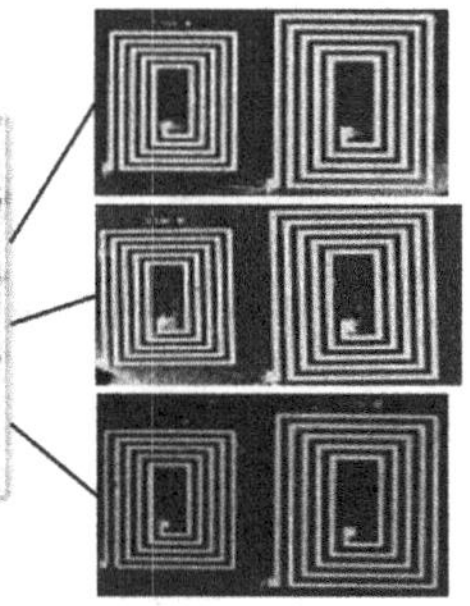

FIGURE 10.4 Electrical conductivities of silver, silver/platinum, and copper fired on PZT green sheets.

transducers, though it is popularly used for making off-resonance positioners. For the high-power ultrasonic transducers, pure silver or Cu electrodes are suitable.

10.3.2 LOW-TEMPERATURE SINTERABLE HARD PZT

In order to use pure Ag or Cu for the internal electrode, ML components should be sintered at 900°C or below. Based on our high-power piezoelectric ceramics, the Sb, Li, and Mn-substituted $0.8Pb(Zr_{0.5}Ti_{0.5})O_3 - 0.16Pb(Zn_{1/3}Nb_{2/3})O_3 - 0.04Pb(Ni_{1/3}Nb_{2/3})O_3$, we further modified them by adding CuO and Bi_2O_3 in order to lower the sintering temperature of the ceramics.[4,5] Since CuO and Bi_2O_3 doping seem to create a melting phase around the piezo-ceramic grain boundaries, a sort of "liquid sintering" decreases the sintering temperature. Table 10.1 summarizes piezoelectric properties of semi-hard piezoelectric ceramics based on $Pb(Zn_{1/3}Nb_{2/3})O_3 - Pb(Ni_{1/3}Nb_{2/3})O_3 - Pb(Zr_{0.5}Ti_{0.5})O_3$, sinterable at 900°C. Under a sintering condition of 900°C for 2 hours, the properties were as: $k_p = 0.56$, Q_m (k_{31}-mode) = 1,023, $d_{33} = 294$ pC/N, $\varepsilon_{33}^X/\varepsilon_0 = 1,326$, and $\tan\delta' = 0.59\%$, when CuO and Bi_2O_3 were added 0.5 wt% each. The maximum vibration velocity of this composition was 0.41 m/s. Figure 10.5 shows the maximum vibration velocity vs. applied field change with various amount of CuO. Note that with increasing CuO content, the applied voltage required for obtaining the same vibration velocity is reduced drastically. In practice, for the composition with 0.4 wt% or higher of CuO, only 6–8 V_{rms}/mm is required for obtaining the $v_0 = 0.48$ m/s.

The composition $0.8Pb(Zr_dTi_{1-d})O_3 - 0.2Pb[(1-c)\{(1-b)(Zn_{0.8}Ni_{0.2})_{1/3}(Nb_{1-a}Sb_a)_{2/3} - b(Li_{1/4}(Nb_{1-a}Sb_a)_{3/4})\} - c(Mn_{1/3}(Nb_{1-a}Sb_a)_{2/3})]O_3 (a = 0.1, b = 0.3, c = 0.3$ and $d = 0.5)$ showed the value of $k_p = 0.56$, $Q_m = 1,951$ (planar mode), $d_{33} = 239$ pC/N, $\varepsilon_{33}^X/\varepsilon_0 = 739$, and the maximum vibration velocity = 0.6 m/s at k_{31} mode. By adjusting the Zr/Ti ratio, compromised properties of $k_p = 0.57$, $Q_m = 1,502$ (planar mode), $d_{33} = 330$ pC/N, $\varepsilon_{33}^X/\varepsilon_0 = 1,653$, and the maximum k_{31}-mode vibration velocity = 0.58 m/s were obtained when Zr/Ti = 0.48/0.52 (Table 10.1). These compositions are suitable for piezoelectric transformers and transducers.

10.3.3 HIGH-POWER PIEZOELECTRIC TRANSFORMERS

Using the above low-temperature-sinterable "hard" PZT, ICAT/the Penn State developed Cu and pure Ag-embedded ML ring dot-type piezo-transformers (PT) (Figure 10.6b), which were sintered at 900°C in a reduced atmosphere with N_2, as illustrated in Figure 10.6a.[6] Because PTs operate at resonance, their impedance is ideally purely resistive and rather small (less than 10 Ω when we use high Q_m material). Thus, the resistance of the internal electrode should be much smaller than the piezo-component's impedance. Figure 10.7 shows (a) equivalent circuit of the ML ring-dot transformer (input side with no load), and (b) their parameters determined for different internal electrode specimens, obtained with HP 4192 analyzer. Copper electrodes exhibit lowest resistance at its resonance, thus reducing ohmic losses and becoming the favorable electrode for the PT application.

TABLE 10.1

Piezoelectric Properties of Semi-hard Piezoelectric Ceramics Based on $Pb(Zn_{1/3}Nb_{2/3})$ $O_3-Pb(Ni_{1/3}Nb_{2/3})O_3-Pb(Zr_{0.5}Ti_{0.5})O_3$

Composition	Sintering condition	Q_m (planar)	Q_m (31-mode)	$\varepsilon_3^T/\varepsilon_0$	d_{33}	k_p	k_{31}	T_c (°C)	v_0 (m/s)
HP-HT-6-2	1,200/2 hours	1,951	1,815	739	239	0.56	0.3	285.6	0.6
HP-HT-12-4	1,200/2 hours	1,502	1,404	1,653	330	0.573	0.33	289.58	0.58
HP-LT-17-3	900/2 hours	1,282	1,023	1,326	294	0.56	–	–	0.41

HT and LT, high- and low-temperature sintering, respectively.
HP-HT-6-2, Pb(Zr,Ti)–Pb(Zn,Ni)Nb with Sb, Li, and Mn substitution.
HP-HT-12-4, Further modification on the HP-HT-6-2.
HP-LT-17-3, Low-temperature sintering of the HP-HT-12-4 with CuO and Bi_2O_3.

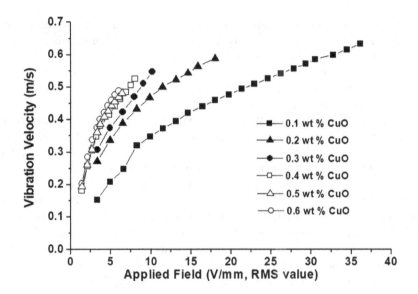

FIGURE 10.5 Vibration velocity variation with applied field in $0.8Pb(Zr,Ti)O_3-0.2Pb[0.7\{0.7(Zn,Ni)_{1/3}-(Nb,Sb)_{2/3}-0.3Li_{1/4}(Nb,Sb)_{3/4}\}-0.3Mn_{1/3}(Nb,Sb)_{2/3}]O_3$ for various x wt% of CuO added to the ceramic.

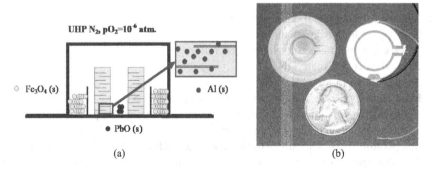

FIGURE 10.6 (a) Experimental setup for sintering Cu-electrode-embedded ML transformers in a reduced N_2 atmosphere. (b) ML cofired transformer with hard PZT and Cu (left) or pure Ag (right) electrode, sintered at 900°C [Penn State trial products].[6]

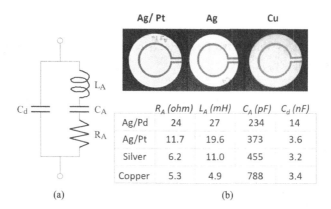

	R_A (ohm)	L_A (mH)	C_A (pF)	C_d (nF)
Ag/Pd	24	27	234	14
Ag/Pt	11.7	19.6	373	3.6
Silver	6.2	11.0	455	3.2
Copper	5.3	4.9	788	3.4

(a)　　　　　　　　　　　　　　　　　(b)

FIGURE 10.7 (a) Equivalent circuit of the ML ring-dot transformers, and (b) their parameters determined for different internal electrode specimens, obtained with HP 4192 analyzer.

Three-layer cofired transformers with copper and silver electrodes were successfully prepared (Figures 10.6b and 10.7). Ring-dot disk ML types (outer dia = 27, center dot dia = 14 mm) with Cu and Ag/Pd (or Ag/Pt) (as references) revealed the maximum power density (at 20°C temperature rise) 42 and 30 W/cm^3, respectively. This big difference comes from the poor electric conductivity of Pd or Pt, compared to Cu or pure Ag. Note that the power density depends not only on the piezo-ceramic composition but also on the electrode species.

ICAT and Center of Dielectric and Piezoelectric Studies at the Penn State together with Solid State Ceramics, Inc., PA, developed recently various Ag/Pd cofired ML PTs using a similar low-temperature sinterable "hard" PZT.[7] Ring-dot disk step-down ML types (OD = 19, center dot D = 8 mm) cofired at 900°C revealed power density of 30, 45, and 60 W/cm^3 at a temperature rise of 20°C, 40°C, and 80°C, respectively. The value for 20°C (maximum vibration velocity definition) matches to the above data by Ural et al.[6] This design was further tested with Cu cofiring in a reduced atmosphere firing furnace by completely avoiding oxidation of Cu. Initial measurements showed that Cu cofired transformers possess about 40% reduction in resistance in identical designs. Since high-power applications are typically driven at high currents because of high admittance of the ML device at the resonance, the difference in the electrode resistance can become critical due to simple Joule heating in the electrodes. Hence, by incorporating Cu cofiring, this effect can be significantly reduced, and improved performance can be observed as well as reduction in overall manufacturing cost (compared with expensive Pd usage).

CHAPTER ESSENTIALS

1. Subsidiary problems in ML piezoelectric devices:
   1. Cracking in long ML actuators
   2. Surface area/volume ratio vs. heat generation
   3. Internal electrode materials.
2. Cracking problem in long ML piezo-actuators:
   1. Manufacture piezo ML units with the length less than 5 mm
   2. Then, laminate the units by using epoxy resin to create a long piezo-ML actuator.
3. Heat generation in piezoelectric transducers:
   Temperature rise is determined by the subtraction of heat dissipation from heat generation.
   1. The heat generation is primarily originated from the hysteresis in polarization-electric field relation (dielectric loss) under an off-resonance frequency, while from the hysteresis in strain–stress relation (elastic loss) under a resonance frequency, which is proportional to the device volume Ve.

2. The heat dissipation is proportional to the device surface area A.
3. Thus, the temperature rise is proportional to Ve/A.
   Hexagon configuration – a flat plate design rather than a cubical shape
   Cylindrical configuration – a tube design rather than a solid rod shape
4. Electrode material selection:
   1. Off-resonance application – Ag/Pd is popular, because of the Ag migration suppression under a large electric field.
   2. Resonance application – High conductivity is essential, pure Ag or Cu is preferred. Ag/Pd, Ag/Pt degrade the resonance performance significantly, including higher temperature rise.
   3. To use Cu, Ag internal electrode, low-temperature sinterable "Hard" PZT ceramic development is essential.
   4. ML fabrication with Cu, Ag electrodes requires "reduced atmosphere" sintering technique.

CHECK POINT

1. The length of "Standard" multilayer (ML) piezoelectric actuators is usually 5–10 mm. Most of the longer MLs (such as 50 mm) are laminated structures. One of the reasons is related with manufacturing cost. What is another technically important reason?
2. There is a copper wire (resistivity = 17 n Ωm) with 50 μm in diameter and 50 cm in length to make an electromagnetic motor coil. What is the resistance of this wire, which will generate heat via Joule heat when voltage is increased? 5 mΩ, 0.5, 5, or 50 Ω.
3. (T/F) In order to suppress the temperature rise of piezoelectric transducers, we had better make the same volume piezo-device in a cubical structure. True or False?
4. (T/F) The internal electrode of high-power ML piezo-transformers should be Ag/Pd, because Pd addition suppresses Ag migration into PZT ceramics. True or False?
5. Which metal electrode exhibits the highest conductivity among Pt, Au, Ag, Ag/Pd, Cu, and Ni? How about the second?

CHAPTER PROBLEMS

1. Review the ML piezoelectric actuator development with Cu internal electrode in these several years. Then, summarize the current technologies from the following aspects:
   a. Low-temperature sinterable "Hard" PZT composition studies – Why "Hard" is difficult for low-temperature sintering from the scientific mechanism?
   b. Low-temperature sinterable "Hard" Pb-free piezoelectric composition.
   c. Manufacturing process of Cu electrode-ML PZT devices. Reducing atmosphere will need to be utilized to maintain metallic copper. The amount of partial pressure of oxygen is delicate for firing PZT in the presence of metallic copper. Kingon et al.[8] showed that the firing atmosphere has to comply the envelope marked as "region C" in the Figure 10.8a. Oxygen partial pressure above the upper line will result in oxidation of copper. On the contrary end, if the atmosphere is too reducing ("region D" in the Figure 10.8a), lead oxide will be reduced and the PZT will decompose. Copper cofiring has been performed on a trial bases[6], and in case of high O_2 pressure, it has oxidized copper, and in case of low O_2 partial pressure, a reduced lead-zirconate-titanate tape was observed (see Figure 10.8b). Thus, cofiring of copper electrodes with PZT requires extensive control of the ambient atmosphere during sintering. Figure 10.6a shows a special technique of magnetite (Fe_3O_4) utilization as a self-regulating buffer oxide in micro-atmosphere (in crucible control) condition. Figure 10.8b shows the experimental results for verifying the above arguments.

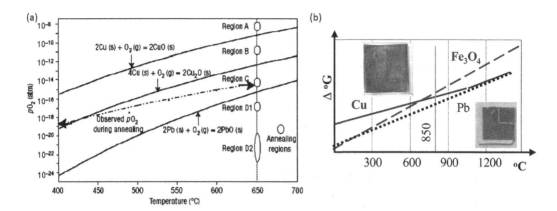

FIGURE 10.8 (a) Metallic copper in the presence of lead requires a tightly controlled PO_2 pressure as temperature is increased during cofiring.[8] (b) Magnetite (Fe_3O_4) can be utilized as a self-regulating buffer oxide in micro-atmosphere (in crucible control) condition.[6]

REFERENCES

1. K. Reichmann, *107th Annual Meeting of the American Ceramic Society*, April 10–13, Baltimore, MD (2005).
2. A. Fujii, *Proceedings of Smart Actuators/Sensors Study Committee*, JTTAS, Dec. 2, Tokyo (2005).
3. J. Zheng, S. Takahashi, S. Yoshikawa, K. Uchino, J.W.C. de Vries, Heat generation in multilayer piezoelectric actuators, *Journal of the American Ceramic Society*. **79**, 3193–3198 (1996).
4. S.–H. Park, S. Ural, C.-W. Ahn, S. Nahm, K. Uchino, Piezoelectric properties of the Sb-, Li- and Mn-substituted $Pb(ZrxTi1-x)O_3$-$Pb(Zn_{1/3}Nb_{2/3})O_3$-$Pb(Ni1/3Nb2/3)O_3$ ceramics for high power applications, *Japanese Journal of Applied Physics*. **45**, 2667–2673 (2006).
5. S.–H. Park, Y.–D. Kim, J. Harris, S. Tuncdemir, R. Eitel, A. Baker, C. Randall, K. Uchino, Low Temperature Co-Fired Ceramic (LTCC) Compatible Ultrasonic Motors, *Proceedings of the 10th International Conference on New Actuators*, Bremen, Germany, June 14–16, 2006, B7.3, pp. 432–435 (2006).
6. S. Ural, S.–H. Park, S. Priya, K. Uchino, High Power Piezoelectric Transformers with Cofired Copper Electrodes—Part I, *Proceedings of the 10th International Conference on New Actuators*, Bremen, Germany, June 14–16, 2006, P23, pp. 556–558 (2006).
7. A. Erkan Gurdal, S. Tuncdemir, K. Uchino, C.A. Randall, Low temperature co-fired multilayer piezoelectric transformers for high power applications, *Materials and Design*. **132**, 512–517 (2017).
8. A. Kingon, S. Srinivasan, Lead zirconate titanate thin films directly on copper electrodes for ferroelectric, dielectric and piezoelectric applications, *Nature Materials*. **4**, 233–237 (2005).
9. K. Uchino, *Micromechatronics*, 2nd Edition, (CRC Press, Boca Raton, FL, 2020). ISBN-13: 978-0-367-20231-6.

11 Concluding Remarks and Future Perspectives

ABSTRACT

Being motivated by multiple junior engineers' request, Uchino tried to elucidate a comprehensive introduction to piezoelectric losses in this book. The author is rather confident that I could provide the reader the basic knowledge on (1) the importance of three losses (dielectric, elastic, and piezoelectric losses), (2) how to measure these three losses by using various characterization techniques, (3) how to develop high-power piezoelectric materials, and (4) how to design the drive and power supply. However, the author is not satisfactory with the clarification of the microscopic loss mechanism in piezoelectrics. Though the determination of real parameters such as permittivity, elastic compliance, and piezoelectric constant is reasonably established via phenomenology, the estimation of the imaginary parameters (dielectric, elastic, and piezoelectric losses) is still very primitive. We still have many research topics remaining for the future generation. This final chapter is devoted to the reader's future activities on this high-power piezoelectric area.

11.1 CONTENT SUMMARY OF THIS TEXTBOOK

This book focused on high-power density actuator/transducer developments, not on sensor applications. Started from the loss phenomenology, we proposed typical high-power characterization methods. Then, current high-power piezoelectric materials and drive/operation schemes were introduced. In the latter part, semi-microscopic loss mechanisms were discussed, based on the domain dynamics models, followed by some know-how on commercialization. Note that the "high-power piezoelectrics" in this book does not mean high absolute power devices (kW, MW), but "high-power density", 10–100 MW/m^3 in practice.

In Chapter 1, we started from "Background of High-Power Piezoelectrics" on the necessity of loss mechanism clarification in order to develop high-energy density piezoelectric components for the portable electronic devices. The developments of multilayer piezoelectric components and high power supply were the key to the high-power piezoelectric components.

For the reader who is not familiar with the fundamentals on piezoelectrics, a brief review of piezoelectrics in Chapter 2 "Foundations of Piezoelectrics", and the concept of losses were introduced in Chapter 3 "Fundamentals of Losses".

The main content starts from Chapter 4 "Phenomenological Approach to Piezoelectric Losses", in which the loss phenomenology in piezoelectrics, including three losses, *dielectric*, *elastic*, and *piezoelectric losses*, was introduced and interrelationships among these losses were discussed. The concept on the difference among "intensive" and "extensive" losses was described in terms of "depolarization field". Based on the piezoelectric constitutive and dynamic equations, Chapter 5 "Equivalent Circuits with Piezo Losses" expanded the equivalent circuit approach in order to facilitate the experimental analysis easier. Actual heat generation analysis was discussed in piezoelectric materials for pseudo-DC and AC drive conditions in Chapter 6 "Heat Generation in Piezoelectrics". Heat generation at off-resonance is attributed mainly to intensive dielectric loss tanδ', while the heat generation at resonance is mainly originated from the intensive elastic loss tanϕ'. You also learned that heat generation profile (i.e., thermal analysis) can determine the mechanical quality factor Q_m (mechanical performance).

Chapter 7 "High-Power Piezo Characterization System" (HiPoCS™) introduced various characterization methodologies of loss factors chronologically, including (1) static P–E, x–X, x–E, and $P–X$ measurements; (2) continuous admittance/impedance spectrum measurement; (3) burst/transient response method; and so on. Mechanical quality factors (Q_A at resonance and Q_B at antiresonance) are primarily measured as a function of vibration velocity. Three losses can be determined from the precise measurements on the mechanical quality factors Q_A and Q_B: for example, k_{31} specimen

- Resonance: $Q_{A,31} = \dfrac{1}{\tan \phi'_{11}}$

- Antiresonance: $\dfrac{1}{Q_{B,31}} = \dfrac{1}{Q_{A,31}} - \dfrac{2}{1 + \left(\dfrac{1}{k_{31}} - k_{31}\right)^2 \Omega_{B,31}^2} \left(2 \tan \theta'_{31} - \tan \delta'_{33} - \tan \phi'_{11}\right)$

where $\tan \delta'_{33}$, $\tan \phi'_{11}$, and $\tan \theta'_{31}$ are intensive loss factors for ε_{33}^T, s_{11}^E, and d_{31}, and $\Omega_{B,31}$ is the normalized antiresonance frequency given by

$$\Omega_{A,31} = \frac{\omega_a l}{2 v_{11}^E} = \frac{\pi}{2} \quad \left[v_{11}^E = 1/\sqrt{\rho s_{11}^E}\right], \quad \Omega_{B,31} = \frac{\omega_b l}{2 v_{11}^E}, \quad 1 - k_{31}^2 + k_{31}^2 \frac{\tan \Omega_B}{\Omega_B} = 0$$

Relating with Chapter 7, Chapter 8 "Drive Schemes of Piezoelectric Transducers" treated the piezo-device driving methods in order to minimize the losses and maximize the transducer efficiency, in particular, at off-resonance capacitive region, at resonance or antiresonance frequency, and at inductive region between the resonance and antiresonance frequencies. Impedance converters, inductive (negative capacitance) and capacitive components, have been introduced for off-resonance and for inductive (between resonance and antiresonance frequencies) region drive, respectively.

High-power piezoelectric materials' development, based on the loss mechanisms, was introduced in Chapter 9 "Loss Mechanisms in Piezoelectrics". Practical high-power "hard" Pb(Zr,Ti)O$_3$ (PZT)-based materials are initially described, which exhibit vibration velocities close to 1 m/s (rms), leading to the power density capability ten times that of the commercially available "hard" PZTs. Based on the macroscopic phenomenological study results, we consider microscopic crystallographic and semi-microscopic domain wall dynamic models to understand the experimental results for various "acceptor" or "donor" ion doping. We proposed an internal bias field model to explain the low loss and high-power origin of these materials. The discussion was expanded to Pb-free piezoelectrics, which have been developed recently. We also introduced the polarization angle dependence of losses (sample geometry) and DC bias electric field and stress dependence of losses (sample driving technique) for the high-power applications. The latter part of this chapter is devoted to the phenomenological handling of the domain wall dynamics, based on Ginzburg–Landau functional:

$$F(P(x),T) = \int dx \left[\frac{\alpha}{2} P(x)^2 + \frac{\beta}{4} P(x)^4 + \frac{\kappa}{2} (\nabla P(x))^2\right]$$

Domain wall structure and wall energy can be calculated as:

$$P(x) = P_0 \tan h\left(\frac{x}{\delta}\right), \delta = \frac{1}{P_0}\sqrt{\frac{2\kappa}{\beta}} = \sqrt{\frac{2\kappa}{-\alpha}}, \gamma_0 = -\frac{2}{3}\alpha P_0^2 \delta = \frac{4}{3} P_0^2 \left(\frac{\kappa}{\delta}\right)$$

Time-dependent/dynamic domain wall equations have been introduced to derive the loss formula in piezoelectrics.

Chapter 10 "High-Power Piezoelectrics for Practical Applications" was devoted particularly to actual high-power piezo-device developers, since I disclosed a sort of know-how from the company executive viewpoint. The reader learned how to suppress the device temperature rise and how to enhance the power density from the viewpoints of multilayer (ML) piezo-device configuration and ML internal electrode selection.

Final Chapter 11 "Concluding Remarks and Future Perspective" aimed to encourage the following generation to conduct the researches on the "imaginary" parameters in order to suppress the heat generation, to enhance the power density and efficiency for the further device miniaturization.

11.2 PRACTICAL DEVELOPMENT CASE STUDIES

The author introduces here two case studies comprehensively on how we developed high-power piezoelectric devices, though the individual contents are overlapped in the previous various chapters: (1) diesel injection valve for an off-resonance application and (2) piezo-transformer (PT) for a resonance device.

11.2.1 DIESEL PIEZO INJECTION VALVE

Diesel engines are recommended rather than regular gasoline cars from the energy conservation and global warming viewpoint. However, as well known, the conventional diesel engine generates toxic exhaust gases such as SO_x and NO_x. In order to solve this problem, injection valves should make very fine mist of diesel fuel in order to burn it completely for minimizing the generation of "particulate matter (PM)" dust as well as toxic gases. Fine mist generation requires high-pressure fuel and quick injection control in so-called "Common Rail System". Figure 11.1a shows the concept on how to create fine diesel fuel mist. Two key issues: (1) high commonrail back pressure and (2) quick multiple injection are required. As shown in the injection timing chart of Figure 11.1b, five (Pilot, pre, main, after, and post) injections are required during one engine cycle (typically 60 Hz). Thus, the actuator response should be maximum 100 μs (or quicker), which the electromagnetic actuators cannot satisfy but merely piezoelectric MLs can do. Note that the ML piezo-actuator is driven basically by an on–off voltage or more precisely a trapezoidal voltage. The voltage rise and fall period should be adjusted exactly to the actuator system resonance period. Otherwise, the displacement overshoot and following ringing create a significant problem for stable fuel injection. The piezoelectric actuator is a key component to increase the burning efficiency and minimize the toxic exhaust gas elements.

However, the highest reliability of the actuator component at an elevated temperature (150°C) for a long period (10 years) is required, which took a long research period. Of course, just from the environmental viewpoint, electric cars with fuel cells or rechargeable batteries will be the best. However, because very high investment is required to increase the mileage rate, electric cars are still not popular. The short-range target may be, in the author's opinion, hybrid cars with a fuel cell and a diesel engine (not the present gasoline engine!). This is a good case study on off-resonance ML piezo-actuators used under harsh environmental conditions.

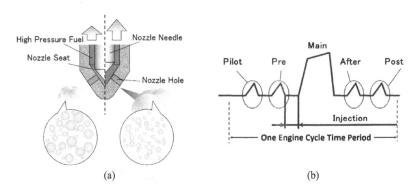

(a)   (b)

FIGURE 11.1 (a) Schematic illustration on how to create fine diesel fuel mist; (b) multiple injection timing chart for the fine mist generation.

11.2.1.1 Piezo-Actuator Material Development

Required actuator specifications for diesel injection valve applications are summarized as:

- Displacement: 50 μm/ length 50 mm (including inactive part)
  - Strain level: 0.14% at 2 kV/mm (effective $d_{33} \approx 700$ pm/V)
- Temperature stability: ±5% for −40°C to 150°C
  - Curie temperature higher than 335°C
- Life time: longer than 10^9 cycles
  - Under a severe drive condition: switching time = 50 μs
  - Electric field = 2 kV/mm
- Heat generation
- Manufacturing price
  - Tape casting cofiring technique is required.

The primary target was to develop high d_{33} "Soft" PZT material usable around 150°C. Thus, Curie temperature should be higher than 300°C. However, because of high responsivity of the ML actuator, additional vibration ringing was observed after the trapezoidal voltage application. It is noteworthy that the vibration ringing increases the heat generation significantly (see Section 6.4). When we drive a piezoelectric device under a pulse drive with some vibration overshoot and ringing, in addition to the normal dielectric loss under a high-electric field drive, the resonating ringing generates more heat via the elastic loss. Figure 11.2 shows the correlation among mechanical quality factor Q_m and piezoelectric constant d_{33} for various PZT ceramics (see Section 9.4.1.3). You can find two major categories in this figure: (1) hard PZT with high $Q_m > 1,000$ and low piezoelectric constant $d_{33} < 300$ pC/N, and (2) soft PZT with low $Q_m < 100$ and high piezoelectric constant $400 < d_{33} < 700$ pC/N. Though soft PZTs are preferably used for off-resonance actuator applications, in general, because of their small Q_m (<100), additional heat generation associated with the vibration ringing could not be neglected. To the contrary, though hard PZTs with high Q_m are good from the heat generation viewpoint, the vibration ringing is larger due to the vibration amplification by the factor Q_m. In conclusion, compromised semi-hard PZT compositions with intermediate Q_m (300–800) and intermediate d_{33} (300–400 pC/N) were chosen for this practical application (circled area on Figure 11.2).

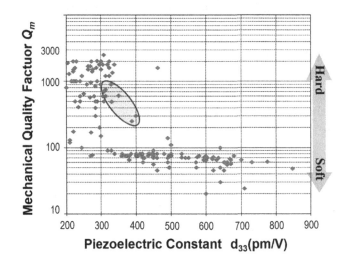

FIGURE 11.2 Correlation among mechanical quality factor and piezoelectric constant for various PZT ceramics.

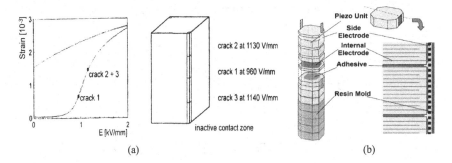

FIGURE 11.3   (a) Side crack initiation processes in an ML piezoelectric actuator. (b) Multi-stacking of ML actuators for diesel injection valves.

Though the detailed compositions are the "trade secret", they are PZT-based with the Curie temperature around 300°C, the low-voltage $d_{33} \approx 350$ pm/V, and mechanical quality factor ≈ 500. Current research directions of the actuator materials include: (1) lower-temperature sinterable PZTs, aiming at Cu internal electrode ink usage and manufacturing energy saving, and (2) Pb-free piezo-electric ceramics, for overcoming the social regulations such as RoHS (Restrictions on the use of certain Hazardous Substances).

11.2.1.2   Multilayer Design

The internal electrode design is not a conventional "interdigital" type for the automobile applications but still keeps the unelectroded part at the corner point of each electrode layer in order to make the external side electrode inexpensively. With increasing the actuator length up to 50 mm, these unelectroded inactive parts start to accumulate significant tensile stress under the electric field application. Figure 11.3a shows experimentally obtained results by Siemens (see Section 10.1). With increasing the applied electric field on the long ML actuator, the first crack occurred at 960 V/mm at the length center of this actuator stack corner. Then, still an increase in the field, the cracks two and three occurred at almost the same fields 1,130 and 1,140 V/mm, now at the length quarter of the stack. Since the length center part already released the stress concentration once the first crack happened, the second crack happened again at the center of the segmented 1/2 length. More cracks can be expected to occur when we apply the electric field up to the maximum 2 kV/mm.

According to the above results, Denso Corporation decided to laminate 2-mm-thick piezo-ML units by using epoxy resin to create a 50-mm-long piezo-ML actuator, as shown in Figure 11.3b. Each 2-mm piezo unit consists of 80-μm PZT thin films (25 layers), which do not generate any cracks during operation up to 2 kV/mm; 80-μm thickness was selected by the economic reason not by the technological restriction. The resin mold design can reduce the unnecessary tensile stress accumulation from the stack actuators.

According to the experimental results (Figure 11.4) on heat generation at an off-resonance frequency from various configurations of ML-type piezoelectric ceramic (soft PZT) actuators, the linear relation was found of the temperature rise with V_e/A, which is reasonable because the volume V_e generates the heat and this heat is dissipated through the area A (see Section 6.1). If we can manufacture a tube design with a center hole, and bolt-clamped, much lower temperature generation is expected.

11.2.1.3   Actuator Encapsulation

The ML actuator was encapsulated because of two reasons: (1) protection from the contamination such as humidity and (2) compressive bias stress application. Complete seal of the capsule with N_2 gas inclusion exhibited a problem first; that is, Ag ion migration was accelerated in the Ag/Pd internal electrode under a large electric field probably due to a significant increase in atmospheric gas

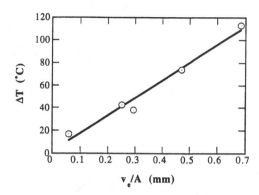

FIGURE 11.4   Temperature rise at off-resonance vs. V_e/A in PZT ML actuators, where V_e and A are the volume and surface area, respectively.

pressure at 150°C (from room temperature). Thus, we created a small pinhole on the capsule not to increase the gas pressure, but still to protect from humidity and contamination. There are also two reasons for the bias stress: (1) output energy efficiency will become maximum, when the compressive bias stress is adjusted to a half of the maximum generative stress (i.e., blocking stress). Refer to *energy transmission coefficient* introduced in Ref. [1] and Example Problem 11.1. (2) Compressive bias stress decreases both dielectric and elastic losses, which significantly suppress the temperature rise (see Section 9.7.2.3). Typically 20 MPa bias stress is used because the blocking stress of PZT actuators is around 40 MPa.

Example Problem 11.1

In a piezoelectric actuator, not all the converted/stored energy k^2 from the input electric energy can be actually used, and the actual mechanical work done depends on the mechanical load under an electric field application. With zero mechanical load or a complete clamp (no strain), no output work is done, even a strain is generated under an electric field. Remember that the work is given by [Force] × [Displacement].

Let us consider the case where an electric field E is applied to a piezoelectric under constant external stress X (<0, because a compressive stress is necessary to work to the outside). This corresponds to the situation that a mass is put suddenly on the actuator, as shown in Figure 11.5a. Figure 11.5b shows two electric field vs. induced strain curves, corresponding to two conditions; under the mass load and no load. Because the area on the field-strain domain does not mean the energy, we should use the stress–strain and field-polarization domains in order to discuss the mechanical and electrical energy, respectively.

- a. Referring to Figure 11.5c, calculate the mechanical energy. Note that the mass shrinks the actuator first by sX (s: piezo-material's compliance, and $X < 0$).
- b. What load can provide the maximum output mechanical energy?
- c. Referring to Figure 11.5d, calculate the actually spent electrical energy. Note that mass load generates the electric energy, which is the loan energy from the mass to the actuator.
- d. The energy transmission coefficient is defined by

$$\lambda_{max} = \left(\text{Output mechanical energy/Input electrical energy}\right)_{max} \qquad \text{(P11.1.1)}$$

Taking the formulas in (a) and (c) into account, consider the maximization and obtain the expression:

$$\lambda_{max} = \left[(1/k) - \sqrt{(1/k^2) - 1}\right]^2 = \left[(1/k) + \sqrt{(1/k^2) - 1}\right]^{-2} \qquad \text{(P11.1.2)}$$

Solution

a. This mechanical energy sX^2 is a sort of "loan" of the actuator credited from the mass, which should be subtracted later (i.e., "paying back"). This energy corresponds to the bottom hatched area in Figure 11.5c. By applying the step electric field, the actuator expands by the strain level dE under a constant stress condition. This is the mechanical energy provided from the actuator to the mass, which corresponds to $|dEX|$. Like paying back the initial "loan", the output work (from the actuator to the mass) can be calculated as the area subtraction (shown by the top dotted area in Figure 11.5c)

$$\int (-X)\, dx = -(dE + sX)X. \tag{P11.1.3}$$

b. Let us consider to maximize the output mechanical energy. The maximum output energy can be obtained when the top dotted area in Figure 11.5c becomes maximum under the constraint of the rectangular corner point tracing on the line (from dE on the vertical axis to $-dE/s$ on the horizontal axis). Since the work energy Eq. (P11.1.3) can be transformed as

$$-s\left(X + \frac{dE}{2s}\right)^2 + \frac{(dE)^2}{4s}, \tag{P11.1.4}$$

$X = -\dfrac{dE}{2s}$ provides the maximum mechanical energy of $\dfrac{(dE)^2}{4s}$. Therefore, the load should be a half of the maximum generative stress when we apply the maximum electric field E.

c. Figure 11.5d illustrates how to calculate the electrical energy spent. The mass load X generates the "loan" electrical energy by inducing $P = dX$ (see the bottom hatched area in Figure 11.5d). By applying a sudden electric field E, the actuator (like a capacitor)

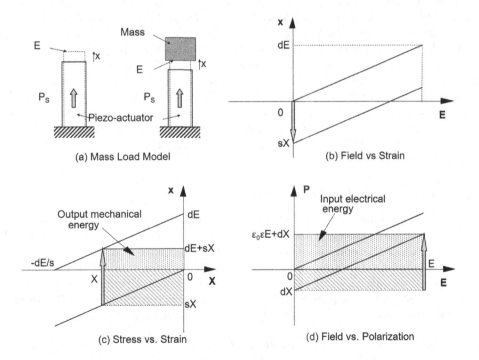

(a) Mass Load Model

(b) Field vs Strain

(c) Stress vs. Strain

(d) Field vs. Polarization

FIGURE 11.5 Calculation of the input electrical and output mechanical energy: (a) load mass model for the calculation, (b) electric field vs. induced strain curve, (c) stress vs. strain curve, and (d) electric field vs. polarization curve.

receives the electrical energy of $\varepsilon_0 \varepsilon E^2$. Thus, the total energy is given by the area subtraction (shown by the top dotted area in Figure 11.5d)

$$\int (E)\, dP = (\varepsilon_0 \varepsilon E + dX) E. \qquad \text{(P11.1.5)}$$

d. Above mechanical energy maximization condition is not the best condition from the *energy transmission coefficient viewpoint*; that is, input electric energy is overspent. We need to choose a proper load now to maximize the *energy transmission coefficient*. From the maximum condition of

$$\lambda = -x \cdot X / P \cdot E$$

$$= -(dE + sX) X / (\varepsilon_0 \varepsilon E + dX) E.$$

$$= -\left[d(X/E) + s(X/E)^2 \right] / \left[\varepsilon_0 \varepsilon + d(X/E) \right] \qquad \text{(P11.1.6)}$$

Letting $y = X/E$, then the maximum point of y is obtained as

$$y_0 = (\varepsilon_0 \varepsilon / d) \left[-1 + \sqrt{(1 - k^2)} \right]. \qquad \text{(P11.1.7)}$$

Finally inserting Eq. (P11.1.7) into Eq. (P11.1.6), we can obtain

$$\lambda_{max} = \left[(1-k) - \sqrt{(1 - k^2 - 1)} \right]^2 = \left[(1-k) + \sqrt{(1 - k^2 - 1)} \right]^{-2}. \qquad \text{(P11.1.8)}$$

11.2.1.4   Pulse Drive Technique

As Ref. [1] describes the Laplace transform analysis, when a piezo-actuator is driven by a step or pulse voltage, the displacement overshoot and vibration ringings are usually associated. If the ML in the injection valve generates the ringings, a serious control problem happens. Figure 11.6 demonstrates how to solve this problem by using a bimorph experiment, where a pseudo-step electric field is applied on the bimorph, and its tip displacement ΔL is plotted as a function of time lapse. The quantity n is a time scale based on a half of the resonance period ($=T/2$) of the piezoelectric actuator. It is noteworthy that ΔL does not exhibit overshoot nor ringing for $n = 2$ (i.e., when the rise time is exactly adjusted to the resonance period). But, for $n = 1$ and 3, the overshoot and ringing follow continuously.

Using this pulse/step drive scheme, the ML actuator in the diesel injection valve was controlled without generating unnecessary ringing nor additional heat generation. Notice that the rise and fall time of the voltage pulse in Figure 11.1b matches exactly to the resonance period of the encapsulated ML piezoelectric actuator.

11.2.2   RING-DOT ML PIEZO-TRANSFORMER

We now provide a development case study on a resonant piezoelectric device; piezoelectric transformer, for example. Different from the off-resonance actuators, the resonance-type devices exhibit very low impedance at the operation mode, which adds the additional development necessity on the electrode materials. If the electric lead shows a comparable or larger resistance than the piezoelectric transducer's impedance, a serious degradation of the transformer performance, including heat generation, is observed. Thus, pure Ag or Cu internal electrode is required in an ML structure, by replacing the conventional Ag/Pd or Ag/Pt (higher resistivity) electrode (see Section 10.3).

11.2.2.1   Low-Temperature Sinterable Hard PZT

In order to use pure Ag or Cu for the internal electrode, ML components should be sintered at 900°C or below. Based on our high-power piezoelectric ceramics, the Sb, Li, and Mn-substituted $0.8Pb(Zr_{0.5}Ti_{0.5})O_3$-$0.16Pb(Zn_{1/3}Nb_{2/3})O_3$-$0.04Pb(Ni_{1/3}Nb_{2/3})O_3$, we further modified them by adding

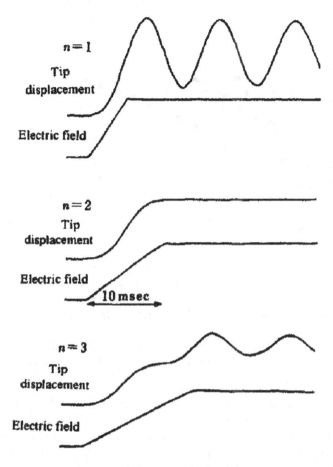

FIGURE 11.6 Bimorph tip displacement produced by a pseudo-step voltage. time scale n is based on ½ of the resonance period T.

CuO and Bi_2O_3 in order to lower the sintering temperature of the ceramics (refer to Section 10.3). Note initially that low-temperature sintering of "Hard" PZT is rather difficult in general. Table 10.1 in Chapter 10 summarizes piezoelectric properties of semi-hard piezoelectric ceramics based on $Pb(Zn_{1/3}Nb_{2/3})O_3$-$Pb(Ni_{1/3}Nb_{2/3})O_3$-$Pb(Zr_{0.5}Ti_{0.5})O_3$, sinterable at 900°C. Under a sintering condition of 900°C for 2 hours, the properties were as: $k_p = 0.56$, Q_m (k_{31}-mode) = 1,023, $d_{33} = 294$ pC/N, $\varepsilon_{33}^X/\varepsilon_0 = 1,326$, and $\tan\delta' = 0.59\%$, when CuO and Bi_2O_3 were added 0.5 wt% each. The maximum vibration velocity of this composition was 0.41 m/s. Figure 11.7 shows the maximum vibration velocity vs. applied field change with various amount of CuO. Note that with the increasing CuO content, the applied voltage required for obtaining the same vibration velocity is reduced drastically. In practice, for the composition with 0.4 wt% or higher, only 6–8 V_{rms}/mm is required for obtaining the $v_0 = 0.48$ m/s. The composition $0.8Pb(Zr_dTi_{1-d})O_3$-$0.2Pb[(1 - c)\{(1 - b) (Zn_{0.8} Ni_{0.2})_{1/3}$ $(Nb_{1-a} Sb_a)_{2/3} - b$ $(Li_{1/4} (Nb_{1-a} Sb_a)_{3/4})\} - c$ $(Mn_{1/3} (Nb_{1-a} Sb_a)_{2/3})]O_3$ ($a = 0.1$, $b = 0.3$, $c = 0.3$, and $d = 0.5$) showed the value of $k_p = 0.56$, $Q_m = 1,951$ (planar mode), $d_{33} = 239$ pC/N, $\varepsilon_{33}^X/\varepsilon_0 = 739$, and the maximum vibration velocity = 0.6 m/s at k_{31} mode. By adjusting the Zr/Ti ratio, compromised properties of $k_p = 0.57$, $Q_m = 1,502$ (planar mode), $d_{33} = 330$ pC/N, $\varepsilon_{33}^X/\varepsilon_0 = 1,653$, and the maximum k_{31}-mode vibration velocity = 0.58 m/s were obtained when Zr/Ti = 0.48/0.52 (Table 10.1). These compositions are suitable for piezoelectric transformers and transducers.[2]

ICAT and Center of Dielectric and Piezoelectric Studies at the Penn State together with Solid State Ceramics, Inc., PA, developed recently various Ag/Pd cofired ML PTs using a similar

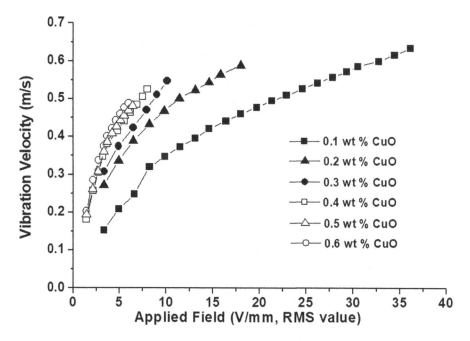

FIGURE 11.7 Vibration velocity variation with applied field in $0.8Pb(Zr,Ti)O_3-0.2Pb[0.7\{0.7(Zn,Ni)_{1/3}-(Nb,Sb)_{2/3}-0.3Li_{1/4}(Nb,Sb)_{3/4}\}-0.3Mn_{1/3}(Nb,Sb)_{2/3}]O_3$ for various x wt% of CuO added to the ceramic.[2]

low-temperature-sinterable "hard" PZT.[3] A commercially available MnO_2- and Nb_2O_5-modified $Pb(Zr,Ti)O_3$ (APC 841, APC Int'l, Mackeyville, PA, USA) was used as the base "hard" piezo-ceramic composition. Metal oxides (CuO, ZnO, and PbO) and carbonates (Li_2CO_3) with purities N 99.9% were mixed according to the formula APC 841 + 0.5 wt% PbO + 1.1 wt% ZnO + 0.2 wt% CuO or Li_2CO_3 for 24 hours in Nalgene jars with ethanol and yttria-stabilized zirconia (YSZ) milling media. These pellets could be sintered at reasonably low temperature between 850°C and 1,050°C.

11.2.2.2   Ag, Cu Internal Electrode ML

Using the above low-temperature-sinterable "hard" PZT, ICAT/the Penn State developed Cu and pure-Ag-embedded ML ring-dot-type PTs (Figure 11.9b). Because PTs operate at resonance, their impedance is ideally purely resistive and rather small (less than 10 Ω when we use high Q_m material). Thus, the resistance of the internal electrode should be much smaller than the piezo-component's impedance. Figure 11.8a shows equivalent circuit of the ML ring-dot transformers, and Figure 11.8b summarizes their parameters determined for different internal electrode specimens, obtained with HP 4192 analyzer. Copper electrodes exhibit lowest resistance at resonance, thus reducing ohmic losses and becoming the favorable electrode for the PT application.

11.2.2.3   Reduced Atmosphere Sintering

Manufacturing process of Cu electrode-ML PZT devices requires reducing atmosphere to maintain metallic copper. The amount of partial pressure of oxygen is delicate for firing PZT in the presence of metallic copper. Kingon et al.[4] showed that the firing atmosphere has to comply the envelope marked as "region C" in the Figure 11.10a. Oxygen partial pressure above the upper line will result in oxidation of copper. On the contrary end, if the atmosphere is too reducing ("region D" in Figure 11.10a), lead oxide will be reduced and the PZT will decompose. Copper cofiring has been performed on a trial basis,[2] and in case of high O_2 pressure, it has oxidized copper, and in case of low O_2 partial pressure, a reduced PZT tape was observed (Figure 11.10b top picture). Thus, cofiring of copper electrodes with PZT requires extensive control of the ambient atmosphere during sintering.

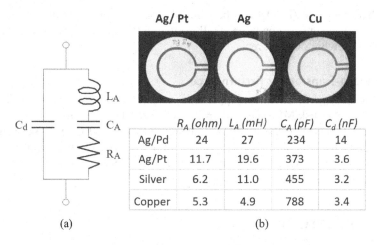

FIGURE 11.8 Equivalent circuit of the ML ring-dot transformers (a), and their parameters determined for different internal electrode specimens, obtained with HP 4192 analyzer (b).

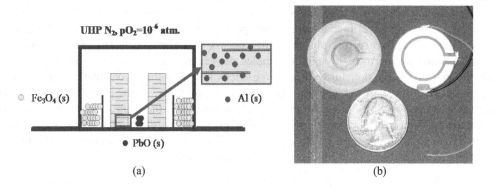

FIGURE 11.9 (a) Experimental setup for sintering Cu-electrode-embedded ML transformers in a reduced N_2 atmosphere. (b) ML cofired transformer with hard PZT and Cu (left) or pure Ag (right) electrode, sintered at 900°C [Penn State trial products].[2]

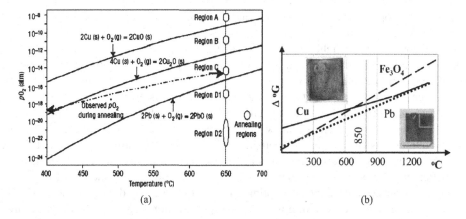

FIGURE 11.10 (a) Metallic copper in the presence of lead requires a tightly controlled PO_2 pressure as temperature is increased during cofiring.[4] (b) Magnetite (Fe_3O_4) can be utilized as a self-regulating buffer oxide in micro-atmosphere (in crucible control) condition.

We found that magnetite (Fe_3O_4) can be utilized as a self-regulating buffer oxide in micro-atmosphere (in crucible control) condition, as illustrated in Figure 11.9a. The resultant electrode is shown in Figure 11.10b bottom picture.

11.2.2.4 ML Piezo-Transformer Performance

Three-layer cofired transformers with copper and silver electrodes were successfully prepared (Figure 11.9b). Ring-dot disk ML types (OD = 27, center dot D = 14 mm) with Cu and Ag/Pd (or Ag/Pt) (as references) revealed the maximum power density (at 20°C temperature rise) 42 and 30 W/cm^3, respectively. This big difference comes from the poor electric conductivity of Pd or Pt, compared to Cu or pure Ag. Note that the power density depends not only on the piezoelectric ceramic composition but also on the electrode species.

11.3 FUTURE PERSPECTIVES

The author is proud to say that we have contributed lots in commercialization of high-power density piezoelectric devices in these 45 years, including high-power "Soft" and "Hard" PZT piezoelectric materials, ML fabrication technologies, power supplies suitable to capacitive load, in addition to compact piezo-motors and transformers. However, I am not yet satisfactory with the clarification of the microscopic loss mechanism in piezoelectrics. Losses are considered to consist of four portions: (1) domain wall motion, (2) fundamental lattice portion, which should also occur in domain-free monocrystals, (3) microstructure portion, which occurs typically in polycrystalline samples, and (4) conductivity portion in highly-ohmic samples. However, in the typical piezoelectric ceramic case, the loss due to the domain wall motion exceeds the other three contributions significantly. Though the determination of real parameters such as permittivity, elastic compliance, and piezo-electric constant is reasonably established via phenomenology, the estimation of the imaginary parameters (dielectric, elastic, and piezoelectric losses) is still very primitive. We still have many research topics remaining for the future generation. This final section is devoted to the reader's future research activities on this high-power piezoelectric area.

11.3.1 WHAT IS "PIEZOELECTRIC" LOSS?

We related the microscopic loss origins only with the domain wall motion-related losses, though it seems to be too simplified. Taking into account the fact that the polarization change is primarily attributed to 180° domain wall motion, while the strain is attributed to 90° (or non-180°) domain wall motion, we supposed that the extensive dielectric and mechanical losses are originated from 180° and 90° domain wall motions, respectively, as illustrated in Figure 11.11 (Section 9.2.3). The dielectric loss comes from the hysteresis during the 180° polarization P reversal under E, while the elastic loss comes from the strain x hysteresis during the 90° polarization reorientation under X. Regarding the piezoelectric loss, we presume that it is originated from *Gauss Law, div (D)* = σ (charge). When we apply a tensile stress on the piezoelectric crystal, the vertically elongated cells will transform into horizontally elongated cells, so that the 90° domain wall will move rightward. However, this "ferroelastic" domain wall will not generate charges because the rightward and left-ward polarizations may compensate each other. Without having migrating charges σ in this crystal, *div (P)* = 0, leading to the polarization alignment *head-to-tail*, rather than "head-to-head" or "tail-to-tail". After the "ferroelastic" transformation, this polarization realignment will need an additional time lag (via electromechanical coupling effect), which we may define the piezoelectric loss. Superposing the ferroelastic domain alignment and polarization alignment (via *Gauss Law*) can generate actual charge under stress application. In this model, the intensive (observable) piezoelectric loss is explained by the 90° polarization reorientation under E, which can be realized by superimposing the 90° polarization reorientation under X and the 180° polarization reversal under E.

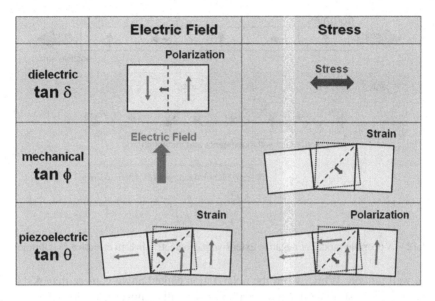

FIGURE 11.11 Polarization reversal/reorientation model for explaining "extensive" dielectric, elastic and piezoelectric losses.

This is the primary reason why we cannot measure the piezoelectric loss independently from the elastic or dielectric losses experimentally.

This domain wall model was employed to explain the loss anisotropy observed in piezoelectric Nb-doped PZT ceramics (PNZT) on polarization angle dependence of losses (Section 9.3.2). Considering three-dimensionally clamped condition, extensive loss parameters were additionally obtained for effective k_{31} and k_{15} vibration mode using "K matrix". Considering an open-circuit condition, due to the different charge development in different modes, the effective dielectric loss varies ($\tan\delta_{11}(\gamma) \neq \tan\delta_{33}(\pi/2-\gamma)$; γ is the cant angle of remnant polarization. Note that in the short-circuited condition where the charge could be well distributed, the effective dielectric loss is mode-independent $\left(\tan\delta'_{11}(\gamma) = \tan\delta'_{33}(\pi/2-\gamma)\right)$. A similar phenomenon is observed for the elastic loss factor. The dielectric and elastic losses are mostly higher when the polarization is angled from the standard structure of each resonator, except for the shear mode in Rhomb PNZT. The piezoelectric loss is more related with the angle between the polarization and the applied electric field than dielectric or elastic loss. The piezoelectric loss is smaller when the angle is larger, meaning the compensation of dielectric and elastic loss is smaller when the polarization is canted from the applied field, from the electromechanical coupling loss perspective. This phenomenon could be a very important point for further study on domain dynamics. The changes are least in MPB structure which is most isotropic. In the effective k_{31} mode, the change of extensive piezoelectric loss is largest in Tet PNZT, which is least isotropic. In the effective k_{15} mode, the change of extensive piezoelectric loss is largest in Rhomb PNZT which has the strongest relative shear property. It is interesting to note that the extensive piezoelectric loss becomes "negative" in Tet PNZT when the polarization is strongly canted with respect to the applied electric field. The negative extensive piezoelectric loss in Tet structure could be obtained from both k_{31} and k_{15} vibration modes.

A possible domain wall dynamic model of the *negative extensive loss* in tetragonal k_{31} vibration mode is proposed in Figure 11.12. When the polarization canted angle γ is less than 45°, the positive strain will facilitate 90° domain wall to move leftward. (Red thick line is expected shift of the 90° domain wall.) Thus, a transient positive field is generated locally. Note the extensive loss parameters are considered under constant D condition which is electrically open-circuited without electrode hypothetically. When the polarization canted angle is 45°, the positive strain will not facilitate

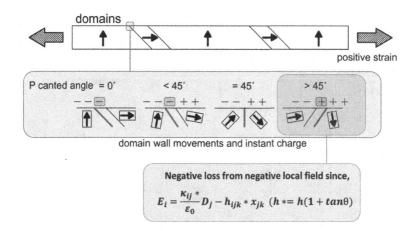

FIGURE 11.12   A possible model of negative extensive piezoelectric loss in effective k_{31} vibration mode.

90° domain wall motion. While, when the polarization canted angle is more than 45° (right-hand side figure), the positive strain will facilitate 90° domain wall to move rightward, leading to generate transient negative electric field locally. In other words, positive and negative extensive piezoelectric loss tan θ are anticipated for the lower and higher angled PNZT samples. Quantitative approach will be required for the detailed discussion.

11.3.2   Do Losses Compensate Each Other?

In a conventional scientific sense, losses are always added; that is, if the material possesses both dielectric and elastic losses, this material exhibits superposed heat generation originated from both dielectric and elastic hysteresis. However, the situation seems to be different in a coupled effect such as piezoelectric effect. Heat generation occurs in the sample uniformly under an off-resonance mainly due to the "intensive dielectric" loss, while heat is generated primarily at the vibration nodal points via the "intensive elastic" loss under the resonance for k_{31} mode. In a piezoelectric PZT, the mechanical quality factor Q_B for the antiresonance (B-type, a subsidiary "coupled" resonance) mode is higher than Q_A for the resonance (A-type) mode. Mechanical quality factors at resonance and antiresonance frequencies can be expressed for k_{31} mode:

$$Q_{A,31} = \frac{1}{\tan\phi'_{11}}; \quad \frac{1}{Q_{B,31}} = \frac{1}{Q_{A,31}} - \frac{2}{1+\left(\dfrac{1}{k_{31}}-k_{31}\right)^2 \Omega^2_{B,31}}(2\tan\theta'_{31}-\tan\delta'_{33}-\tan\phi'_{11}) \quad (11.1)$$

$Q_A < Q_B$ in PZT and other Pb-based perovskite piezoelectrics indicate that intensive piezoelectric loss tanθ' is larger than the average of the elastic tanϕ' and dielectric tanδ'. Since the maximum vibration velocity is also higher for the antiresonance mode than for the resonance mode, we proposed the antiresonance usage for the motor and transducer applications.

The key is to notice the piezoelectric loss factor's contribution. Without the piezoelectric loss, $Q_A > Q_B$ should always be derived from only two dielectric and elastic losses. Because a large piezoelectric loss contributes in *subtraction* of the other two losses, the antiresonance mode becomes much higher Q_B, leading to lower heat generation, in practice.

The relation between mechanical quality factor and real electrical power and mechanical vibration is given by the following equation, which allows the calculation of the mechanical quality factor at any frequency from the real electrical power (P_d) spent and the mechanical energy stored in the sample (evaluated from the tip rms vibration velocity (V_{rms}) measurements) for a longitudinally vibrating piezoelectric resonator (k_t, k_{33}, k_{31}):

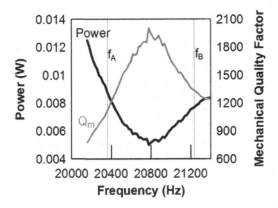

FIGURE 11.13 Mechanical quality factor measured using real electrical power (including the phase lag) for a hard PZT APC 851 k_{31} plate.

$$Q_{m,l} = 2\pi f \frac{\frac{1}{2}\rho V_{rms}^2}{P_d/Lwb} \tag{11.2}$$

The change in mechanical quality factor in a PZT (APC 851) ceramic plate (k_{31}) under constant vibration condition of 100 mm/s RMS tip vibration velocity (i.e., stored mechanical energy constant) is shown in Figure 11.13 (see Section 7.2.4). Very interestingly, the mechanical quality factor reaches a maximum value between the resonance and antiresonance frequencies, the point of which may suggest the optimum condition for the transducer operation merely from an efficiency viewpoint [though the piezo-specimen has "inductive" characteristic at this frequency] and also for understanding the behavior of piezoelectric material properties under high-power excitation. The maximum Q_m or minimum total loss may be realized by the better compensation of the piezoelectric loss and other dielectric and elastic losses, but the microscopic origin of these three-loss compensation is not clear yet.

Another noteworthy fact is that most of Pb-free piezoelectric perovskites exhibit $Q_A > Q_B$ or $Q_A \approx Q_B$, which suggests that the contribution of the intensive piezoelectric loss $\tan\theta'$ is not large in Pb-free piezoelectrics, in comparison with Pb-based piezoelectrics. Detailed microscopic and crystallographic explanation may be required (see Section 9.4.2.2).

11.3.3 HOW CAN WE CALCULATE THE DOMAIN WALL MASS AND INTERNAL FRICTION?

The author is deeply indebted to A. S. Sidorkin via his book "*Domain Structure in Ferroelectrics and Related Materials*" (Viva Books, New Delhi, India, 2017, ISBN 978-81-309-3154-8) for the discussion on domain wall dynamics. Our phenomenological discussion starts from Ginzburg–Landau functional:

$$F(P(x),T) = \int dx \left[\frac{\alpha}{2} P(x)^2 + \frac{\beta}{4} P(x)^4 + \frac{\kappa}{2} (\nabla P(x))^2 \right] \tag{11.3}$$

We can derive static domain wall structure; polarization distribution $P(x)$ and domain wall thickness δ, and the wall energy density γ_0 as

$$P(x) = P_0 \tanh\left(\frac{x}{\delta}\right), \quad \delta = \frac{1}{P_0}\sqrt{\frac{2\kappa}{\beta}} = \sqrt{\frac{2\kappa}{-\alpha}}, \tag{11.4}$$

$$\gamma_0 = -\frac{2}{3}\alpha P_0^2 \delta = \frac{4}{3} P_0^2 \left(\frac{\kappa}{\delta}\right). \tag{11.5}$$

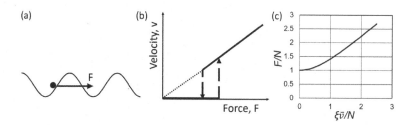

FIGURE 11.14 (a) Prandtl–Tomlinson model: A point mass in a periodic potential. (b) Macroscopic law of friction: dependence of the average velocity on the force. (c) Friction force of the Prandtl–Tomlinson model vs. average velocity in the overdamped case.

Though α and β (and γ in the first-order phase transition case) can easily be determined experimentally, we do not find a suitable measuring technique on the "inter-coupling" κ parameter.

Loss factor evaluation needs a sort of damping factor or time-delay introduction. When we consider the dissipation Γ and the presence of the external electric field E, which are actually essential to discuss the dielectric loss, we can expand the formula as follows:

$$\mu \frac{\partial^2 P}{\partial t^2} + \Gamma \frac{\partial P}{\partial t} - \kappa \frac{\partial^2 P}{\partial x^2} + \alpha P + \beta P^3 = E \tag{11.6}$$

Here, α and β are the polarization "intra-coupling" coefficient initially introduced in a Landau free energy for a monodomain single crystal specimen, which can easily be determined from various experimental methods as mentioned above: (1) inverse permittivity vs. temperature measurement (Curie and Curie–Weiss temperature, and Cuie–Weiss constant) and (2) spontaneous polarization, piezoelectric constant via the relation: $d_{33} = 2\varepsilon_0 \varepsilon Q_{33} P_S$. The parameter κ is introduced as the polarization "inter-coupling" (i.e., polarization cluster–cluster interaction) in Ginzburg–Landau functional for discussing a polydomain single crystal (static structure). How we can simulate this κ is a remaining issue. Can we obtain κ experimentally?

A new parameter μ is a sort of mass proportional to the material's density normalized by the effective charge, which is verified as the *mobility* of the domain wall and is provided by

$$\mu = \rho a^6 / e^{*2}. \tag{11.7}$$

and another parameter Γ is the dissipation factor or "viscous friction" constant of the domain wall. μ and Γ are correlated by

$$\mu = \frac{3}{2} \frac{\delta}{P_0} \frac{1}{\Gamma}. \tag{11.8}$$

Again, how we can simulate/calculate Γ theoretically has not been solved. Can we expand Prandtl–Tomlinson model, shown in Figure 11.14? Is the domain wall velocity obtained experimentally related with Figure 11.14b, or is the periodic potential in Figure 11.14a correlated with the domain wall energy γ_0? Once the Γ value is evaluated, the dielectric loss is represented as

$$\tan \Phi \approx \Gamma \omega / (-2\alpha). \tag{11.9}$$

In this textbook, we derived only the dielectric loss in Eq. (11.9), though the result is not very satisfactory. How we can evaluate this Γ value is still remaining, even after tracing the present domain wall dynamic model. How we can integrate the elastic and piezoelectric losses are still remaining for the readers' generation.

CHAPTER ESSENTIALS

1. Case Study 1: Pulse drive actuator – diesel piezo-injection valve
   a. Piezo-actuator material – semi-soft PZT usable at 150°C, pulse drive requires intermediate Q_m
   b. Long ML design – manufacturing unit length is a couple of mm (less than 10 mm)
   c. Actuator encapsulation – DC bias stress (20 MPa) to stabilize the performance
   d. Pulse drive scheme – rise/fall time of the pseudo-step voltage adjusted to the exact resonance period.
2. Case Study 2: Resonance drive device – ring-dot ML PT
   a. Piezo-material – very hard PZT, sinterable at 900°C
   b. Ag, Cu internal electrode – note that popular Ag/Pd, Ag/Pt electrode ink reduces the performance
   c. Reduced atmosphere sintering – ambient atmosphere control is essential
   d. ML PT performance – using the same PZT composition, by changing the internal electrode from Ag/Pd to Cu, the maximum power density increases by 40% from 30 to 42 W/cm².
3. Future perspective:
   a. What is "piezoelectric" loss?
   b. Do losses compensate each other?
   c. How can we calculate the domain wall mass and internal friction?

REVIEW CHECK POINT

1. There is a copper wire (resistivity = 17 n Ωm) with 50 µm in diameter and 50 cm in length to make an electromagnetic motor coil. What is the resistance of this wire, which will generate heat via Joule heat when voltage is increased? 5 mΩ, 0.5, 5, or 50 Ω.
2. (T/F) Displacement (mechanical) vs. electric field hysteresis is one of the origins for heat generation in piezoelectric actuators. True or False?
3. (T/F) The IEEE Standard on Piezoelectricity, ANSI/IEEE Std. 176–1987 suggests that the mechanical quality factors at the resonance Q_A and at the antiresonance Q_B are the same. True or False?
4. (T/F) A polycrystalline piezoelectric PZT has three independent piezoelectric d matrix components, d_{33}, d_{13}, and d_{15}. True or False?
5. (T/F) The following two force configurations are equivalent mathematically. True or False?

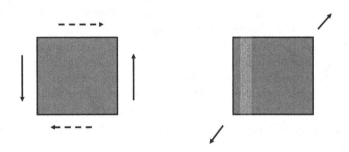

6. Elastic Gibbs energy (in 1D expression) is given by:

$$G_1 = (1/2)\alpha P^2 + (1/4)\beta P^4 + (1/6)\gamma P^6 + \cdots - (1/2)sX^2 - QP^2X.$$

Why don't we include the "odd-number" power terms of polarization P? Answer simply.

7. Elastic Gibbs energy is given by:

$$G_1 = (1/2)\alpha P^2 + (1/4)\beta P^4 + (1/6)\gamma P^6 + \cdots - (1/2)sX^2 - QP^2X.$$

   Which parameter can you find the primary expansion term in respect of "temperature"? Answer simply.

8. (T/F) The phenomenology suggests that the spontaneous polarization of a ferroelectric changes with temperature linearly. True or False?

9. (T/F) The phenomenology suggests that the spontaneous strain in a ferroelectric changes with temperature linearly. True or False?

10. Provide the inverse relative permittivity formula in a paraelectric phase in terms of temperature T.

11. (T/F) The phenomenology suggests that the piezoelectric constant of a ferroelectric material exhibits the maximum just below its Curie temperature. True or False?

12. (T/F) The phenomenology suggests that the Curie–Weiss temperature of a ferroelectric material is always higher than (or equal to) the Curie temperature. True or False?

13. How is the piezoelectric coefficient d related with the electrostrictive coefficient Q, spontaneous polarization P_S, and relative permittivity ε in the Devonshire phenomenology? Using the air permittivity ε_0, provide a simplest formula.

14. (T/F) There is a highly resistive (no electric carrier/impurity in a crystal) piezoelectric single crystal (spontaneous polarization P_S) with a monodomain state without surface electrode. The *depolarization electric field* in the crystal is given by $E = -\left(\dfrac{P_S}{\varepsilon_0}\right)$. True or False?

15. What is a typical number for k_{33} in soft PZT ceramics? 1%, 10%, 50%, or 70%?

16. There is a PZT ML actuator with a cross-section area $5\times5\,mm^2$. Provide a generative (blocking force) roughly. 1, 10, 100 N, or 1 kN?

17. How large displacement can be generated in a soft PZT ML actuator with a length of 10 mm? 100 nm, 1, 10, 100 μm?

18. What is the fundamental longitudinal resonance frequency of a PZT ML actuator with a length 10 mm? 1, 10, 100 kHz, or 1 MHz?

19. What is the MPB composition of the PZT system (MPB: morphotropic phase boundary) at room temperature? $PbZrO_3{:}PbTiO_3$ = (48:52), (50:50), (52:48), or (none of these)?

20. (T/F) The MPB composition of the PZT system exhibits the maximum electromechanical coupling k, piezoelectric coefficient d, and the minimum permittivity ε. True or False?

21. (T/F) When we add "Coulomb" damper to a mass-spring system under free vibration (no external force), vibration amplitude decreases exponentially. True or False?

22. (T/F) When we add "viscous damper" to a mass-spring system under free vibration (no external force), vibration amplitude decreases exponentially. True or False?

23. (T/F) The Laplace transform for $\sin(at)$ is $1/(s^2 + a^2)$. True or False?

24. What is the Laplace transform for the Dirac Impulse (delta) function?

25. What is the Laplace transform for the Heaviside Step function?

26. (T/F) Mechanical quality factor and the damping ratio ζ are related as $Q_m = \dfrac{\omega_0}{\Delta\omega} = 1/\zeta$. True or False?

27. (T/F) Complex spring constant is equivalent to the viscoelastic damping model. True or False?

28. (T/F) The high-frequency portion of the Bode plot for the second-order system of a piezoelectric device and the displacement is approximated with an asymptotic straight line having a negative slope of 20 dB/decade. True or False?

29. (T/F) Because the polarization is induced after the electric field is applied (time delay), the P vs. E hysteresis loop should show the clockwise rotation. True or False?

30. (T/F) The hysteresis area of the strain x vs. electric field E corresponds directly to the piezoelectric loss factor $\tan\theta'$. True or False?

31. (T/F) The permittivity under mechanically clamped condition is smaller than that under mechanically free condition. True or False?

32. (T/F) The elastic compliance under open-circuit condition is larger than that under short-circuit condition. True or False?

33. (T/F) The piezoelectric resonance is only the mechanical resonance mode, and the anti-resonance is not the mechanical resonance. True or False?

34. (T/F) The fundamental resonance mode of the k_{33} mode has an exact half-wave length vibration on the rod specimen. True or False?

35. (T/F) The fundamental resonance mode of the k_{31} mode has an exact half-wave length vibration on the plate specimen. True or False?

36. The extensive losses are interrelated with the intensive losses in terms of the **K**-matrix in the k_{31} mode.

$$\begin{bmatrix} \tan\delta \\ \tan\phi \\ \tan\theta \end{bmatrix} = [\mathbf{K}] \begin{bmatrix} \tan\delta' \\ \tan\phi' \\ \tan\theta' \end{bmatrix}, \text{ where } [\mathbf{K}] = \frac{1}{1-k^2} \begin{bmatrix} 1 & k^2 & -2k^2 \\ k^2 & 1 & -2k^2 \\ 1 & 1 & -1-k^2 \end{bmatrix}.$$

Obtain the inverse **K**-matrix, in this case.

37. Provide the relationship between the mechanical quality factor Q_M at the resonance frequency and the intensive elastic loss in the k_{31}-type specimen.

38. Provide the relationship between the mechanical quality factor Q_M at the antiresonance frequency and the extensive elastic loss in the k_t-type specimen.

39. (T/F) The strain distribution in a high k_{33} rod specimen is more uniform at the antiresonance mode than that at the resonance mode. True or False?

40. (T/F) The strain distribution in a high k_{33} rod specimen is more uniform at the resonance mode than that at the antiresonance mode. True or False?

41. (T/F) The strain distribution in a high k_{31} rod specimen is sinusoidal at the resonance mode. True or False?

42. (T/F) The strain distribution in a high k_{31} rod specimen at the antiresonance mode indicates that the nodal lines are located slightly outside from the both plate edges. True or False?

43. When $(\tan\delta'_{33} + \tan\phi'_{11})/2 < \tan\theta'_{31}$ is satisfied, which is larger Q_A or Q_B for the k_{31}-type specimen?

44. (T/F) When we consider an equivalent electric circuit of a mechanical system in terms of LCR series connection, the spring constant c is directly proportional to capacitance C (in proportion to). True or False?

45. (T/F) Because of the damped capacitance in the equivalent circuit of a k_{31}-type piezo-plate oscillator, the antiresonance mode comes out, in addition to the resonance mode. True or False?

46. How can you describe the quality factor Q in a series-connected LCR circuit?

47. (T/F) When we consider an off-resonance drive of a ML piezo-actuator, heat generation comes primarily from the elastic loss $\tan\phi'$. True or False?

48. (T/F) When we consider a resonance drive of a k_{31}-type piezo-plate-actuator, heat generation comes primarily from the elastic loss $\tan\phi'$. True or False?

49. (T/F) The heat generation is originated from three losses (dielectric, elastic, and piezoelectric) in piezoelectrics. These three losses are always added to produce the final temperature rise. True or False?

50. When we excite the fundamental resonance mode on a k_{31}-type rectangular PZT plate under a high-voltage drive, which part of the specimen generates the maximum temperature rise? Answer simply.

51. (T/F) In order to suppress the temperature rise in piezoelectric actuators/transducers, a tube design is better than solid rod design. True or False?

52. (T/F) To generate the same vibration velocity of a piezoelectric transducer, the resonance drive is the most efficient, rather than antiresonance drive in PZT-based materials. True or False?

53. (T/F) When we drive the piezoelectric k_{31} plate mechanically under a sinusoidal force from the plate ends, the resonance frequency should be observed at the same frequency under both short-circuit and open-circuit conditions of the specimen. True or False?

54. (T/F) When we drive the piezoelectric k_{31} plate electrically under constant high-voltage condition, the admittance spectrum shows a significant skew distortion at the antiresonance frequency. True or False?

55. (T/F) The mechanical quality factors Q_A (at resonance) and Q_B (at antiresonance) should be the same in a piezoelectric PZT k_{33} rod specimen. True or False?

56. (T/F) The mechanical quality factor Q_A at resonance is larger than Q_B at antiresonance in a PZT piezoelectric k_{31} plate specimen. True or False?

57. Can you find more energy efficient frequency for driving a PZT piezoelectric k_{31} plate specimen than its resonance or antiresonance frequencies? If yes, where do you find this most efficient frequency?

58. (T/F) The off-resonance dielectric permittivity and loss values are usually used for evaluating the piezoelectric constant and its loss factor in the admittance spectrum method, because it is theoretically impossible to measure the dielectric performance at the resonance mode. True or False?

59. (T/F) The voltage supply has a small output impedance, while the current supply has a large output impedance. True or False?

60. When we measure the admittance spectrum on a k_{31}-type piezoelectric plate, the phase changes from +90° to −90° around the resonance point with an increase in drive frequency. What is the phase lag at the resonance point?

61. When we measure the admittance spectrum on a k_{31}-type piezoelectric plate, the phase changes from +90° to −90° around the resonance point with an increase in drive frequency. How do you call the frequencies that provide the phase ±45°?

62. Describe the resonance frequency f for the following electrical circuit:

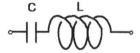

63. (T/F) When a piezoelectric actuator is driven by a rectangular pulse voltage, the mechanical ringing is completely suppressed when the pulse width is adjusted exactly to a half of the resonance period of the sample. True or False?

64. When a piezoelectric actuator is driven under a pseudo-DC condition by a switching amplifier, the efficiency is very low. What electric component (impedance converter) is recommended to be coupled in the driving circuit in order to increase the efficiency?

65. (T/F) Negative capacitance is utilized to drive a "capacitive" actuator device efficiently. True or False?

66. Which is a "Soft" PZT result, left- or right-hand side figure?

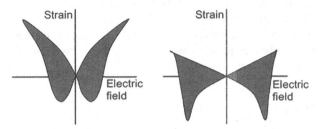

67. In a polycrystalline PZT, we observed $P_S = 27$ µC/cm². Estimate the single crystal value roughly with the Uchida–Ikeda model.

68. (T/F) The tetragonal symmetry Fe-doped PZT ceramics shows the largest mechanical quality factor under vibration velocity 0.05 m/s. Thus, it is the best specimen to be used for high-power density (vibration velocity 0.5 m/s) applications. True or False?

69. (T/F) For the diesel injection valve application, the largest displacement or the largest piezoelectric d_{33} material is the most suitable candidate. Thus, very soft PZT composition should be utilized for this application. True or False?

70. (T/F) The origin of the multidomain structure in a thin ferroelectric crystal without electrodes is to reduce the "depolarization energy" in the specimen. True or False?

71. (T/F) The P–E hysteresis loop is observed in a PZT specimen as below. We can conclude that this sample possesses a "positive" internal bias field. True or False?

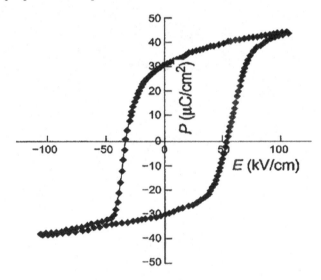

72. (T/F) The domain wall thickness in a ferroelectric crystal increases with increasing temperature up to the Curie temperature. True or False?

73. (T/F) Ferroelectrics exhibit the maximum permittivity and maximum dielectric loss just around the Curie temperature. True or False?

74. (T/F) In order to suppress the temperature rise of piezoelectric transducers, we had better make the same volume piezo-device in a cubical structure. True or False?

75. (T/F) The internal electrode of high-power ML PTs should be Ag/Pd, because Pd addition suppresses Ag migration into PZT ceramics. True or False?

76. The length of "Standard" ML piezoelectric actuators is usually 5–10 mm. Most of the longer MLs (such as 50 mm) are laminated structures of the thin unit MLs. One of the reasons is related with manufacturing cost. What is another technically important reason?

77. Which has higher electric conductivity between Ag and Ag/Pd?
78. Which has higher electric conductivity between Ni and Cu?
79. Which has higher electric conductivity between Ag and Cu?
80. A k_{31}-type piezo-ceramic plate with permittivity ε_{33}^X is completely clamped along the length direction. Supposing the electromechanical factor k_{31}, provides the longitudinally clamped permittivity ε_{33}^{x1} in terms of ε_{33}^X and k_{31}.

CHAPTER PROBLEMS

11.1   Search the recent papers on the following contents, then write a summary article as a new research proposal:
1. High-power piezoelectric materials
2. ML manufacturing technologies, in particular with Cu, pure Ag internal electrodes
3. Loss phenomenology
4. High-power characterization methods
5. Domain wall dynamics and loss mechanisms.

11.2   Even though both k_t and k_{33} modes piezoelectric specimens (shown in **below** figure) exhibit basically the elastic compliance c_{33}^D and s_{33}^D, under D-constant condition, respectively, the loss values are different owing to the boundary condition difference: $x_1 = x_2 = 0$ (strain zero), or $X_1 = X_2 = 0$ (stress zero). Taking into account the exact definition of "extensive loss" $\tan\phi_{33}$ for the c_{33}^D in the k_t mode, consider the difference of $\tan\phi_{33}'''$ for the s_{33}^D in the k_{33} mode.

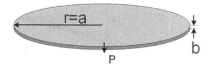

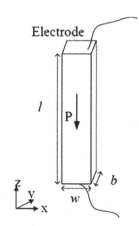

a.   Verify the following relations:

$$c_{33}^E = c_{33}^D(1 - k_t^2) \ (k_t \text{ mode})$$

$$s_{33}^D = s_{33}^E(1 - k_{33}^2) \ (k_{33} \text{ mode})$$

b. Derive the elastic compliance and stiffness relations in the ∞ mm symmetry (equivalent to 6 mm), referring to Section 4.2: $\dfrac{1}{s_{33}} = c_{33} - \dfrac{2c_{13}^2}{c_{11}+c_{12}}$ and $\dfrac{1}{c_{33}} = s_{33} - \dfrac{2s_{13}^2}{s_{11}+s_{12}}$.

c. The mechanical quality factor Q_B at the antiresonance is provided as follows.
   - Why is Q_B in the k_t mode given directly by the extensive loss $\tan\phi_{33}$?
   - Taking $c_{33}^{D*} = c_{33}^D(1+j\tan\phi_{33})$ for k_t and $s_{33}^{D*} = s_{33}^D(1+j\tan\phi_{33}''')$ for k_{33}, discuss the relation between $\tan\phi_{33}$ and $\tan\phi_{33}'''$.

k_t mode:

$$Q_{B,t} = \frac{1}{\tan\phi_{33}}, \quad \frac{1}{Q_{A,t}} = \frac{1}{Q_{B,t}} - \frac{2}{k_t^2 - 1 + \Omega_{A,t}^2/k_t^2}\left(2\tan\theta_{33} - \tan\delta_{33} - \tan\phi_{33}\right)$$

$$\Omega_{B,t} = \frac{\omega_b l}{2v_{33}^D} = \frac{\pi}{2} \quad \left[v_{33}^D = 1/\sqrt{\rho/c_{33}^D}\right], \quad \Omega_{A,t} = \frac{\omega_a l}{2v_{33}^D}, \quad \Omega_{A,t} = k_t^2 \tan\Omega_{A,t}$$

k_{33} mode:

$$Q_{B,33} = \frac{1}{\tan\phi_{33}'} = \frac{1-k_{33}^2}{\tan\phi_{33}' - k_{33}^2(2\tan\theta_{33}' - \tan\delta_{33}')}$$

$$\frac{1}{Q_{A,33}} = \frac{1}{Q_{B,33}} + \frac{2}{k_{33}^2 - 1 + \Omega_A^2/k_{33}^2}\left(2\tan\theta_{33}' - \tan\delta_{33}' - \tan\phi_{33}'\right)$$

$$\Omega_{B,33} = \frac{\omega_b l}{2v_{33}^D} = \frac{\pi}{2} \quad \left[v_{33}^D = 1/\sqrt{\rho s_{33}^D}\right], \quad \Omega_{A,33} = \frac{\omega_a l}{2v_{33}^D}, \quad \Omega_{A,33} = k_{33}^2 \tan\Omega_{A,33}$$

REFERENCES

1. K. Uchino, *Micromechatronics*, 2nd Edition, (CRC/Dekker Press, New York, NY, 2019).
2. S. Ural, S.–H. Park, S. Priya, K. Uchino, High Power Piezoelectric Transformers with Cofired Copper Electrodes- Part I, *Proceedings of the 10th International Conference on New Actuators*, Bremen, Germany, June 14–16, 2006, P23, pp. 556–558 (2006).
3. A. Erkan Gurdal, S. Tuncdemir, K. Uchino, C.A. Randall, Low temperature co-fired multilayer piezoelectric transformers for high power applications, *Materials and Design*. **132**, 512–517 (2017).
4. A. Kingon, S. Srinivasan, Lead zirconate titanate thin films directly on copper electrodes for ferroelectric, dielectric and piezoelectric applications, *Nature Materials*. **4**, 233–237 (2005).

Appendix
Answers to "Check Point"

CHAPTER 1

1) 5Ω 2) True   3) True   4) True

CHAPTER 2

1) Lorentz factor   2) True   3) False (k_{31}!)   4) False   5) $G_1(-P)$ should not change the energy.

6) α term   7) False   8) True   9) True   10) True

11) True   12) F (Q is constant)   13) False   14) $d = 2\,\varepsilon_0\varepsilon\,QP_S$ 15) False

16) 70%   17) 1 kN   18) 100 kHz   19) 52:48   20) False

CHAPTER 3

1) False   2) 1   3) $1/s$ 4) False   5) True

6) False   7) True   8) False

CHAPTER 4

1) False   2) False   3) True   4) False   5) True

6) False   7) False   8) True   9) $Q_M = 1/\tan\phi'_{11}$ 10) $Q_M = 1/\tan\phi_{33}$

11) False   12) Q_B

CHAPTER 5

1) True   2) True   3) False   4) False   5) $Q = \sqrt{L/C}\,/R$

6) $[K]^{-1} = [K]$ 7) False   8) True   9) True   10) $Q_M = 1/\tan\phi'_{11}$

11) $\dfrac{1}{Q_{A,t}} = \tan\phi_{33} - \dfrac{2}{k_t^2 - 1 + \Omega_{A,t}^2/k_t^2}$ 12) True   13) Q_B

$$\times\left(2\tan\theta_{33} - \tan\delta_{33} - \tan\phi_{33}\right)$$

CHAPTER 6

1) False   2) True   3) False   4) Node at the plate center

5) True   6) False   7) True

CHAPTER 7

1) False   2) False   3) False   4) False   5) False
6) A frequency in-between the resonance and antiresonance frequencies 7) False

CHAPTER 8

1) Counterclockwise   2) False   3) True   4) 0°
5) Quadrantal frequencies   6) $f = 1/2\pi\sqrt{LC}$ 7) False 8) Inductor
9) True   10) Pulse width   11) True   12) In-between resonance and
                                 modulation   antiresonance frequencies
13) Capacitor

CHAPTER 9

1) Left   2) 32 µC/cm²   3) False   4) False   5) True
6) True   7) True   8) False   9) False

CHAPTER 10

1) To escape from the crack generation in the piezo-inactive edge region of an ML actuator
2) 5 Ω   3) False   4) False   5) #1 Cu, #2 Ag

CHAPTER 11

1) 5Ω   2) True   3) True   4) False (k_{31}!) 5) False
6) In order not to change the energy by flip over P to $-P$; that is, $G_1(P) = G_1(-P)$. 7) α term
8) False   9) True   10) $1/\varepsilon = (T{-}T_0)/C$ 11) True   12) False
13) $d = 2\varepsilon_0\varepsilon QP_S$ 14) False ($-P_S/\varepsilon_0\varepsilon$!) 15) 70%   16) 1 kN   17) 10 µm
18) 100 kHz   19) (52:48)   20) False   21) False   22) True
23) False [correct answer: $a/(s^2 + a^2)$] 24) 1   25) $1/s$ 26) False (1/2!)
27) True   28) False (-40) 29) False   30) False   31) True
32) False   33) False   34) False   35) True   36) $[K]^{-1} = [K]$
37) $Q_M = 1/\tan\phi'_{11}$ 38) $Q_M = 1/\tan\phi_{33}$ 39) False 40) True 41) True
42) False   43) Q_B 44) False (1/c!)   45) True   46) $Q = \sqrt{L/C}/R$
47) False   48) True   49) False   50) Plate center (i.e., nodal line)
51) True   52) False   53) False   54) False   55) False
56) False   57) Yes, a frequency between the resonance and antiresonance frequencies.
58) False   59) True   60) 0°   61) Quadrantal frequencies
62) $f = 1/2\pi\sqrt{LC}$ 63) False (exact resonance period!) 64) Inductor 65) True
66) Left-hand side 67) 32 µC/cm²   68) False   69) False   70) True
71) False   72) True   73) True   74) False   75) False
76) To escape from the crack generation in the piezo-inactive edge region of an ML actuator
77) Ag   78) Cu   79) Cu   80) $\varepsilon_{33}^{x_1} = \varepsilon_{33}^X(1 - k_{31}^2)$

Index

Printed in the United States
By Bookmasters